Cocoa Production and Processing Technology

Cocoa Production and Processing Technology

Emmanuel Ohene Afoakwa

CRC Press is an imprint of the
Taylor & Francis Group, an **informa** business

CRC Press
Taylor & Francis Group
6000 Broken Sound Parkway NW, Suite 300
Boca Raton, FL 33487-2742

First issued in paperback 2016

Version Date: 20140108

ISBN 13: 978-1-138-03382-5 (pbk)
ISBN 13: 978-1-4665-9823-2 (hbk)

Visit the Taylor & Francis Web site at
http://www.taylorandfrancis.com

and the CRC Press Web site at
http://www.crcpress.com

This book is dedicated to my dear wife, Ellen and our lovely children Cita, Nana Afra, Maame Agyeiwaa and Kwabena Ohene-Afoakwa (Junior) whose wisdom, prayers and support have helped me achieve great success in my career and life.

Contents

Preface

Over the past few decades, cocoa has increasingly gained spectacular attention on the global market as it continues to become one of the most lucrative and heavily traded food commodities in the world. This has led to interesting continuous increases in cocoa production across the world, most especially by the four main growing countries in West Africa—Côte d'Ivoire, Ghana, Nigeria and Cameroon—now together providing ~75% of the global cocoa market. Coupled with these and the recent expansion of cocoa production from Southeast Asia—Indonesia, Malaysia and Vietnam—has raised questions by various stakeholders in the cocoa business and processors in the confectionery industry over the quality of cocoa that enters the international market. That notwithstanding, the cocoa market has become far more sophisticated than it was in the 1990s and despite the challenges it faces it is still one of the largest food commodities exported from the developing countries to the rest of the world.

Many questions, however, continue to be raised by various organisations involved in the cocoa business as well as manufacturers and consuming countries on the quality, sustainability and traceability of cocoa. Such concerns are not new, but have led to several discussions over the past decades which laid the foundations for the quality assessment of cocoa beans used today. Recent developments on the emergence of Southeast Asia as a new block in the cocoa market and the continuously increasing production capacities by the old players, together with cocoa processors and the consuming public wanting even higher standards, have regenerated these concerns. It is thus important for cocoa producers across the globe to understand the factors that can bridge the gap in the sustainable production of high-quality cocoa beans for the international market. Many of these concerns stem from the fact that the major cocoa-producing countries use far different production and post-harvest practices and strategies which are inconsistent and nonharmonised. This is because the factors leading to sustainable production of high-quality cocoa beans including cocoa genotype, environmental conditions and post-harvest treatments are not well understood.

This book provides overviews of up-to-date scientific and technical explanations of the technologies and approaches to modern cocoa production practices, global production and consumption trends as well as principles of cocoa processing and chocolate manufacture. Principally, it provides detailed information on the origin, history and taxonomy of cocoa, as well as fairtrade and organic cocoa industries and their influence on the livelihoods and cultural practices of smallholder farmers. Other important aspects cover factors that promote production, sustainability and traceability of high-quality cocoa beans for the global confectionery industry.

The chapters cover the entirety of the cocoa cultivation, harvesting and post-harvest treatments with special emphasis on cocoa bean composition, genotypic variations in the bean and their influence on flavour quality, post-harvest pre-treatments (pulp pre-conditioning by pod storage, mechanical and enzymatic depulping, and bean spreading), fermentation techniques, drying, storage and transportation. Details of the cocoa fermentation processes as well as the biochemical and microbiological changes involved and how these influence flavour formation and development during industrial processing are discussed. Much attention is also given to the cocoa trading systems, bean selection and quality criteria. Other important aspects covered include scientific and technological explanations of the various processes involved in industrial processing of fermented and dried cocoa beans into liquor, cake, butter and powder, and these include cleaning and sorting, winnowing, sterilisation, roasting, alkalisation, grinding, liquor pressing into butter, deodorisation and cocoa powder production. It also covers the general principles of industrial chocolate manufacture, with detailed scientific explanations of the various stages of chocolate manufacturing processes including mixing, refining, conching and tempering/fat pre-crystallisation systems. The discussions also cover the factors that influence the quality characteristics of finished chocolates, quality parameters, post-processing defects and preventive strategies for avoiding post-processing quality defects in chocolate. These in tandem with the earlier discussions provide innovative techniques related to sustainability and traceability in high-quality cocoa production as well as new product development with significance for cost reduction and improved cocoa bean and chocolate product quality.

The ideas and explanations provided in this book evolved from my research activities on cocoa and the various interactions I have had with cocoa farmers in Ghana, who produce bulk cocoa beans with the highest quality worldwide, and those from many other countries across the world as well as other stakeholders engaged in the production, storage, marketing, processing and manufacturing of cocoa and chocolate products. It contains detailed explanations of the technologies that could be employed to assure sustainable production of high-quality and safe cocoa beans for the global confectionery industry. With opportunities for improvements in quality possible through improved production practices and more transparent supply chain management, plant breeding strategies and new product development associated with fairtrade, organic and the development of niche premium quality products, there is a need for greater understanding of the variables as well as the science and technologies involved.

It is hoped that this book will be a valuable resource for academic and research institutions around the world, and as a training manual on the science and technology of cocoa production and processing, and chocolate manufacture. It is aimed at cocoa producers, traders and businesses as well as confectionery and chocolate scientists in industry and academia, general practising food scientists and technologists, and food engineers. The chapters on research developments are intended to help generate ideas for new research activities relating to process improvements, product quality control and assurance, as well as development of new niche/premium cocoa and chocolate products.

It is my vision that this book will inspire all bulk cocoa-producing countries across the world to strive to produce high-quality cocoa beans, similar to those of Ghana beans, for the global cocoa market and as well inspire many local and multinational cocoa-processing industries in their quest for adding value to the many raw materials that are produced within these countries, especially cocoa.

Acknowledgements

I am sincerely grateful to my parents—the late Mr Joseph Ohene Afoakwa (Esq.) and Mrs Margaret Ohene Afoakwa—for ensuring I obtained the best education in spite of the numerous challenges they faced during some periods of their lives. Their profound love, prayers, support and advice strengthened me from my childhood, giving birth to the many dreams and aspirations which have all become realities in my life today. As well, I am thankful to the government of Ghana and to all cocoa farmers in Ghana whose toil and sweat were used to fund my education through the Ghana Cocoa Board Scholarship Scheme, which I earned all throughout my secondary education, without which I could not have remained in school to make it to the university level. I am indeed grateful for the support received from the government and the people of Ghana throughout my education.

My gratitude and appreciation also go to the management of the Nestlé Product Technology Centre (York, UK) for providing the funding and support for my training in chocolate technology at the Nestlé Product Technology Centre York, UK; and also to Dr Alistair Paterson, Centre for Food Quality, University of Strathclyde, Glasgow, UK; Mr Mark Fowler, head of the Applied Science Department of the Nestlé Product Technology Centre (York, UK) and Dr Steve Beckett (retired confectionery expert) for their support, encouragement, patience and friendliness during the period of my doctoral training in York. Many thanks also go to Joselio Vieira, Angela Ryan, John Rasburn, Peter Cooke, Philip Gonus, Angel Manéz, Jan Kuendigar, Ramana Sundara and Sylvia Coquerel of the Nestlé Product Technology Centre York, UK and to Dr Jeremy Hargreaves (Nestlé Head Office, Vevey, Switzerland) whose advice, guidance and support enhanced my understanding of the science and technology of chocolates.

My sincere thanks also go to the many friends and colleagues around the world who have mentored, encouraged and inspired me in various ways throughout my career including Professor Samuel Sefa-Dedeh, Professor George Sodah Ayernor, Professor Ebenezer Asibey-Berko, Professor Anna Lartey, Professor Esther Sakyi-Dawson, Professor Kwaku Tano-Debrah, Dr Agnes Simpson Budu, Dr William Bruce Owusu, Dr George Amponsah Annor, Dr Fred Vuvor, Dr Esi Colecraft and Dr Gloria Otoo, all of the Department of Nutrition and Food Science, University of Ghana, Legon-Accra, Ghana.

I especially want to express my sincere appreciation to my graduate students including Ms Jennifer Quao, Mr Evans Akomanyi, Mr Edem John Kongor, Mr Eric Ofosu-Ansah and Daniel Tetteh Amanquah who conducted aspects of the research components included in this book. My appreciation also goes to my doctorate students, teaching and research assistants including Roger Phillips Aidoo, Bobby Antan Caiquo and Prince Kelly Anyomitse for helping me with the typing and editing of various aspects of the manuscripts. Many thanks also go to Professor Linley Chiwona-Karltun of the Swedish University of Agricultural Sciences, Uppsala, Sweden.

Finally, my profound appreciation and love go to my siblings Sammy, Juliana and Regina for their prayers and support throughout my education, and again to my dear wife, Ellen and our children Cita, Nana Afra, Maame Agyeiwaa and Kwabena Ohene-Afoakwa (Junior) for supporting me and most importantly providing the much needed love, encouragement and affection that have strengthened me throughout my career.

About the Author

Emmanuel Ohene Afoakwa, PhD, is an associate professor in food science in the Department of Nutrition and Food Science, University of Ghana. He holds a PhD degree in food science from the University of Strathclyde, Glasgow, UK and MPhil and BSc (Honours) degrees in food science from the University of Ghana, Legon-Accra, Ghana. He also holds a certificate in international food laws and regulations from Michigan State University, East Lansing, Michigan, USA. In addition, he holds a post-graduate certificate in food quality management systems from the International Agricultural Centre of Wageningen University, Wageningen, the Netherlands. He is also a trained and licensed food auditor of the World Food Safety Organization, UK.

Dr Afoakwa has vast relevant experience in food science and technology and international food laws and regulations. He is a member of several professional bodies including the Institute of Food Technologists (IFT), Food Science and Nutrition Network for Africa (FOSNNA), Information Technology for the Advancement of Nutrition in Africa (ITANA), The African Network for School Feeding Programmes (ANSFEP), the Ghana Institute of Nutrition and Food Technology (GINFT) and the Ghana Science Association (GSA). He has authored and co-authored 160 publications (including 70 peer-reviewed journal publications, 4 books, 4 book chapters, 2 encyclopaedia chapters and 80 conference presentations with published abstracts) in food science and technology, food and nutrition security, and school feeding programmes.

In the pursuance of his duties as a food technologist, he has travelled to 34 different countries across the globe where he has gained high international recognition for his work. He is a member of the board of directors of the Global Child Nutrition Foundation (GCNF) in Washington, DC, USA; the executive secretary to the African Network for School Feeding Programmes; the executive secretary to the Ghana Institute of Nutrition and Food Technology, and the scientific secretary to the Society on Information Technology for the Advancement of Nutrition in Africa (ITANA). He also serves as a member of editorial boards of several international journals as well as a technical reviewer for more than 10 international peer-reviewed journals around the world. In addition, he is a technical advisor to the International Foundation for Science (IFS) within the area of food science and nutrition. As well, he is a consultant trainer in scientific writing and grant proposal development with the African Women in Agricultural Research and Development (AWARD). He has vast experience in food technology and nutrition, and translates his research findings through process and product development into industrial production towards the achievement of the UN Millennium Development Goals (MDGs) mainly on food and nutrition security, and sustainable agricultural development.

Dr Afoakwa is an expert in cocoa and chocolate technology and has published extensively and given several presentations at international conferences around the world including the Annual Meeting of International Food Technologists (IFT) in the United States, the International Conferences of Food Science and Technology, World Congress of Food Science and Technology (IUFoST Bi-annual Congresses) in France, China and Brazil, and the ZDS Chocolate Technology International Congress by ZDS Solingen in Cologne, Germany.

1 Introduction to the World Cocoa Economy

1.1 INTRODUCTION

Cocoa (*Theobroma cacao* L.), generally known to have originated from Central and South America, is an important agricultural export commodity in the world and forms the backbone of the economies of some countries in West Africa, such as Côte d'Ivoire and Ghana. It is the leading foreign exchange earner and a great source of income for many families in most of the world's developing countries. In Ghana, cocoa is the second foreign exchange earner and many farmers and their families depend on it for their livelihood (Afoakwa 2010). The World Cocoa Foundation estimates the number of cocoa farmers worldwide currently to be 5–6 million and the number of people who depend upon cocoa for their livelihood, worldwide, 40–50 million. Hence the economic importance of cocoa cannot be overemphasised and current global market value of the annual cocoa crop is US $5.1 billion (Ghana Cocoa Board 2010; World Cocoa Foundation 2010).

Cocoa continues to be an important source of export earnings for many producing countries, in particular in Africa. Africa's heavy dependence on cocoa as well as on other primary commodities as a source of export earnings has been vulnerable to market developments, in particular price volatility, and weather conditions. However, in some circumstances, real exchange rates, domestic marketing arrangements and government intervention have acted to buffer price movements for cocoa producers. Cocoa was the second source of export earnings in Ghana in 2010, after gold, generating US $2.2 billion. In Côte d'Ivoire, dependence on cocoa exports has been declining in recent years, with export revenues from crude oil and petroleum products increasing significantly over the same period. These are estimated to have surpassed revenues from cocoa in 2005. However, with lower oil prices in 2009 and the price of cocoa surging, cocoa-derived export revenues increased significantly in 2009, surpassing oil revenues and reaching a total of US $3.7 billion in 2009 and US $3.8 billion in 2010 (ICCO 2012a).

The African region accounts for approximately 75% of net world exports of cocoa, and is by far the largest supplier of cocoa to the world markets, followed by Asia and Oceania (16%) and the Americas (6%). The cocoa market remains highly concentrated, with the top five countries accounting for 87% of world net exports, whereas over 98% originated from the top 10 countries during the five-year period from 2006/2007 to 2010/2011. Côte d'Ivoire is the world's leading exporter of cocoa, representing 37% of global net exports, followed by Ghana (22%) and Indonesia (15%). With increased processing at the origin, cocoa products now represent a slightly higher proportion of total cocoa exports in most cocoa-producing countries (ICCO 2012b).

1.2 MAJOR CHANGES IN THE WORLD COCOA TRADE

Major changes have taken place in the world cocoa economy over the last 10 years up to the current 2012/2013 season. These include, among others, the development of supply and demand of and for cocoa, cocoa farm gate prices, trade flows of cocoa beans between regions, past and present price developments, the reliance of cocoa-producing countries on the cocoa sector in terms of export revenues and recent developments concerning chocolate consumption.

World cocoa production rose from nearly 3.2 million tonnes in the 2002/2003 cocoa season to an estimated 4 million tonnes forecast for the 2012 season. This represents an average annual growth rate of 3.3%, using a three-year moving average to smooth out the effect of weather-related aberrations. Annual production levels have deviated considerably from the trend value, mainly arising from the influence of climatic factors. Although production suffered in the 2006/2007 season, declining by nearly 10% and resulting in the record deficit of nearly 280,000 tonnes, an all-time record output of over 4.3 million tonnes was achieved during the 2010/2011 cocoa year, arising from excellent weather conditions favouring crop development across Africa, the world's largest cocoa-producing region.

World cocoa consumption, as measured by grindings of cocoa beans by the industry, also increased on average by 2.9% per annum over the review period. Grindings have shown a more consistent trend than production, rising from nearly 3.1 million tonnes in 2002/2003 to over 3.9 million tonnes in 2010/2011 with a forecast of nearly 4 million tonnes for 2011/2012. The review period witnessed only one decline, albeit notable, in 2008/2009 when consumer demand fell in the midst of global economic woes and the steady increase in cocoa bean prices.

Taking the period 2002/2003 to 2011/2012 as a whole, production surpluses occurred in five out of the last ten seasons and production deficits in the other five of the last ten seasons. Total end-of-season stocks rose from 1.395 million tonnes in 2002/2003 to an estimated 1.732 million tonnes at the end of the current season. However, as a result of increased grindings, the ratio of world cocoa bean stocks to grindings is estimated to have declined from 46% in 2002/2003 to 43% at the end of the 2011/2012 crop year.

There has been an increased demand for cocoa beans in Asia, Eastern Europe and Latin America, which reflects the increasing consumption of chocolate in these countries. Over the past decade, cocoa consumption, as measured by grindings, has increased by 2.5% from the 3,608,000 tonnes in 2006/2007 to 4,008,000 tonnes in 2012/13. Despite a relative slowdown during that 2006/2007 season, the cocoa market has been characterised over the last five years by a sustained demand for cocoa, rising by 3.8% per annum (based on a three-year moving average). This was supported by a strong demand for cocoa butter to rebuild stocks, as well as by rising chocolate consumption in emerging and newly-industrialised markets and changes in chocolate consumption behaviour in mature markets towards higher cocoa content chocolate products (Afoakwa 2010; ICCO 2012b). Other market trends such as growing interest in 'ethically' produced chocolates (organic, Fairtrade, rainforest) have marginally increased demand for beans produced according to specific requirements. These trends suggest an increased demand for cocoa beans produced under more controlled conditions, whether for quality or certification (organic, Fairtrade) purposes (ICCO 2012b).

Fairtrade is concerned with ensuring a fair price and fair working conditions for producers and suppliers, promoting equitable international trading agreements. Over the past decade Fairtrade has experienced considerable growth in the food sector with direct influence on Fairtrade cocoa sourcing and supply. This growth has been significantly aided by labelling and certification through the Fairtrade Foundation mark and its availability in the mainstream cocoa marketing system. Sales in Fairtrade cocoa have increased remarkably over the past decade with annual purchases increasing progressively from 1996 to 2012, and almost doubling between 2005 and 2012. In 2006, Fairtrade cocoa attracted a relatively larger market worldwide with annual purchases of 10,919 MT representing about a 93% increase in sales of 2005, an indication that the sustained Fairtrade certification process has been a viable strategy to achieving the objectives of ethical trading. Paying premium prices means that Fairtrade cocoa in niche markets is positioned as premium-priced produce in the market. Further progress made within the Fairtrade and organic cocoa industries would be examined. As well, the entire Fairtrade labelling and certification systems provides an overview of trends in world sourcing, marketing systems and supply chain management of Fairtrade cocoa over the past decade.

Projections for the next five years predict that cocoa prices will remain steady, with both supply and demand increasing by about 3% per year. Africa has been and is projected to remain the principal cocoa producer with 70% market share, assisted by recent improvements to political and social conditions in Côte d'Ivoire. Another predicted growth factor is the continued increase of chocolate consumption in Asian markets (ICCO 2012a,b). Furthermore, the market share held by dark and specialty chocolate is expected to continue to increase, thus also increasing demand for quality cocoa beans. At the same time, concerns have been raised over the impact of climate change, the international economic downturn and a growing awareness of child labour on cocoa production and prices. These all have the potential to reduce supply, or decrease prices gained at market (COPAL 2008).

1.3 POST-HARVEST TREATMENTS AND COCOA BEAN QUALITY

Market trends have fuelled the overall demand for cocoa beans; at the same time, much greater attention is being paid to the quality of the cocoa beans being produced worldwide. Over the past 50 years, much of the research into cocoa bean fermentation, drying and processing has been aimed at solving certain quality or flavour problems. This book also outlines the progress that has been made in improving cocoa quality, focussing on the role of fermentation and to a lesser extent, drying.

The impact of post-harvest treatment on fresh cocoa beans and the effects of these treatments on fermentation and final bean quality have been investigated. Three basic processes have been evaluated for the treatment of fresh cocoa beans prior to fermentation: pod storage, mechanical depulping and enzymatic depulping. All three of these treatments were developed or investigated in attempts to reduce the problem of acidity in dried fermented cocoa beans. Over-acidity in processed cocoa beans has been linked to the production of high levels of lactic and acetic acid during fermentation. By removing a portion of the pulp, or reducing the fermentable sugar content of the beans, it has been shown that less acid is produced during fermentation, leading to less acidic beans (Duncan et al. 1989; Sanagi, Hung, and Yasir 1997). Removal of up to 20% of the cocoa pulp from fresh Brazilian cocoa beans significantly improved the flavour quality of the beans produced (Schwan and Wheals 2004). Methods for mechanically depulping fresh cocoa beans include presses (Rohan 1963; Wood and Lass 1985), centrifuges (Schwan, Rose, and Board 1995) or simply spreading beans onto a flat surface for several hours prior to fermentation, causing a significant increase in the sweating produced in the first 24 hours of fermentation. In addition to reducing acidity, benefits of depulping include shorter fermentations and increased efficiency and the ability to use the excess pulp in the manufacture of jams, marmalade, pulp juice, wine or cocoa soft drinks (Buamah, Dzogbefia, and Oldham 1997; Schwan and Wheals 2004; Dias et al. 2007; Afoakwa 2010).

Storage of cocoa pods before the beans are removed for fermentation can also be beneficial to fermentation outcomes (Sanagi et al. 1997). It has been shown that upon storage, the pulp volume per seed decreases, due to water evaporation and inversion of sucrose (Biehl et al. 1989) and the total sugar content is diminished, reducing acid production during fermentation. The flavour quality of Malaysian beans was improved by pod storage for up to 21 days prior to fermentation (Barel et al. 1987; Duncan et al. 1989; Aroyeun, Ogunbayo, and Olaiya 2006). Findings from our recent work on Ghanaian cocoa also revealed that storage of cocoa pods for five days after harvest enhances the fermentative quality of the beans and as well reduces the fermentation time from six days to four days (Afoakwa et al. 2012).

Generally, quality may be considered as a specification or set of specifications which are to be met within given tolerances or limits. However, in the context of cocoa quality, it is used to include not just the all-important aspects of flavour and purity, but also physical characteristics that have direct bearing on manufacturing performance, especially yield of cocoa nib (Biscuit, Cake, Chocolate and Confectionery Alliance, BCCCA 1996). The different aspects or specifications of quality in cocoa therefore include: flavour, purity or wholesomeness, consistency, yield of edible material and cocoa butter characteristics.

The quality of cocoa beans is an important trade parameter because the quality of chocolate depends to a large extent upon the quality of the cocoa beans used to make the chocolate. After cocoa is harvested, the beans have to be fermented and dried, a process which enables them to develop the characteristic cocoa flavour after they have been roasted. Nearly all exported cocoa is sold on the international markets in London, New York and Paris. Inasmuch as chocolate is sold in a very competitive market, manufacturing companies would like to buy the best quality cocoa. Fine and flavour cocoas have distinctive aroma and flavour characteristics and are therefore sought after by chocolate manufacturers but they represent only 5% of global cocoa production. Generally to make good quality chocolate, cocoa beans must have cocoa flavour potential; be free from off flavours such as smoky and mouldy flavours; should not be excessively acidic, bitter or astringent; should have uniform sizes and on the average weigh 1 gram, should be well fermented, thoroughly dry with a moisture content of between 6 and 8%; have a free fatty acid content of less than 1%; cocoa butter content of 50 to 58%; shell content of less than 11 to 12% and be free from live insects, foreign objects, harmful bacteria, and pesticide residue. In recent times, concerns regarding the chemical and microbial safety of cocoa beans continue to emerge in the global cocoa trade. These largely focus on the presumably high concentrations of pesticide residues and ochratoxins (OTA) in cocoa beans produced within the West African sub-region.

That notwithstanding, questions continue to be raised by various organisations involved in the cocoa business as well as manufacturers and consuming countries on the quality, sustainability and traceability of cocoa. Such concerns are not new, but have led to several discussions over the past decades which laid the foundations for the quality assessment of cocoa beans used today. Recent developments on the emergence of Southeast Asia as a new block in the cocoa market and the continuous increasing production capacities by the old players, together with cocoa processors and the consuming public wanting even higher standards, have regenerated these concerns. It is thus important for cocoa producers across the globe to understand the factors that can bridge the gap in the sustainable production of high-quality cocoa beans for the international market. Many of these concerns stem from the fact that the major cocoa-producing countries use far different production and post-harvest practices and strategies which are inconsistent and unharmonised. This is because the factors leading to sustainable production of high-quality cocoa beans including cocoa genotype, environmental conditions and post-harvest treatments are less well understood.

1.4 CONCEPT OF THIS BOOK

This book provides overviews of up-to-date scientific and technical explanations of the technologies and approaches to modern cocoa production practices, global production, grinding, stocks, surplus and consumption trends as well as principles of cocoa processing and chocolate manufacture. Principally, it provides detailed information on the origin, history and taxonomy of cocoa, as well as Fairtrade and organic cocoa industries and their influence on the livelihoods and cultural practices of smallholder farmers. It discusses some of the factors that promote production, sustainability and traceability of high-quality cocoa beans for the global confectionery industry.

The chapters broadly cover the traditional and some modern cocoa cultivation practices, growth, pod development, harvesting and post-harvest treatments with special emphasis on cocoa bean composition, genotypic variations in the bean and their influence on flavour quality, post-harvest pre-treatments (pulp pre-conditioning by pod storage, mechanical and enzymatic depulping and bean spreading), fermentation techniques, drying, storage and transportation. Details of the cocoa fermentation processes as well as the biochemical and microbiological changes involved and how these influence flavour formation and development during industrial processing are discussed. Much attention is also given to the cocoa trading systems, bean selection and quality criteria.

Some chapters cover the scientific and technological explanations of the various processes involved in industrial processing of the fermented and dried cocoa beans into liquor, cake, butter and powder. These include cleaning and sorting, winnowing, sterilisation, roasting, alkalisation, grinding, liquor pressing into butter, deodorisation and cocoa powder production. It also covers the general principles of industrial chocolate manufacture, with detailed scientific explanations on the various stages of chocolate manufacturing processes including mixing, refining, conching, tempering/fat pre-crystallisation systems and moulding. The discussions also cover the factors that influence the quality characteristics of finished chocolates, quality parameters, post-processing defects and preventive strategies for avoiding post-processing quality defects in chocolate. These in tandem with the earlier discussions provide innovative techniques related to sustainability and traceability in high-quality cocoa production as well as new product development with significance for cost reduction and improved cocoa bean and chocolate product quality.

The ideas and explanations provided in this book evolved from my research activities on cocoa and the various interactions I have had with cocoa farmers in Ghana, who produce the bulk cocoa beans with the highest quality worldwide, and smallholder farmers from many other countries across the world as well as other stakeholders engaged in the production, storage, marketing, processing and manufacturing of cocoa and chocolate products. It contains detailed explanations of the technologies that could be employed to assure sustainable production of high-quality and safe cocoa beans for the global confectionery industry. With opportunities for improvements in quality possible through improved production practices and more transparent supply chain management, plant breeding strategies and new product development associated with Fairtrade, organic and the development of niche premium quality products, there is a need for greater understanding of the variables as well as the science and technologies involved.

It is hoped that this book will be a valuable resource for academic and research institutions around the world, and as a training manual on the science and technology of cocoa production and processing and chocolate manufacture. It is aimed at cocoa producers, traders and businesses as well as confectionery and chocolate scientists in industry and academia, general practising food scientists and technologists and food engineers. The chapters on research developments are intended to help generate ideas for new research activities relating to process improvements and product quality control and assurance, as well as development of new niche/premium cocoa and chocolate products.

It is my vision that this book will inspire all bulk cocoa-producing countries across the world to strive at producing high-quality cocoa beans, similar to those of Ghana beans, for the global cocoa market and as well inspire many local and multinational cocoa-processing industries in their quest for adding value to the many raw materials that are produced within these countries, especially the magic beans from the golden pod, cocoa.

2 History and Taxonomy of Cocoa

2.1 HISTORY OF COCOA

Cocoa (*Theobroma cacao* L.) is a native species of tropical humid forests on the lower eastern equatorial slopes of the Andes in South America. Allen (1987) reported the centre of genetic diversity of *T. cacao* to be the Amazon Basin region of South America and all the 37 collecting expeditions listed by End, Wadsworth, and Hadley (1990) to seek germplasm of wild cacao were to the Amazon Basin region. The word cacao is derived from the Olmec and the subsequent Mayan languages (*Kakaw*) and the chocolate-related term *cacahuatl* is Nahuatl (Aztec language) derived from Olmec/Mayan etymology (Dillinger et al. 2000). Cocoa was considered divine in origin, and in 1737 the Swedish botanist Carolus Linneaus named the cocoa tree *Theobroma cacao*, now its official botanical name, from the Greek word *ambrosia* (Alvim 1984; Anon 2008). Based on archaeological information, Purdy and Schmidt (1996) reported that the Mayans cultivated cocoa 2,000–4,000 years before Spanish contact. It is recorded that cocoa was domesticated and consumed for the first time by the Maya and Aztecs. The Maya, Olmec, Toltec and Aztecs used the beans of cocoa as both currency and as the base for a bitter drink (Purdy and Schmidt 1996; Nair 2010; Anon 2011).

The name cocoa is a corruption of the word *cacao*, which originated from the Amazons in South America. Its cultivation and value spread in ancient times throughout central and eastern Amazonia and northwards to Central America (Afoakwa 2010). Cocoa was first cultivated by the Aztecs in Mexico, South America, and spread throughout the Caribbean islands. Later in the 1520s, Hernán Cortés, a Spaniard, took cocoa to Spain as a beverage and to Spanish Guinea as a crop. The Spanish not only took cocoa to Europe, they introduced the crop into Fernando Pó in the seventeenth century, and thus laid the foundation of the future economies of many West African countries. Currently, West Africa produces ~75% of world cocoa (ICCO 2012a,b).

The use of cocoa beans dates back at least 1,400 years (Rössner, 1997), when Aztecs and Incas used the beans as currency for trading or to produce the so-called *chocolatl*, a drink made by roasting and grinding cocoa nibs, mashing with water, often adding other ingredients such as vanilla, spices or honey. In the 1520s the drink was introduced to Spain (Minifie 1989) although Coe and Coe (1996) emphasised that the European arrivals in the New World, including Christopher Columbus and Cortés were unimpressed with the Mayan beverage, sweetening it with honey. Nevertheless, conquistadors familiarised the chocolate beverage throughout Europe and being expensive, it was initially reserved for consumption by the highest social classes and only in the seventeenth century did consumption of chocolate spread throughout Europe. After the conquest of Central America in 1521, Hernán Cortés and his conquistadores took a small cargo of cocoa beans to Spain in 1528, together with utensils for making the chocolate drink. By 1580 the drink had been popularized in the country and consignments of cocoa were regularly shipped to Spain. The popularity of chocolate as a drink spread quickly throughout Europe, reaching Italy in 1606, France in 1615, Germany in 1641 and Great Britain in 1657 (Fowler 2009; Afoakwa 2010).

Large-scale cultivation of cocoa was started by the Spanish in the sixteenth century in Central America. It spread to the British, French and Dutch West Indies (Jamaica, Martinique and Surinam) in the seventeenth century and to Brazil in the eighteenth century. From Brazil it was taken to São Tomé and Fernando Pó (now part of Equatorial Guinea) in 1840; and from there to other parts of West Africa, notably the Gold Coast (now Ghana), Nigeria and the Côte d'Ivoire. The cultivation of cocoa later on spread to the Caribbean islands, Asia, and Africa. It is currently grown on a number of Pacific islands, including Papua New Guinea, Fiji, Solomon Islands, Samoa, and Hawaii (Hebbar, Bittenbender, and O'Doherty 2011). In Ghana, available records indicate that the Dutch missionaries planted cocoa in the coastal areas of the then Gold Coast as early as 1815, and in 1857 Basel missionaries also planted cocoa at Aburi (Ghana Cocoa Board 2010). However, these did not result in the spread of cocoa cultivation until Tetteh Quarshie, a native of Osu, Accra, who had travelled to Fernando Pó and worked there as a blacksmith, returned in 1879 with *Amelonado* cocoa pods and established a farm at Akwapim Mampong in the Eastern Region. Farmers bought pods from his farm to plant and cultivation spread from the Akwapim area to other parts of the Eastern Region (Ghana Cocoa Board 2010). In 1886, Sir William Bradford Griffith, the governor, also arranged for cocoa pods to be brought in from São Tomé, from which seedlings were raised at Aburi Botanical Gardens and distributed to farmers. In recognition of the contribution of cocoa to the development of Ghana, the government in 1947 established the Ghana Cocoa Board (COCOBOD) as the main government agency responsible for the development of the industry. Currently, there are six cocoa-growing regions in Ghana, namely Ashanti, Brong Ahafo, Eastern, Volta, Central and Western regions. Ghana is the world's second largest producer of cocoa beans, producing approximately 20% of the world's cocoa (ICCO 2013a).

As the consumption of chocolate became more and more widespread during the eighteenth century, the Spanish monopoly on the production of cocoa soon became untenable and plantations were soon established by the Italians, Dutch and Portuguese. At this point, chocolate was still consumed in liquid form and was mainly sold as pressed blocks of a grainy mass to be dissolved in water or milk to form a foamy chocolate drink. The mass production of these chocolate blocks also began in the eighteenth century when the British Fry family founded the first chocolate factory in 1728 using hydraulic equipment to grind the cocoa beans. The first US factory was built by Dr James Baker outside Boston a few decades later and in 1778 the Frenchman Doret built the first automated machine for grinding cocoa beans. The production of cocoa and chocolate was truly revolutionized in 1828 by the invention of Coenraad Van Houten of a cocoa press which succeeded in separating cocoa solids from cocoa butter. The resulting defatted cocoa powder was much easier to dissolve in water and other liquids and paved the way, in 1848, for the invention of the first real 'eating chocolate', produced from the addition of cocoa butter and sugar to cocoa liquor (Dhoedt 2008).

In the United Kingdom in 1847, Joseph Fry was the first to produce a plain eating chocolate bar, made possible by the introduction of cocoa butter as an ingredient (Beckett 2000). Demand for cocoa then sharply increased, and chocolate processing became mechanised with the development of cocoa presses for production of cocoa butter and cocoa powder by Van Houten in 1828, milk chocolate in 1876 by Daniel Peters, who had the idea of adding milk powder, an invention of Henri Nestlé, a decade earlier. This was followed by the invention of the conching machine in 1880 by Rudolphe Lindt, from where chocolate came to take on the fine taste and creamy texture we now associate with good quality chocolate. It was still very much an exclusive product, however, and it was not until 1900 when the price of chocolate's two main ingredients, cocoa and sugar, dropped considerably that chocolate became accessible to the middle class. By the 1930s and 1940s, new and cheaper supplies of raw materials and more efficient production processes had emerged at the cutting edge of innovation with fast manufacturing technologies and new marketing techniques through research and development by many companies in Europe and the United States, making chocolate affordable for the wider populace. Chocolate confectionery is now ubiquitous with consumption averaging 8.0 kg/person per annum in many European countries.

2.2 TAXONOMY OF COCOA

Cocoa (*Theobroma cacao*) belongs to the genus *Theobroma* and it is classified under the sub-family *Sterculioidea* of the mallow family Malvaceae. *Cacao*, together with the kola nut, was once classified under the now obsolete family *Sterculiaceae*. The name given to the plant provides an indication of how valuable it is; the generic name *Theobroma* is derived from the Greek for 'food of the gods'; from θεος (*theos*), meaning 'god', and βρῶμα (*broma*), meaning 'food' (Wikipedia 2013).

There are 22 known species assigned to the genus *Theobroma*; and out of these, *Theobroma cacao* is the only species widely cultivated outside its native range of distribution (Hebbar et al. 2011) and is reported to be the only species of economic importance. The 22 species are subdivided into six sections based on their morphological characters:

1. Andropetalum (*T. mammosum*)
2. Glossopetalum (T. angustifolium, T. canumanense, T. chocoense, T. cirmolinae, T. gradiflorum, T. hylaeum, T. nemorale, T. obovatum, T. simiarum, T. sinuosum, T. stipulatum, T. subincanum)
3. Oreanthes (T. bernouillii, T. glaucum, T. speciosum, T. sylvestre, T. velutinum)
4. Rhytidocarpus (*T. bicolor*)
5. Telmatocarpus (*T. gileri, T. microcarpum*)
6. Theobroma (*T. cacao*)

(Figueira et al. 2002; Hebbar et al. 2011).

The 22 species are grown in Brazil except those of Andropetalum (*T. mammosum*). *T. grandiflorum, T. obovatum, T. speciosum, T. sylvestre, T. subincanum, T. microcarpum, T. bicolor* and *T. cacao* are native to the Amazon basin of Brazil (Figueira et al. 2002). All these species have at least one fatty acid component similar to that of *T. cacao* (Figueira et al. 2002). The composition of the fatty acids in terms of palmitic acid for the *theobroma* species differs from *T. cacao* whereas at least one of the other fatty acids is similar to that of *T. cacao*. For example, species from the section Glossopetalum have stearic acid content similar to that of *T. cacao* whereas *T. sylvestre* and *T. microcarpum* have oleic acid content similar to that of *T. cacao* (Figueira *et al.* 2002). The chemical composition of the nibs of *T. sylvestre* and *T. speciosum* in terms of fatty acid composition are similar to *T. cacao* (Carpenter et al. 1994; Figueira et al. 2002; Quast, Luccas, and Kieckbusch 2011).

Cupuassu (*Theobroma grandiflorum*) is a fruit native to the Amazon region (Quast et al. 2011; Figueira et al. 2002). Among the *Theobroma* species, *Theobroma grandiflorum* has the largest fruit, with the unfermented seeds containing about 84% moisture and 60% fat on dry weight basis (Quast et al. 2011). *Theobroma grandiflorum* has found applications in the food, pharmaceutical and chemical industries. The fat is found to be an alternative fat substitute for cocoa in chocolate production (Figueira et al. 2002; Lannes, Medeiros, and Gioielli 2003; Medeiros et al., 2006). Just as with cocoa, *Theobroma grandiflorum* seeds are fermented, dried, deshelled and nibs milled to obtain cupuassu liquor, which is used for a Brazilian product called *cupulate*, which has nutritional and sensorial characteristics that are very close to chocolate (Oliveira et al. 2004).

Several other species are cultivated or wild-harvested on a relatively small scale for human consumption. These are *T. bicolor* (mocambo, pataste), *T. grandiflorum* (cupuaçu), and to a lesser extent, *T. speciosum* and *T. subincanum* (Hebbar et al. 2011). Nair (2010) reported that *Theobroma bicolor* Humb. and Bonpl. are cultivated for the edible pulp around the beans, and the beans are used like those of cocoa. The beans of *Theobroma angustifolium* Moc. and Sesse. are mixed with cocoa in Mexico and Costa Rica and the sweet pulp around the beans of *Theobroma grandiflorum* (Wild. ex Spreng.) Schumann is used for making a drink in parts of Brazil and is also eaten.

2.3 MORPHOLOGICAL AND VARIETAL CHARACTERISTICS OF COCOA

2.3.1 Cocoa Plant

The cocoa plant is usually a small tree, 4 to 8 metres tall, although when shaded by large forest trees it may reach up to 10 metres in height. The stem is straight, the wood light and the bark is thin, somewhat smooth and brownish. The fruit (pods) reach up to 15–25 cm in length. The mature fruit or pod consists of a comparatively thick husk containing between 30 to 50 seeds embedded in a thick mucilaginous pulp. All cultivated cocoas show great variability and it is generally agreed that they can be divided within the species. The principal varieties of the cocoa tree *Theobroma cacao* are

1. *Forastero* from the Amazonas region, and grown mainly in West Africa as bulk cocoa
2. *Criollo*, rarely grown because of disease susceptibility
3. *Trinitario*, a hybrid of *Forastero* and *Criollo*
4. *Nacional* with fine flavour, grown in Ecuador

Forastero varieties form most of the 'bulk' or 'basic' cocoa market. World annual cocoa bean production is approximately 4.25 million metric tonnes and major producers are the Côte d'Ivoire, Ghana, Indonesia, Brazil, Nigeria, Cameroon and Ecuador. There are also a number of smaller producers, particularly of 'fine' cocoa, which forms less than 5% of world trade. Currently in 2013, West Africa alone produces ~75% of global production with Côte d'Ivoire and Ghana producing approximately 38% and 22%, respectively, totalling about 60%.

2.3.1.1 Forastero Cocoa

Forastero means 'foreigner' in Spanish and refers to any cocoa trees that are not *Criollo* or a hybrid and usually produce deep purple seeds (Hebbar et al. 2011). *Forastero* is native to the Amazon region and largely grown in West Africa and Southeast Asia. It forms 95% or the 'bulk' of the world production of cocoa (Afoakwa 2010; Fowler 2009; Delonga et al. 2009; Afoakwa et al. 2012) and is the most widely used due to its higher yield than the *Criollo* variety. *Forastero* varieties exhibit greater variability both in tree and fruit morphology. The pods when ripe are hard, yellow and have a more rounded shape like a melon (Figures 2.1 and 2.2) containing 30 or more pale to deep purple beans. This variety is generally more vigorous and less susceptible to diseases such as swollen shoot, mottle leaf, yellow mosaic, cocoa necrosis, witches broom and black pod (Dzahini-Obiatey, Domfah, and Amoah 2010; Afoakwa 2010) as well as pests such as capsids and cocoa pod borer (*Conopomorpha cramerella*; Fowler 2009; Afoakwa 2010; ICCO 2012a,b) than the *Criollo* variety. *Forastero* cocoa beans are

FIGURE 2.1 Typical unripe *Forastero* cocoa pods.

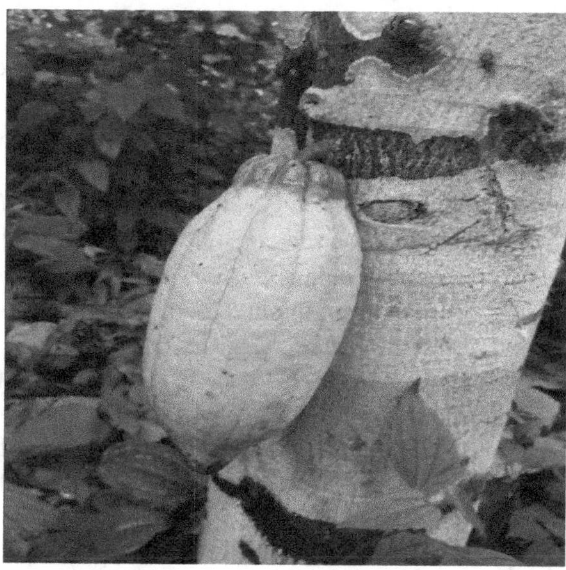

FIGURE 2.2 Typical ripened *Forastero* cocoa pod.

characterized by darker brown cotyledons which are slightly bitter but have the strongest flavour. Chocolate produced from these beans is rich in chocolate flavour but low in complex or fruity flavour notes (DeZaan Cocoa Manual 2009; Hebbar et al. 2011).

Several cultivars of *Forastero* are grown in Ghana. The main percentage of cultivars are *Amelonado* (13.3%) and *Amazonica* (34.4%) including a new hybrid, the mixed hybrid (52.3%). The farmers in Ghana locally call the mixed hybrid variety *akokora bedi* which literally means 'the aged will surely enjoy'. This is due to the short duration it takes to begin bearing fruits. *Amelonado* is the *Forastero* variety widely grown in West Africa (Fowler 2009; Hebbar et al. 2011) with the varieties including *Comum* in Brazil, *West African Amelonado* in Africa, *Cacao Nacional* in Ecuador and Matina or *Ceylan* in Costa Rica and Mexico (ICCO 2012b).

The *Forastero* type of cocoa now forms the greater part of all cocoa grown and is hardy and vigorous, producing beans with the strongest flavour. It is a much more plentiful variety of high-quality cocoa, representing most of the cocoa grown in the world. Grown mainly in Brazil and Africa, it is hardier, higher yielding and easier to cultivate than *Criollo* and is used in just about every blend of chocolate that is made. The pods are short, yellow, smooth without warts, with shallow furrows and have 30 or more pale to deep purple beans.

2.3.1.2 Criollo Cocoa

Criollo refers to a group of genetically similar trees that produce lightly pigmented seeds and share several other morphological traits (Hebbar et al. 2011). This variety exhibits symptoms of inbreeding depression and has a history of low vigour, poor productivity, high susceptibility to diseases, insects and stress attack; hence, it is less cultivated (Afoakwa 2010; Hebbar et al. 2011). This type is now very rare and only found in old plantations in Venezuela, Central America, Madagascar, Sri Lanka and Samoa (Fowler 2009). Because it is of the highest quality cocoa bean, it is less bitter and more aromatic and thus has a milder and nuttier cocoa flavour (Fowler 2009; DeZaan Cocoa Manual 2009; Rusconi and Conti 2010) than any other bean, hence, it is highly priced. The yield of *Criollo* cocoa plantation is lower than that of a *Forastero* plantation of the same size. The fruits of the *Criollo* variety typically have a soft thin husk or pod with a textured surface and usually have some degree of red pigmentation with 20–30 white or faint purple beans. When *Criollo* pods are ripe, they are long, yellow or red, with deep furrows and big warts (Figure 2.3).

2.3.1.3 Trinitario Cocoa

The *Trinitario* cocoa variety is a hybrid between the *Criollo* and *Forastero* varieties. *Trinitario* was developed in Trinidad (Willson 1999; DeZaan Cocoa Manual 2009) hence the name, but later spread to Venezuela, Ecuador, Cameroon, Samoa, Sri Lanka, Java and Papua New Guinea (ICCO 2012b). This variety is of much higher quality than the *Criollo* and is of higher yield and more resistant to diseases than the former *Forastero* (Mossu 1992; Afoakwa 2010; Dand 1996). Some *Trinitario* varieties produce cocoa beans with special flavours. They have mostly hard pods and

FIGURE 2.3 Typical *Criollo* cocoa.

(a) (b)

FIGURE 2.4 Typical *Trinitario* cocoa pods.

are variable in colour, could be long or short and they contain 30 or more beans of variable colour (Figure 2.4) but white beans are rare.

2.3.1.4 Nacional Cocoa

The *Nacional* cocoa variety grows in Ecuador. It is believed to have originated from the Amazonian area of Ecuador (Fowler 2009; Afoakwa 2010; DeZaan Cocoa Manual 2009) and has a distinctive aroma and flavour characteristics (Afoakwa 2010; Hebbar et al. 2011) but is less cultivated and hence contributes about 5% of global cocoa production. Currently, pure *Nacional* cocoa varieties are rare. The ones with *Arriba* flavour in Ecuador are hybrids between *Nacional* and *Trinitario* (Fowler 2009). Typical *Nacional* cocoa pods are shown in Figure 2.5. Research conducted in 2008 in Latin America suggested a new classification of cacao germplasm into 10 major groups. These are *Marañon*, *Curaray*, *Criollo*, *Iquitos*, *Nanay*, *Contamana*, *Amelonado*, *Purús*, *Nacional* and *Gulana* (DeZaan Cocoa Manual, 2009; Motamayor et al. 2008). This new classification reflects much more accurately the genetic diversity of cacao. Important characteristic differences between typical *Criollos*, *Forastero* and *Trinitario* cocoas are presented in Table 2.1.

2.4 VARIETAL EFFECT ON COCOA BEAN FLAVOURS

The different cocoa bean genotypes or varieties discussed above influence both flavour quality and intensity in chocolate during manufacturing

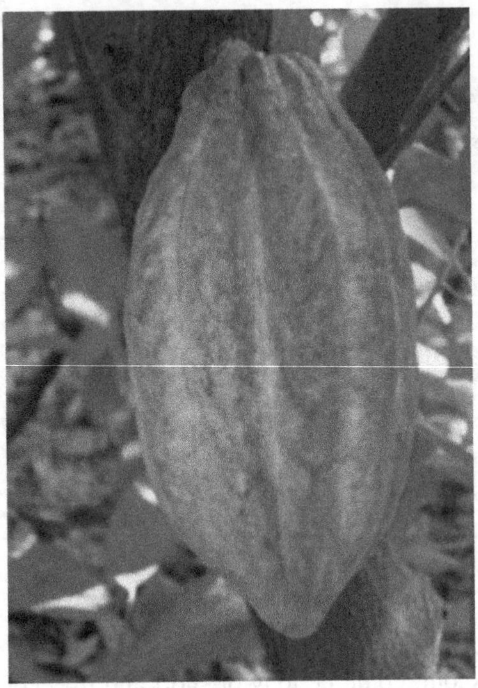

FIGURE 2.5 Typical *Nacional* cocoa pod.

(Taylor 2002, Luna et al. 2002; Counet et al. 2004). The differences are largely due to the wide differential in chemical composition of the derived beans, likely determining the quantities of flavour precursors and activity of enzymes, and thus contributions to flavour formation. Reineccius (2006) concluded that varietal differences were primarily due to quantitative (as opposed to qualitative) differences in flavour precursor and polyphenol contents. Contents of sugars and enzymic breakdown of polysaccharides form an important source of precursors. However, post-harvest processes (fermentation and drying), and roasting have a strong influence on final flavours (Kattenberg and Kemming 1993; Clapperton et al. 1994; Luna et al. 2002; Counet and Collin 2003). Three primary cocoa types: *Forastero* (bulk grade), *Criollo* (fine grade), and hybrid, *Trinitario* (fine grade) show wide variations in final flavour (Beckett 2000; Awua 2002; Amoye 2006). *Nacional* cacao is viewed as a third fine variety: producing the well-known *Arriba* beans with distinctive floral and spicy flavour notes (Despreaux 1998; Luna et al. 2002; Counet et al. 2004). These differences in flavour can be ascribed to bean composition variation from botanical origin, location of growth and farming conditions. Bulk varieties dominate blends and fine grades, used in lesser quantities, are selected to make specific contributions to overall flavour profile.

TABLE 2.1
Characteristics of the Different Cocoa Varieties

Characteristics		*Criollo*	*Forastero*	*Trinitario*
Pod husk	Texture	Soft, crinkly	Hard, Smooth	Mostly hard
	Colour	Red occurs	Green	Variable
Beans	Average no. per pod	20 to 30	30 or more	30 or more
	Colour of cotyledons	White, ivory or very pale purple	Pale to deep purple	Variable; white beans rarely
Agronomic	Tree vigour	Low	Vigorous	Intermediate
	Pest and disease susceptibility	Susceptible	Moderate	Intermediate
Quality	Fermentation need	1 to 3 days maximum	Normally 5 days	4 to 5 days
	Flavour	Weak chocolate; mild and nutty	Good chocolate	Good chocolate; full cocoa
	Fat content	Low	High	Medium
	Bean Size (g/100 beans)	85	94	91

Each bean variety has a unique potential flavour character. But growing conditions such as climate, amount and time of sunshine and rainfall, soil conditions, ripening, time of harvesting, and time between harvesting and bean fermentation all contribute to variations in final flavour formation. Table 2.2 summarises how differences in genetic origin, cocoa variety and duration of fermentation influence flavour profile but different conditions may lead to significant differences in flavour from a single cocoa variety. A good example is the difference in flavour profile between a single *Forastero* variety produced originally in Ghana and now grown in Malaysia (Clapperton 1994), arising possibly through geographic climatic conditions and duration or method of fermentation.

Bulk cocoas typically show strong flavour characters; fine cocoas are perceived as aromatic or smoother (Kattenberg and Kemming 1993; Jinap, Dimick, and Hollender 1995; Luna et al. 2002). Clapperton et al. (1994) noted consistent differences in flavour attribute specifically overall cocoa flavour intensity, acidity, sourness, bitterness and astringency. Bean origins include the West African *Amelonado* variety (AML), four Upper Amazon clones [*Iquitos Mixed Calabacillo* 67 (IMC67), *Nanay* 33 (NA33), *Parinari* 7 (PA7), *Scavina* 12 (SCA12)] and *Unidentified*

TABLE 2.2
Origin, Cocoa Variety and Fermentation Duration Effects on Flavour Character

Origin	Cocoa Type	Duration (Days)		Special Flavour Character
Ecuador	*Nacional (Arriba)*	2	Short	Aromatic, floral, spicy, green
Ecuador	*Criollo (CCN51)*	2		Acidic, harsh, low cocoa
Ceylon	*Trinitario*	1.5		Floral, fruity, acidic
Venezuela	*Trinitario*	2		Low cocoa, acidic
Venezuela	*Criollo*	2		fruity, nutty
Zanzibar	*Criollo*	6	Medium	Floral, fruity
Venezuela	*Forastero*	5		Fruity, raisin, caramel
Ghana	*Forastero*	5		Strong basic cocoa, fruity notes
Malaysia	*Forastero/ Trinitario*	6		Acidic, phenolic
Trinidad	*Trinitario*	7-8	Long	Winy, raisin, molasses
Grenada	*Trinitario*	8-10		Acidic, fruity, molasses
Congo	*Criollo/Forastero*	7-10		Acidic, strong cocoa
Papua New Guinea	*Trinitario*	7-8		Fruity, acidic

Source: Reprinted from E.O. Afoakwa et al. *Critical Reviews in Food Science and Nutrition* 48:840–857, 2008a. With permission.

Trinitario (UIT1) grown in Sabah, Malaysia. Flavour characters in UIT1 differ from West African *Amelonado*, characterised by intense bitterness and astringency associated with caffeine and polyphenol contents. Fermented beans from Southeast Asia and the South Pacific are characterised by a higher acidity (more lactic and acetic acids) than West African beans (Clapperton et al. 1994) due to varietal differences, box fermentation and rapid artificial drying.

Cocoa liquors differ in sensory character. The West African group (Ghana, Côte d'Ivoire and Nigeria) are generally considered sources of standard (benchmark) cocoa flavour with a balanced but pronounced cocoa character with subtle to moderate nutty undertones. Cameroon liquors are renowned for bitterness, those from Ecuador for floral-spicy notes. American and West Indian varieties range from aromatic and winy notes from Trinidad cocoa to the floral or raisin-fruity notes of Ecuadorian stocks making unique contributions to blends. Asian and Oceanian beans exhibit a range of flavour profiles ranging from subtle cocoa and nutty/sweet notes in Java beans to the intense acid and phenolic notes of Malaysian (De La Cruz et al. 1995).

Counet et al. (2004) reported fine varieties with short fermentation processes had high contents of procyanidins, whereas *Trinitario* from New Guinea and *Forastero* beans were specifically higher in total aroma. Aroma compounds formed during roasting were found to vary quantitatively directly with fermentation time and inversely with procyanidin content of cocoa liquors.

High concentrations of phenol, guaiacol, 2-phenylbutenal and γ-butyrolactone characterise Bahia beans known for typical smoked notes. Also reported are higher contents of 2-methylpropanal and 3-methylbutanal in Caracas (Venezuela) and Trinidad dried fermented beans (Dimick and Hoskin 1999). Of Maillard products, Reineccius (2006) reported roasting yields higher levels of pyrazines in well-fermented beans (Ghana, Bahia) than in less-fermented (*Arriba*) or unfermented from Sanchez (Dominican Republic) or Tabasco (Mexico). Lower in astringency and bitterness imparted by polyphenols, *Criollo* beans, in which anthocyanins are absent, are often less fermented than *Forastero* (Carr, Davies, and Dougan 1979; Clapperton 1994; Clapperton et al. 1994; Luna et al. 2002).

3 World Production, Grinding and Consumption Trends of Cocoa

3.1 INTRODUCTION

Cocoa is grown principally in West Africa, Central and South America and Southeast Asia. In order of annual production size, the eight largest cocoa-producing countries at present are Côte d'Ivoire, Ghana, Indonesia, Nigeria, Cameroon, Brazil, Ecuador and Malaysia (Figure 3.1). These countries together produce about 4.1 million metric tonnes representing ~95% of world production (ICCO 2011). In the early 1970s cocoa production was concentrated in Ghana, Nigeria, Côte d'Ivoire and Brazil, but it has now expanded to areas such as the Pacific region, where countries such as Indonesia have shown spectacular growth rates in production (ICCO 2008).

About 90–95% of world production is grown by smallholders on a low input, low output basis. Typically, family or village labour is used at relatively little cost. In Ghana, cocoa is cultivated on about 1.5 million hectares of land by some 800,000 farm families in six out of the ten regions. The western region produces about 41% of Ghana's cocoa with the Volta region contributing the least (ca. 2%). It is cultivated almost exclusively by smallholder farmers with average farm sizes of about 4.0 hectares (ICCO 2008). Currently, annual cocoa production worldwide is 4.28 million tonnes and the annual increase in demand for cocoa has been 3% per year for the past 100 years (ICCO 2008; Afoakwa 2010).

3.2 WORLD PRODUCTION OF COCOA

Theobroma cacao originated in the Amazon Basin and optimal conditions for growth are 20–30°C (68–86°F), 1,500–2,500 mm of annual rainfall and 2,000 hours of sunshine per year. Table 3.1 shows density of production is centred within West Africa, now accounting for ~75% of world cocoa production in the 2011/2012 growing season. West African countries are ideal in climatic terms for growing cocoa as a cash crop.

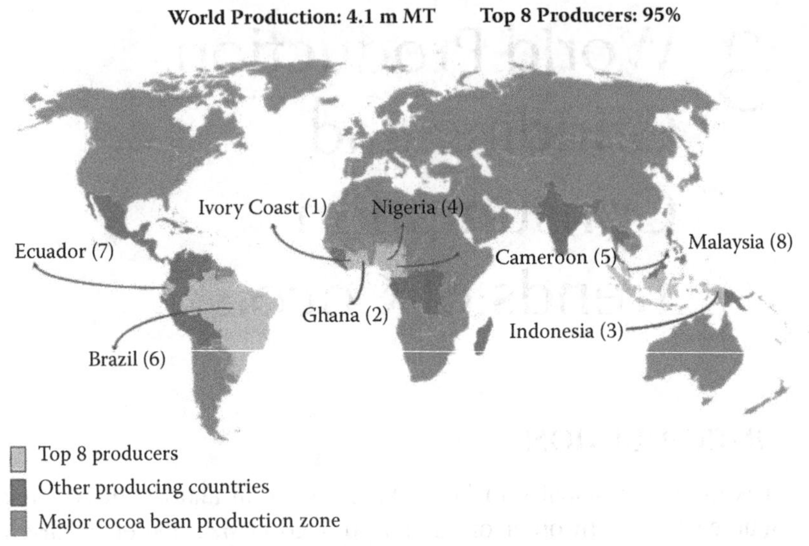

FIGURE 3.1 World leading cocoa producing countries. (Adapted from ICCO 2008. Annual report 2006/2007. International Cocoa Organisation, London UK.)

However, as a consequence, natural or man-made problems have potentially a disproportionately large impact upon the cocoa trade. Smallholders of West Africa have dominated world production since the 1930s. In the 1980s, the emergence of Malaysia and Indonesia gave a more balanced geographical spread of production. However, a period of low prices wiped out Malaysia as a major producer and Brazil as a major exporter, increasing West Africa's share of production. In 2005–2006, 71% of world cocoa came from Africa: Côte d'Ivoire, 37.8% and Ghana,19.9% (ICCO 2008).

In 2006/2007, world production of cocoa beans dropped by almost 9% from the previous season to 3.4 million tonnes, mainly as a consequence of unfavourable weather conditions in many cocoa-producing areas. West Africa, the main cocoa-producing region, was hit by a severe harmattan and its inherent dry weather, which lasted from the end of 2006 to February 2007, had a strong negative impact on production. In Asia and in South America, El Niño-related weather conditions developed in September 2006 and continued until the beginning of 2007. Cocoa production in the two major producing countries was hit severely in 2006/2007. Figure 3.2 shows the world cocoa production trends by continent from 2005/2006 to 2010/2011.

Production in Ghana declined by 17% from the previous season to 614,000 tonnes, resulting mainly from a very poor mid crop. In Côte d'Ivoire, cocoa output reached 1,292,000 tonnes, down by 116,000 tonnes from the 2005/2006 season. As in Ghana, the second harvest of the season proved very disappointing, as the trees did not recover from the poor level

TABLE 3.1
World Cocoa Production between 2004 and 2012 (1,000 tonnes)

	2004/05		2005/06		2006/07		2008/09		2009/10		2010/11		2011/2012	
Africa	2375	70.3%	2642	71.0%	2392	70.4%	2516	70.0%	2486	68.4%	3226	74.9%	2905	71.3
Cameroon	104		166		129		224		209		229		207	
Côte d'Ivoire	1286		1408		1292		1, 23		1242		1511		1486	
Ghana	599		740		614		662		632		1,025		879	
Nigeria	200		200		190		250		235		240		230	
Others	181		128		167		157		168		221		104	
America	445	13.2%	446	12.0%	411	12.1%	478	13.3%	516	14.2%	559	13.0%	639	15.7
Brazil	171		162		126		157		161		200		220	
Ecuador	116		114		114		135		150		161		190	
Others	158		170		171		186		205		199		229	
Asia & Oceania	559	16.5%	636	17.1%	597	17.5%	598	16.6%	633	17.4%	524	12.2%	531	13.0
Indonesia	460		530		490		490		550		440		450	
Papua New Guinea	48		51		50		59		39		47		45	
Others	51		55		57		48		44		37		36	
World Total	3379	100.0%	3724	100.0%	3400	100.0%	3593	100.0%	3635	100.0%	4309	100.0%	4075	100.0%

Source: ICCO, 2008; *Annual report 2006/2007*. International Cocoa Organisation, London UK. ICCO, *Annual Report of the International Cocoa Organization for 2010/2011*. 2012a; ICCO, *ICCO Quarterly Bulletin of Cocoa Statistics*, XXXIX: 1, Cocoa year 2012/13. 2013a. With permission.

Note: Totals may differ from sum of constituents due to rounding.

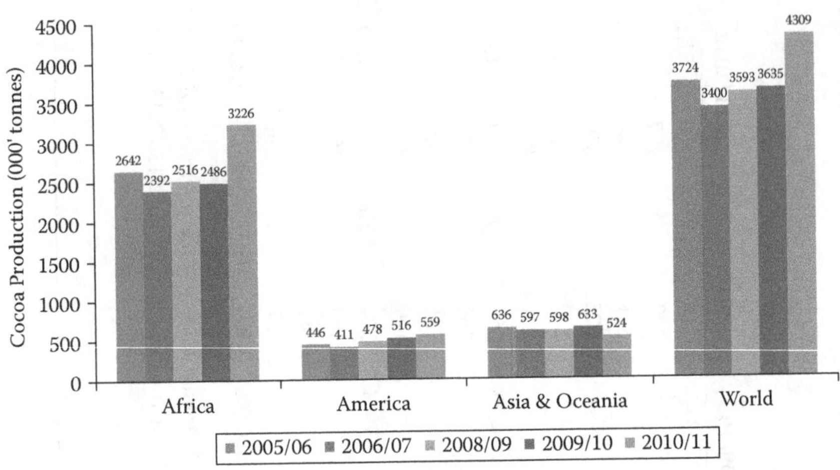

FIGURE 3.2 World cocoa production trends by continent from 2005/2006 to 2010/2011.

of soil moisture and lack of rainfall which lasted until February 2007, causing many developing pods to shrivel. The statistical picture for the mid crop in Côte d'Ivoire could have been worse. Indeed, the 2007/2008 main crop experienced an early and strong start at the end of August: almost 100,000 tonnes of cocoa beans reached Ivorian ports in September 2007. These cocoa beans were statistically counted as part of the 2006/2007 mid crop and, consequently, enhanced the production figures of the 2006/2007 cocoa season, whereas in fact, they were part of the 2007/2008 main crop (ICCO 2008).

3.3 COCOA YIELD IN PRODUCING COUNTRIES

Africa is expected to remain the world's leading cocoa-producing area for many decades. Production in Côte d'Ivoire, the world's largest cocoa bean producer, is expected to grow by 2.0% a year from 1.2 million tonnes of the base period to 1.5 million tonnes in 2014, and account for about 37% of global cocoa production due mainly to changes in foreign direct investment and by market liberalisation. However, yields in Côte d'Ivoire and Ghana are well below levels seen in Asia partly because of less use of agricultural inputs. Figure 3.3 shows the world cocoa yields in major producing countries in 2010/2011 and 2011/2012.

The recent surge in world cocoa prices has made it easier for the growers to use more inputs. If this trend continues, the volume of cocoa produced in Côte d'Ivoire could increase further. Output in Ghana, the second largest cocoa bean producer in Africa and in the world, would grow from 410,000 tonnes in 1998–2000 to 490,000 tonnes in 2010, an annual average growth

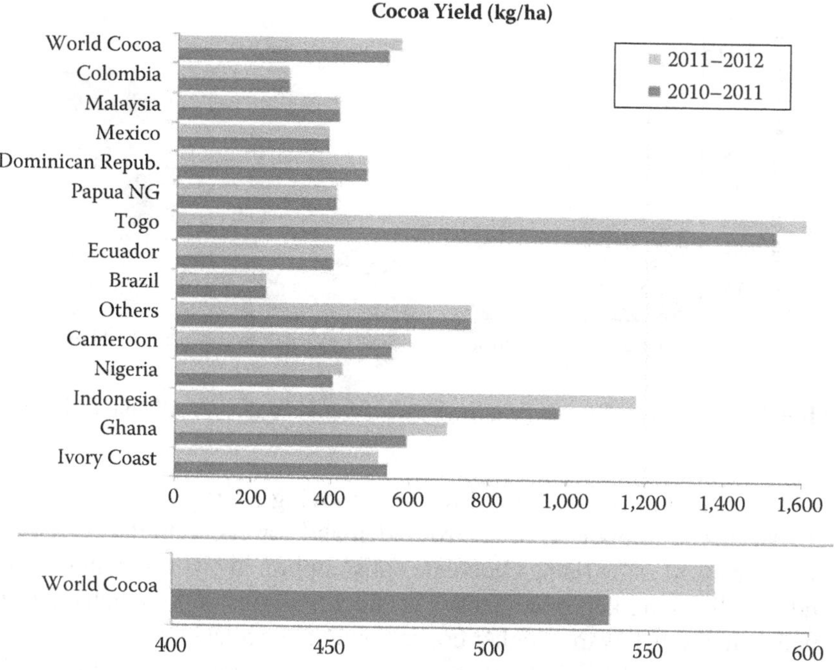

FIGURE 3.3 World cocoa yields in major producing countries in 2010/2011 and 2011/2012. (Adapted from ICCO, *Annual Report of the International Cocoa Organisation for 2010/2011*, 2012a. With permission.)

rate of 1.6%. The corresponding growth rate for the previous decade was 3.3%. The lower projected growth rate over the next decade would result from the outbreak of diseases (such as swollen shoot virus, black pod and mirids), increased competition at the world market and low export prices. Over the same period, Nigeria and Cameroon are projected to increase output by 1.4% (ICCO 2010a; Afoakwa 2010).

3.4 WORLD COCOA GRINDINGS TRENDS BETWEEN 2005/2006 AND 2011/2012

Cocoa is largely produced in developing countries but it is mostly consumed in the developed countries (Figure 3.4). The consuming countries are the countries that import and grind the cocoa into finished and semi-finished products. Few multi-national companies dominate both processing of cocoa and chocolate manufacturing (World Cocoa Foundation 2010). Cocoa consumption, as measured by grindings, increased by 2.5% from the 2005/2006 season to 3,608,000 tonnes in 2006/2007 (Table 3.2). Despite a relative slowdown during that season, the cocoa market was characterised over the last five years by a sustained demand for cocoa, rising by

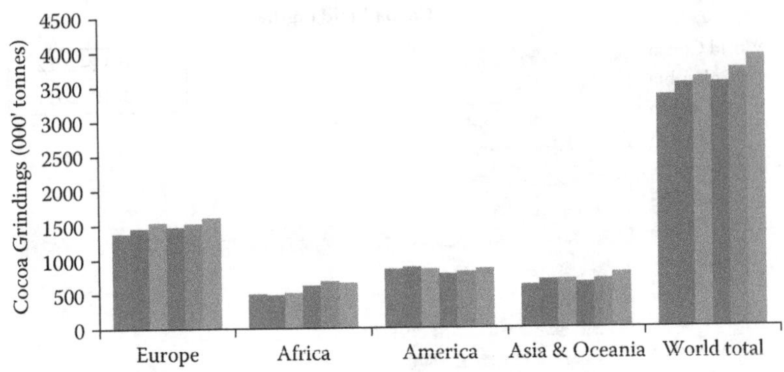

FIGURE 3.4 World cocoa grindings trends between 2004/2005 and 2010/2011.

3.8% per annum (based on a three-year moving average; ICCO 2007a,b; 2010a). It was supported by a strong demand for cocoa butter to rebuild stocks, as well as by rising chocolate consumption in emerging and newly-industrialised markets and changes in chocolate consumption behaviour in mature markets towards higher cocoa content chocolate products.

At the regional level, developments were heterogeneous in 2006/2007, with grindings rising by around 6% in Europe to 1,540,000 tonnes and to 514,000 tonnes in Africa (Table 3.2). Meanwhile, they remained at almost the same level, at 699,000 tonnes in Asia and Oceania and declined by 3% in the Americas to 853,000 tonnes. Processors located in Germany and Ghana contributed to almost half of the increase in world grindings, reflecting the installation of additional capacities in these countries. The Netherlands and the United States remained the major cocoa-processing countries, each with grindings of more than 400,000 tonnes during the year.

The strong increase in grindings recorded in 2009/2010 was due to re-stocking and recovering demand continued, albeit at a slower rate, during the 2010/2011 season. As depicted in Table 3.2, the ICCO Secretariat estimated that global cocoa grindings would increase by 5% in the 2010/2011 season, to 3.923 million tonnes, the highest grindings figure on record (the last highest level of 3.775 million tonnes being recorded in 2007/2008 prior to the start of the world economic crisis).

Cocoa-processing activity increased in most regions: the largest increase of over 12% occurred in Asia and Oceania (up by 87,000 tonnes to 795,000 tonnes) followed by nearly 6% to 1.612 million tonnes in Europe and by 5.5%, to 860,000 tonnes in the Americas. However, grindings declined by 4%, to 657,000 tonnes in Africa, mainly due to the political disruptions that occurred during the first quarter of the year in Côte d'Ivoire when processing plants were not working at full capacity and grindings fell below trend.

TABLE 3.2
Grindings of Cocoa Beans (1,000 tonnes)

	2004/05		2005/06		2006/07		2008/09		2009/10		2010/11		2011/2012 (Estimates)	
Europe	1379	41.4%	1456	41.4%	1540	42.7%	1475	41.8%	1524	40.8%	1612	41.1%	1522	38.5
Germany	235		306		357		342		361		439		407	
Netherlands	460		455		465		490		525		537		500	
Others	684		695		719		643		638		636		615	
Africa	501	14.9%	485	13.8%	514	14.3%	622	17.6%	685	18.3%	657	16.7%	712	18.0
Côte d'Ivoire	364		336		336		419		411		361		430	
Ghana	–		–		–		133		212		230		212	
Others	137		149		179		70		61		67		71	
America	853	25.4%	881	25.0%	853	23.7%	780	22.1%	815	21.8%	860	21.9%	840	21.3
Brazil	209		209		224		216		226		239		242	
United States	419		432		418		361		382		401		387	
Others	225		226		212		203		207		220		210	
Asia & Oceania	622	18.5%	698	19.8%	699	19.4%	655	18.5%	708	19.0%	795	20.3%	874	22.1
Indonesia	115		140		140		120		130		190		268	
Malaysia	249		267		270		278		298		305		297	
Others	258		291		289		256		280		299		309	
World total	3354	100%	3520	100%	3608	100%	3531	100%	3731	100%	3923	100%	3948	100%
Origin grindings	1262	37.6%	1293	36.8%	1325	36.7%	1419	40.2%	1527	40.9%	1598	40.7%	1716	43.5

Source: ICCO, *ICCO Quarterly Bulletin of Cocoa Statistics*, XXXVI: 4, Cocoa year. 2009/2010, 2010a; ICCO, Executive Committee at the 146th Meeting, 2013b. With permission

Note: Totals and differences may differ due to rounding.

The overall increase in world grindings is likely to be attributed to changing global consumption patterns in emerging countries. There are millions of new consumers in these markets including Asia where demand is focused on powder-based products. Traditionally, the higher-value product butter was the driver behind grindings growth but, in recent years, powder demand has outpaced butter demand. In emerging markets, there is a taste for lighter treats, that is, milder products based on cocoa powder, as palates are less used to chocolate. Also, the hot climate and lack of refrigeration in a market like India make it harder to stock butter-based products.

3.5 WORLD STOCKS OF COCOA BEANS

As shown in Table 3.3, world stocks of cocoa beans increased from over 1.432 million tonnes to 1.775 million tonnes at the end of the 2010/2011 cocoa year which is equivalent to 45.2% of estimated annual grindings in 2010/2011. Findings from ICCO (2012b) showed that world stocks of cocoa beans were assessed as at the end of the 2011 crop year and resulted in the following: of the total, 72% were located in cocoa-importing countries, 24% held in cocoa-producing countries and the remaining 4% in other importing countries. The concentration of cocoa storage in a few geographical locations is highlighted by the fact that 53% of global cocoa bean stocks were held in warehouses in entry ports located in Europe and 47% in the main entry ports in the Netherlands, Belgium, Germany and the United Kingdom, near the leading cocoa-processing industries in Western Europe (ICCO 2012a). Table 3.3 shows the world cocoa grindings and stocks trends between 2001/2002 and 2010/2011.

3.6 INTERNATIONAL COCOA PRICE DEVELOPMENTS

Yearly averages of international cocoa prices ranged between US $1,534 and $3,246 per tonne during the period covered. The lowest occurred in 2003/2004 when the world cocoa economy experienced a surplus of 287,000 tonnes, and the highest was reached in 2009/2010 when a deficit of 132,000 was attained (Table 3.3).

In 2002/2003, the first season, the annual average of the ICCO daily price was at US $1,873 per tonne, an increase of 19% over the previous season when it reached US $1,580 per tonne. Cocoa prices increased after an attempted coup on 19 September 2002 in Côte d'Ivoire. Concerns over potential disruptions to the flow of cocoa at the beginning of the following crop year, originating from the continuing political and social crisis in the world's leading cocoa-producing country, pushed international prices to 16-year highs at US $2,436 per tonne in October 2002.

TABLE 3.3
World Cocoa Bean Production, Grindings and Stocks

Crop Year (Oct-Sept)	Gross Crop	Grindings	Surplus/ Deficit	Total End-of-Season	Stocks to Grinding Ratios	ICCO Daily Price (annual average)	
		in 1,000 Tonnes			(%)	$US/tonne	SDRs/Tonne
2001/02	2877	2886	**–29**	1315	**45.6**	1580	1231
2002/03	3179	3077	**+80**	1395	**45.3**	1873	1369
2003/04	3548	3237	**+287**	1682	**52.0**	1534	1047
2004/05	3378	3382	**–38**	1644	**48.6**	1571	1049
2005/06	3808	3522	**+248**	1892	**53.7**	1557	1068
2006/07	3430	3675	**–279**	1613	**43.9**	1854	1226
2007/08	3737	3775	**–75**	1538	**40.7**	2516	1573
2008/09	3593	3531	**+26**	1564	**44.3**	2599	1707
2009/10	3635	3731	**–132**	1432	**38.4**	3246	2115
Estimates							
2010/11	4309	3923	**+343**	1775	**45.2**	3105	1969

Source: ICCO (2010a; ICCO, Executive Committee at the 146th Meeting, 2012b. With permission.

Note: Surplus/deficit is current net world crop (gross crop adjusted for loss in weight) minus grindings. Totals and differences may differ due to rounding.

The remaining prospect of a third successive production deficit that could reduce even further the stocks-to-grindings ratio in 2002/2003, and short-covering by trade, investment funds and speculators, also contributed to an additional increase in cocoa prices in the first half of the 2002/2003 cocoa year. Nevertheless, the harvesting, transportation and commercialisation of cocoa proceeded normally in Côte d'Ivoire, despite the prevailing political and social unrest in the country. Moreover, the higher international cocoa bean prices were closely reflected in a rising trend in farm gate prices in Côte d'Ivoire which prompted higher standards of husbandry, as well as increased sales of pesticides and fertilisers. Thus, rising yields may have helped farmers to offset the impact of civil unrest in the country.

After a three-year period of constant increases, average international cocoa prices recorded a sizeable drop of 18% in the 2003/2004 season, at US $1,534 per tonne. Concerns over potential disruptions to harvesting, evacuation and export of cocoa from Côte d'Ivoire, caused by the political and social unrest in the country, gradually lessened during the 2003/2004 season. The other major underlying factors influencing the movement of prices were weather conditions and the resulting outlook for the crops in West Africa. Although most analysts expected a global production deficit at the beginning of the season, production forecasts were progressively revised upwards towards a large surplus at the end of the season. However, the downward pattern broke sharply at the beginning of July 2004, reflecting concerns among market participants about the weather conditions in West Africa. At the end of August, futures prices reached their highest levels for the 2003/2004 season, at US $1,800 per tonne. In September, larger mid crops in West Africa and improving rain conditions contributed subsequently to a fall in prices.

At US $1,571 per tonne in the 2004/2005 cocoa year, the average international cocoa prices remained at almost the same level as during the previous season. This was mainly explained by the near-balanced supply and demand situation experienced during that campaign. Nonetheless, two rallies occurred during the season.

Despite a large supply surplus recorded in the 2005/2006 cocoa year, average international cocoa prices remained at almost the same level as during the previous season, at US $1,557 per tonne. At the beginning of the 2005/2006 season, strong arrivals of cocoa beans from the two major cocoa-producing countries, Côte d'Ivoire and Ghana, resulted in prices falling to under US $1,400 per tonne. However, with demand for cocoa in Europe remaining strong and the bulk of the main harvest almost completed in Africa by January, worries surfaced of a supply shortage. Subsequently, prices rose to a 10-month high in mid-January, at US $1,653. During the months that followed, the absence of fresh fundamental news did not move the futures market in any precise direction. The situation started to change drastically at the end of June, with international prices surging to US $1,807 in the following month. The scale and speed of the price rally took many market operators by surprise. Prices were supported by a relatively low level of stocks of tenderable cocoa in LIFFE certified warehouses in Europe. This was clearly reflected in an inverted price structure on the futures market, with nearby positions trading at a premium (backwardation) compared to the more usual contango situation. By the end of July when concerns over availability of stocks for short-term supplies eased, terminal prices fell, reaching lower levels at the end of the season than those experienced before the rally.

International prices continued to ease at the beginning of the 2006/2007 cocoa season, falling to US $1,491 in the middle of October 2006. After reaching this low point, both cocoa futures markets embarked on a rising trend. This development was supported by the global supply and demand situation. Indeed, during the 2006/2007 to 2009/2010 seasons, the cocoa market experienced three supply deficits out of four seasons: in 2006/2007 by 279,000 tonnes, in 2007/2008 by 75,000 tonnes and in 2009/2010 by 132,000 tonnes. The small supply surplus of 26,000 tonnes, which occurred in the 2008/2009 cocoa year, resulted from a collapse in global demand for cocoa beans, estimated at almost 7% compared to the level reached in the previous season. Globally, during these four seasons, stocks of cocoa beans declined by 460,000 tonnes. This development resulted mainly from the lack of growth in the cocoa output of Côte d'Ivoire, the world's major cocoa-producing country.

The 2006/2007 season witnessed a record supply deficit of 279,000 tonnes, supporting the increase in average international prices for the season which rose by 19% to US $1,854 per tonne. Weather conditions had been unfavourable since the beginning of the season. Firstly, *El Niño* related weather conditions developed in the tropical Pacific which significantly reduced output in several countries. Secondly, cocoa-producing countries in West Africa had suffered from a severe harmattan and its subsequent dryness that started in December. In addition, the market experienced a strong demand for cocoa beans during this season (up by over 4%) which added pressure on prices. As a result, cocoa futures prices showed a steady rise until early July 2007, reaching levels 49% higher at US $2,215 compared to the lowest level recorded at the beginning of the season. The 2007/2008 season recorded a second consecutive supply deficit, albeit smaller than the previous one, at 75,000 tonnes. In consequence, the average international price for the season rose by 36% to US $2,516 per tonne. In the first two months of the 2007/2008 season, cocoa futures prices moved with no particular direction, influenced by uncertainties regarding the supply and demand situation and by currency movements.

At the end of November 2007, futures prices resumed the upward movement initiated in the middle of October 2007. Despite an expected substantial increase in world cocoa bean production of 9% over the previous season, demand for cocoa beans was still expected to surpass production. However, a correction occurred in March 2008 when concerns over the impact of the US financial crisis associated with the near collapse of Bear Stern, a US investment bank, sparked panic in the financial markets. In reaction, investment funds decided to reduce their risks by taking their profits across all assets, including cocoa. The downward correction was short-lived and cocoa futures prices resumed their upward movement, with the ICCO daily price reaching its highest level for 28 years on 1 July 2008, at US $3,296 per tonne (ICCO 2012b).

The major movements which characterised the evolution of international cocoa prices during the 2007/2008 season, as previously described, were also experienced by most commodities. The fact that prices had followed the same trend across many commodities suggested that there were some common causes in price movements, such as the turbulence in the world's financial markets, the deterioration of global economic growth and, most importantly, the fluctuation of the US dollar against other major currencies. However, in general, the cocoa market showed stronger growth than other commodities in periods of upward movement and was more resilient in periods of downward movement.

A new price correction started in July 2008, with international cocoa prices declining to levels 41% lower, at US $1,956 after the start of the 2008/2009 season, in the middle of November 2008, compared to the peak reached in July. Downward pressure initially originated from the lack of purchasing interest from the processing and manufacturing sector as a result of the relatively high price of cocoa, as well as from news related to a global slowdown in the demand for cocoa beans. The downward trend initiated in the beginning of July was reinforced by the strengthening US dollar against other major currencies from the end of the month until the end of the season. The US dollar gained over 11% against the Pound Sterling and the Euro during this period. In addition, in the first half of September, the intensification of the global financial crisis may have accelerated the declining movement of cocoa futures. Indeed, the turmoil in the US financial market during the first two weeks of the month (bankruptcy of Lehman Brothers, sale of Merrill Lynch to Bank of America and rescue by the US government of American International Group) and its subsequent impact on European markets, prompted non-commercial market participants to reduce their risk exposure in all assets, including cocoa.

After a four-month period of decline, prices bounced back in the middle of November 2008, decoupling from the movement of other commodities. As global cocoa production was expected to decline during the 2008/2009 cocoa year, most analysts were, at the time, forecasting a global cocoa production deficit for the 2008/2009 season. If this had occurred, it would have been the third consecutive supply deficit. However, the extent of the impact of the deterioration of the economic environment on demand for cocoa had been underestimated by most analysts and demand fell by almost 7%, representing the sharpest yearly decline since the 1946/1947 cocoa year, when data were first published by the trading house Gill & Duffus Group Ltd. In the end, the 2008/2009 cocoa year recorded a global supply surplus of 26,000 tonnes. However, the average international price for the 2008/2009 cocoa year increased by 3% compared to the previous season, to US $2,599 per tonne.

At the beginning of the 2009/2010 cocoa year, with most analysts expecting a supply deficit, prices continued to follow a rising trend, reaching their highest level in over 31 years in the middle of December 2009, at US $3,637 per tonne. In January 2010, cocoa prices reversed to a lower level, arising from the publication of lower than expected processing activity in North America and Europe. Moreover, the steady strengthening of the US dollar was another key factor in the decline of cocoa prices as well as of other commodities. Thereafter, international cocoa prices moved sideways.

In total, although the 2010/2011 season experienced a record supply surplus of 343,000 tonnes, the average international price for cocoa declined only from US $3,246 to US $3,105 per tonne, as it was supported by concerns over the political situation in Côte d'Ivoire. The poor global economic growth prospects during 2011 and bearish fundamental news of increased supplies and overstocked warehouses within the industry halted an upward momentum in cocoa prices during the first quarter of the 2011/2012 cocoa season. Cocoa prices relentlessly followed a downward trend during this period and, at the end of the 2011 year, the New York market recorded a 30% decrease over the previous year, and the London market recorded a 31% decrease over the same period.

After hitting such low levels, cocoa futures experienced a significant recovery by the middle of January 2012. The turnaround in cocoa futures was mainly supported by reports of dry weather conditions and waning weekly cocoa arrivals in Côte d'Ivoire. By the end of January, a lower than expected year-on-year increase in grindings data of just 1.8% in West Europe and 1.5% in North America, combined with unfavourable weather conditions in the West African producing areas caused cocoa prices to move sideways. Nevertheless, on 31 January, the Ivorian government initiated its forward sales programme for the 2012/2013 main crop, an initiative which forms part of the country's planned reforms of the cocoa sector. In particular, this programme aims to guarantee fixed prices for cocoa farmers in Côte d'Ivoire.

3.7 COCOA PROCESSING TRENDS

Traditionally, total world grindings have been used to measure global demand, as manufacturers tend to process cocoa beans in accordance with demand for cocoa products (cocoa paste/liquor, cocoa butter, cocoa cake and cocoa powder). Hence any excess of supply over demand became part of the total world stocks of cocoa beans. Until recently, the composition of demand for, and the structure of relative prices of, cocoa products were relatively stable with no frequent large build-up of excess stocks of cocoa products. Hence the level of world grindings closely mirrored the pattern in global demand and consumption of cocoa in finished products over most of the review period (ICCO 2012b).

Between 2002/2003 and 2011/2012, primary cocoa consumption (as measured by total world grindings of cocoa beans) continued along an upward trend, growing at an average rate of 2.9% per annum, representing a total increase of over 915,000 tonnes over the period. World grindings have increased almost every year with the exception of 2008/2009, when they collapsed in the midst of the global economic crisis, by over 6%. The growth in world grindings is estimated to have fallen during the current 2011/2012 year, increasing by just under 2% to reach nearly four million tonnes, but resulting in a record level for grindings.

After a slowdown in grindings at the beginning of the last decade and closure of some non-profitable plants, a substantial decrease in stocks of cocoa butter ensued. The price of cocoa butter subsequently increased, following demand, and the ratio rose to its highest level during the period, to 2.92 in 2004/2005, remaining in the high bracket for the next three seasons. By contrast, the high level of stocks prompted the cocoa powder price ratio to decline from 0.68 in 2004/2005 to its lowest level of 0.55 in 2007/2008. The recovery in the butter price ratio resulted in a combined product ratio, averaging 3.60 during the 2004/2005 season, the highest level recorded in the last 20 years. This contributed to a recovery in world grindings, starting from 2002/2003, as processing margins improved. The cocoa-processing business had been very profitable in recent years up to the 2007/2008 season with butter prices reaching near record levels. However, the deterioration of the global economic environment and the steady increase in the price of cocoa beans since October 2006 were followed by a reduced demand for cocoa, confirmed in 2008/2009 when processing activity declined by almost 7% compared to the previous season: chocolate consumers suffered from a reduction in their income and chocolate manufacturers increased the price of their products or reduced the size of their portions. However, the deterioration of the economic environment did not have such a large negative impact on demand for products based on cocoa powder and the dip in processing activity in 2008/2009 led to a low availability of cocoa powder. As pressure mounted with increased demand for the product, the powder price started to rise, with the ratio averaging 1.92 in the first half of the 2011/2012 cocoa year (ICCO, 2012b).

The main force behind the strong demand for cocoa powder is the change in global consumption patterns in emerging countries. There are many new consumers in these markets, including Asia, where demand is focused on powder-based products, as discussed in Section 3.4.

This contrasting trend in demand for cocoa powder and cocoa butter was clearly reflected in an upward trend in cocoa powder price ratios and parallel declines in cocoa butter ratios over the same period. The cocoa butter price ratio (the price of cocoa butter relative to cocoa beans) declined from 1.92 in 2002/2003, the beginning of the period under review, to 1.48 in 2010/2011 and further to 1.10 during the October 2011 to March 2012 period, whereas powder ratios climbed from 1.50 to 1.71 and 1.92 during the corresponding periods. Similarly, the combined product ratio declined from 3.42 to 3.19 and 3.02.

Europe remained by far the largest cocoa-processing region during the 2003 to 2012 period. The increase in European grindings was estimated at 277,000 tonnes between 2002/2003 and 2011/2012, which corresponded to an average annual growth rate of 2.1%. However, the pace of growth was lower than that recorded for world grindings, estimated at 2.9%. Europe's share has consequently declined (from 43% to 40%) during the 2006/2007 period. In a similar pattern, processing in the American region grew by a meagre 0.5% per annum over the 2011/2012 period, with its share declining from 26% to 21%. In contrast, grindings in the African region increased by an average rate of over 5.7%, its share rising from 14% at the beginning of the 2006/2007 period to an expected 18% in 2011/2012. With an annual growth rate of 5.6%, the largest regional volume increase, of 314,000 tonnes, occurred in Asia and Oceania, mainly as a consequence of a gradual and steady increase of grindings in Indonesia and Malaysia. The share of the region is forecast to rise from 16% at the beginning of the 2006/2007 period to 20% in 2011/2012. Statistical information on trends in cocoa processing in individual countries and by region over the last four cocoa years, and for the current 2011/2012 season are available (ICCO 2012b; ICCO 2013b).

Most cocoa processing continues to be performed in cocoa-importing countries near the major centres of cocoa consumption in Europe and North America, with the Netherlands maintaining its position as the world's leading cocoa-processing country. Germany surpassed the United States towards the end of the review period, realising a very rapid growth in processing during recent years, partly attributed to multi-nationals transferring activity to the country after a series of takeovers. Greater use of bulk shipments and economies of scale from processing large volumes encouraged the expansion of cocoa-processing facilities located near ports in cocoa-importing countries.

Origin grindings have increased and have become more widespread among cocoa-producing countries over the last 10 years, supported, in some countries, by government policies favouring the export of value-added semi-finished products rather than raw cocoa beans. Greater involvement of multi-national companies in up-stream activities, including internal marketing, shipping and local processing in cocoa-producing countries, also resulted in substantial investment in cocoa-processing capacity at origin, most notably in West Africa and Asia. The grindings share of origin countries is expected to rise to 42% in 2011/2012 from around 35% in 2002/2003. Côte d'Ivoire is presently the world's third largest cocoa-processing country, after the Netherlands and Germany. Although it claimed the second position in 2009/2010, grindings slowed down in 2010/2011 arising from the political situation during the season. Following an expansion of its grinding capacity, Malaysia reinforced its position as the leading cocoa-processing country in Asia, ranking fifth in the world. Brazil, Ghana and lately Indonesia were also among the leading cocoa-producing countries which process significant amounts of cocoa beans of around 200,000 tonnes (ICCO 2012b; ICCO 2013b).

3.8 COCOA AND CHOCOLATE CONSUMPTION

3.8.1 Apparent Cocoa Consumption

The total world grindings accurately reflect global demand for cocoa beans over the medium and long term. However, increases in grindings do not necessarily indicate increases in actual consumption at the country or regional level, arising from significant international trade in cocoa and chocolate products. A more appropriate measure of cocoa consumption at the country or regional level is therefore the amount of cocoa beans used in the manufacture of the confectionery, food, beverage or cosmetic products that are actually consumed in the country or region. This 'apparent domestic cocoa consumption' for a country is calculated as grindings plus net imports of cocoa and chocolate and chocolate products, in bean equivalent terms. The cocoa products—cocoa butter, cocoa paste/liquor and cocoa powder/cake—are converted into bean equivalents using standard conversion factors; net trade in chocolate and chocolate-based products is converted to bean equivalents, based on general assumptions about the cocoa content of the chocolate products involved.

The latest estimates of apparent domestic cocoa consumption compiled by the ICCO Secretariat illustrate recent trends in regional and country consumption patterns. The most recent year for which data are available,

2010/2011, indicates that the European region accounted for 48% of total world consumption of cocoa (down from 51% at the beginning of the review period in 2002/2003) followed by a similar pattern in the Americas, at 33% (down from 34%) whereas Asia's share increased from 13% to 15% and Africa's from 2% to 3% (ICCO 2012b; ICCO 2013b).

Between 2002/2003 and 2010/2011, world cocoa consumption expanded by 731,000 tonnes (up by 24%) with most of the increase coming from higher consumption in the traditional cocoa-consuming countries of Europe (up 262,000 tonnes or 17%) and consumption increased by 227,000 tonnes (up by 22%) in the Americas over the same period. The most dynamic regions in terms of cocoa consumption were the Asian region (up by 50% or 188,000 tonnes) and the African region (up by 74% or 54,000 tonnes). In 2010/2011, the leading consumers of cocoa by country were the United States, Germany, France, the United Kingdom, the Russian Federation, Brazil, Japan, Spain, Italy and Canada.

The world *per caput* consumption of cocoa has also witnessed a similar pattern of growth over the review period, rising from 0.54 kg in 2002/2003 to 0.61 kg in 2010/2011. It is worth noting that the only decline during the 2003 to 2012 period occurred in 2008/2009 during the global economic slowdown. Thereafter, cocoa consumption resumed an upward trend, reaching once more, in 2010/2011, its pre-crisis level of 0.61 kg per head.

Most notable, during the review period, were the increases in *per caput* consumption levels recorded by Germany (from 3.40 kg to 3.96 kg) and Slovenia (from 2.15 kg to 3.07 kg) in the European Union and the Russian Federation and Brazil from the BRIC countries (Brazil, the Russian Federation, India and China). Brazil expanded their consumption from 0.55 kg per head in 2002/2003 to 0.92 kg per head in 2010/2011. Although India and China posted increases, the large populations of the two countries do not reflect the growth accordingly. Consumption levels in most Eastern European and Central Asian countries also rose from the low levels witnessed at the beginning of the period under review. A combination of high economic growth rates, successful promotional activities and low prices stimulated cocoa consumption. It should be noted that, in some mature markets, consumption has remained either stable (e.g., France, Japan and the United States) or has even declined, as in Italy (ICCO 2012b; ICCO 2013b).

3.8.2 Chocolate Consumption

Cocoa is mainly consumed as chocolate confectionery, chocolate-coated products (biscuits, ice creams), or in other food products containing cocoa powder, including beverages, cakes, snacks, and the like. The principal ingredients in chocolate are cocoa paste, which imparts the basic chocolate flavour, cocoa butter which provides the characteristic mouth feel, sugar and a flavouring agent. Milk or milk powder is added to produce milk chocolate; nuts, biscuits and other fillings are added to make filled chocolates. Cocoa powder is used in a wide range of food products and beverages. The growth in cocoa consumption in the Far East and in Eastern Europe was largely attributed to an increase in demand for products containing cocoa powder. Relatively small amounts of cocoa butter are used in cosmetic products and, more recently, new products are being manufactured from cocoa by-products in some cocoa-producing countries (Afoakwa and Paterson 2010; ICCO 2012b).

However, information on consumption of products containing cocoa is only published for leading consuming countries, and often after a considerable delay, making it difficult to assess or interpret trends in global consumption. Data published by the Association of the Chocolate, Biscuit & Confectionery Industries of the European Union (CAOBISCO) in July 2012 show that consumption of all chocolate confectionery products in the 19 countries for which statistics are available for the 2002–2010 period (which include most of the traditional leading cocoa-consuming countries) increased by 10%, an average annual growth of only 1.2%. During the review period, the average year-on-year growth ranged between 0.4% and 4.5%, except in 2009, when consumption shrank by 2.4% arising from the global economic crisis (ICCO 2012b).

3.9 CHOCOLATE MARKET

The chocolate market is still dominated by consumers from Western Europe and North America. Even in the face of the economic hardship of recent years, chocolate is still regarded as an affordable luxury. Although the volume of chocolate consumed has increased at only a low pace, the volume of cocoa consumed has increased more rapidly as taste has changed over the years. On the high end of the mature market, chocolate aficionados are asking for single estate and origin 'high cocoa content' products, with their own distinctive flavours. Innovation is also being used to appeal to consumers in the saturated markets: new flavours, new packaging and new sizes for health-conscious consumers. Sustainable sourcing is growing in importance: demand for cocoa grown in a responsible manner is rising as companies respond to consumer preferences.

On the other side of the market, the youthful populations of the BRIC countries with their disposable incomes, are a major driving force behind the growth in chocolate consumption. Manufacturers are catering to specific consumer tastes: Cadbury India reported its highest-ever sales and net profit in 2011, after ramping up distribution and adding new products from its portfolio. In China, where the chocolate confectionery market is very young and products are often bought as gifts, the market is currently experiencing very rapid growth, mainly thanks to the growing middle class of more than 300 million people. With the Indonesian market also expanding at a very rapid pace, the Asian market is expected to hold a 20% share of the global market by 2016 (ICCO 2012b; ICCO 2013b).

4 Fairtrade Cocoa Industry

4.1 INTRODUCTION

Fairtrade is a trading partnership that aims for sustainable development of excluded and disadvantaged producers, seeking greater equity in international trade by offering better trading conditions to, and securing the rights of marginalised producers and workers, especially in the southern hemisphere. Fairtrade organisations (backed by consumers) are actively engaged in supporting producers, raising awareness and campaigning for changes in the rules and practice of conventional international trade, with regulated terms of trade which ensure that farmers and workers in some of the poorest countries in the world are adequately protected and can build a more sustainable future (Fairtrade Federation 1999; EFTA 2005; FLO 2006).

The concept of "Fairtrade" has existed since the early 1960s. It was founded by a society of importers and non-profit retailers in the wealthy, northern European countries and small-scale producers in developing countries, who, while fighting against low market prices and high dependence on brokers, were seeking a more direct type of trade with the European market. It is united in the view that conventional trading relations between the southern and northern hemispheres are unfair and unsustainable and that this issue can be addressed through a different approach to the trading system. Its goal is to tackle poverty in developing countries through trade, and its pragmatic approach is one of the key reasons for its success. However, the diversity of the movement, its lack of structure and economies of production scale were impediments to its sustainability. Since the early 1990s, the Fairtrade movement has become more organised to address the challenges it faces and has thus become one of the fastest growing markets in the world, generating about US $200 million annually through sales (Brown 1993; Damiani 2002; FLO 2004; EFTA 2006; Kilian et al. 2006). The harmonisation of definitions, the increased professionalism and emphasis on quality assurance, the direct marketing through supermarkets and the establishment of working relations with mainstream businesses to enable economies of scale have further secured the steady growth of Fairtrade.

Viewed positively, the globalisation of world trade, currently totalling £3.5 trillion per annum, has helped to lift more than 400 million people out of poverty from the Tiger economies of East Asia and elsewhere (Geographical 2004). However, although international trade can be powerful redistributors of global wealth, it brings with it other problems in its wake. Such problems make for an imbalance of economic power between what producers are paid at subsistence and below subsistence wages in third-world countries compared with the profits made by rich retailers and distributors in the food supply chain, predominantly in the West (Denny and Elliott 2003; Wright and Heaton 2006). The Fairtrade cocoa system secures better livelihoods for farmers by modernising cocoa farming through productivity improvements, introduction of systems of best-known farm practices and improvements in living and working conditions. Through codes of good farm practices, containing guidelines for sustainable production, farmers benefit from better market access for Fairtrade cocoa and chocolate products. This is very important in view of new requirements from consumers as demand for Fairtrade cocoa and chocolate products continues to increase.

Consumers increasingly want products which provide a decent living for the farmers, are produced in a socially acceptable way, not harming the environment and which are safe and healthy to enjoy (FLO 2005). Delivering such products is in the direct interest of the farmers, the cocoa processors, the traders/exporters and the chocolate manufacturers. The benefits resulting for the farmers and the other stakeholders in the chain in delivering 'Fairtrade cocoa' are enhanced livelihood for farmers and their families, improved market access and sustainable increases of production and consumption.

Currently, over 4.5 million people in developing countries are benefiting from increasing sales of Fairtrade cocoa in 20 national markets across Europe, North America, Japan and Mexico. The Fairtrade mark appears on a range of cocoa products including chocolate bars, chocolate sauces, hot chocolate drinks, chocolate-coated snack bars and biscuits. This range of products is growing progressively and standards for new categories have been introduced on a regular basis over the past few years. Since 1997, retail sales of Fairtrade cocoa certified products in the United Kingdom have been growing by an average of 50% per annum and are now running at over £195 million. In recent years, many studies have confirmed the rise of a so-called 'ethical consumerism' and their systematic influences on global food trade. According to these studies, consumers' values have shifted from pragmatic, price and value-driven imperatives to a new focus on ethical values and on the story behind the products (Shaw and Clarke, 1999; Bourn and Prescott 2002; Anon 2002; Carrigan and Attalla 2001; Kilian et al. 2006; Wright and Heaton, 2006; Poelman et al. 2007). This chapter examines the entire Fairtrade labelling and certification systems, and seeks to make a significant contribution to understanding current trends in world sourcing, marketing systems and supply chain management of Fairtrade cocoa over the past decade.

4.2 FAIRTRADE COCOA LABELLING, STANDARDS AND CERTIFICATION CONCEPTS

The roots of the Fairtrade movement can be traced back for at least 40 years and from that time it has developed into a worldwide network of organisations seeking to relieve the causes of poverty by tackling some of the injustices of world trade. Traditionally such organisations have been development projects or community enterprises that were set up to assist people in disadvantaged communities earn a more sustainable livelihood through export trade. Because different people have different problems, these organisations developed a variety of operating models to achieve their objectives but all share common principles of empowering producers to improve their own lives by working through democratic organisations, and enabling consumers to use their purchasing power to change the way world trade is managed.

Fairtrade labelling builds on the work of these pioneering 'alternative traders' by defining standards and operating a certification scheme that enables all supplying and retailing businesses to participate in Fairtrade. The first labelling initiative was set up for coffee by the Max Havelaar Foundation in the Netherlands in 1988 (FLO 2005). Other national labels were set up over the next few years and standards were developed for other products. In 1997, seventeen of these organisations joined together to form Fairtrade Labelling Organizations International (FLO) as an umbrella organisation to control the standards and to certify producer organisations. The Fairtrade Foundation is a member of Fairtrade Labelling Organizations International, the umbrella organisation for Fairtrade labelling which works with over 508 producer organisations in 58 developing countries representing 5 million people: farmers, workers and their dependents (Fairtrade Foundation 2004). It sets international Fairtrade Producer standards and the terms of trade for Fairtrade certified products; facilitates and develops Fairtrade business; liaises with Fairtrade certified producers to assist in strengthening their organisations and improve their production and market access; and conducts lobbying and advocacy, promoting the case for trade justice. In the interests of ensuring efficiency and transparency, all certification is carried out by FLO Certification Limited, a separate legal entity. FLO Cert is the world's largest social certifier and is responsible for inspecting and certifying production according to the

TABLE 4.1
Members of the Fairtrade Labelling Organization (FLO)

Country	National Initiative
	Europe
Austria	Fairtrade
Belgium	Max Havelaar
Denmark	Max Havelaar
Finland	Reilum kaupan edistämisyhdistys ry.
France	Max Havelaar
Germany	TransFair
Italy	TransFair
Ireland	Fairtrade Mark
Luxembourg	TransFair Minka
Netherlands	Stichting Max Havelaar
Norway	Max Havelaar
Sweden	Rättvisemarkt
Switzerland	Max Havelaar Stiftung
United Kingdom	Fairtrade Foundation
	North America
Canada	TransFair
Mexico	Comericio Justo
United States	TransFair
	Asia and Oceania
Australia	Fairtrade Association of Australia and New Zealand
Japan	Fairtrade Label
New Zealand	Fairtrade Association of Australia and New Zealand

Source: ICCO. *Consultative Board Report on the World Cocoa Economy. Fifth Meeting*, London, 2005. With permission.)

defined Fairtrade standards, monitoring the trade in primary and semi-finished products to ensure compliance.

Fairtrade labels are established in 20 countries (Table 4.1) including Austria, Australia and New Zealand, Belgium, Canada, Denmark, Finland, France, Germany, Ireland, Italy, Japan, Luxembourg, Mexico, the Netherlands, Norway, Sweden, Switzerland, the United Kingdom and the United States.

The national labelling organisations license Fairtrade labels for use on specific products, monitor the supply chains of certified cocoa products, help companies develop new Fairtrade certified cocoa products and work with others to raise consumer awareness and support. Producer standards apply to organisations representing farmers and workers and ensure that such organisations are democratic and accountable to their members, capable of ensuring compliance with social and economic criteria and committed to a programme of ongoing improvement. However, being able to apply Fairtrade premiums for the benefit of members, producer standards also ensure respect for basic human rights as defined in ILO conventions, decent wages and employment conditions for employed workers, ongoing improvements in worker health and safety and environmental protection by control and reduction of chemical inputs (Fairtrade Foundation 2006).

To receive FLO certification all producers, merchants, processors, wholesalers and retailers must adhere to the determined standards. There are two types of generic standards: one for small producers and another for workers on plantations and in factories. The first type applies to small property owners organised into cooperatives or other types of organisations with a democratic and participatory structure. The second applies to organised workers whose employers pay decent salaries, guarantee them the right to join unions and provide lodging when relevant. The plantations and factories must comply with minimum health, safety and environmental standards, without using child or forced labor.

The established norms distinguish between 'minimum requirements', which must be met to obtain Fairtrade certification, and 'progress requirements' that encourage producers to improve constantly labor conditions and product quality, foster environmentally friendly practices and invest in the organisation and its associates. Business standards stipulate that merchants must:

1. Pay a price to producers that covers the costs of sustainable production and housing.
2. Pay a premium that producers can invest in development.
3. Make a partial pre-payment when producers ask for it.
4. Sign long-term contracts that enable better planning and encourage sustainable production practices.

Fairtrade cocoa standards require buyers to help develop more direct and transparent supply chains, and commit to long-term relationships with producers. Additionally, some standards apply to specific products and determine minimum prices and quality, as well as processing requirements.

4.3 WORLD SOURCING OF FAIRTRADE COCOA

Fairtrade cocoa is sourced from many countries that are typified by low or very low GDP per head, often with poor infrastructure and communications. It is characterised by the dominance of small farmers with an estimated 95% of annual world cocoa production derived from smallholdings in the size range of one to three hectares. The sector is further characterised by price volatility, a declining trend in real prices, concentration of production in a limited number of countries, low productivity and huge crop losses due to pests and diseases. However, with growth in population and rising GDP, global demand for Fairtrade chocolate has continued to grow, with expanding markets for Fairtrade cocoa. Cocoa sold with the Fairtrade label captures a very small share of the cocoa market (~1%). However, based on the steady growth of Fairtrade and the support of public opinion and governments, some Fairtrade participants claim that the idea will move beyond niche markets and become more mainstream, and the increasing popularity of Fairtrade Products can be taken as a proof. The recent strong growth in premium chocolate is also driven by increased consumer concern over poverty in less-developed countries, and about a trading system that is considered as unfavourable to these countries. Consequently, purchases of labelled Fairtrade cocoa increased from 700 tonnes in 1996 to 5,657 tonnes in 2005, as shown in Table 4.3, representing an average annual growth of 23% during the period 2004–2005. In 2005, 83% of the total sales of Fairtrade cocoa were realised in only six countries, the United Kingdom (40%), Germany and France (13% each), Austria, Italy and Switzerland (6% each). However, as shown in Table 4.4, the estimated market share of Fairtrade labelled cocoa in each country is still less than 2%, with the highest shares in Luxembourg (1.55%), Austria (1.12%) and the United Kingdom (1.02%; ICCO 2005).

In 2006, consumers worldwide bought €1.6 billion worth of Fairtrade Certified Products, 42% more than sales in 2005. Likewise, sales in Fairtrade cocoa have been increasing progressively over the past decade with annual purchases almost doubling between 2005 and 2006. In 2006, Fairtrade cocoa attracted a relatively larger market worldwide with annual purchases of 10,919 MT representing about 93% increase in sales of 2005 (Figure 4.1). This growth in sales was accompanied by more licensees (companies that sell the final packaged Fairtrade Products) joining the scheme. The number of licensees in 2006 reached 1,954, increasing by 29% in comparison to figures observed in 2005.

Despite this number sounding impressive, Fairtrade Certified Producers are still far away from being able to sell all their produce under Fairtrade terms. In fact, FLO estimates that approximately 20% of the total production of Fairtrade Certified Producers is sold under Fairtrade terms. This is one reason why despite the impressive growth in the last 10 years, the Fairtrade market still has plenty of room for growth. FLO and its member Labelling Initiatives are working to open new markets and identify new business opportunities for cocoa producers. Fortunately, they are not alone

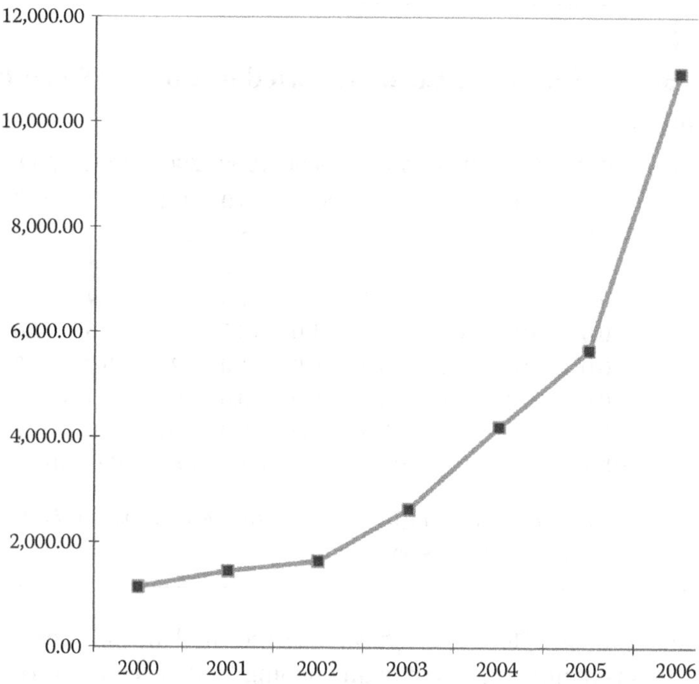

FIGURE 4.1 Fairtrade cocoa exported between 2000 and 2006 (MT). (Adapted from ICCO, *Consultative Board Report on the World Cocoa Economy. Fifth Meeting*, 2005. With permission.)

in this endeavour, as millions of consumers and supporters are helping out. FLO-registered cocoa producer associations and National Initiatives have to report to FLO, respectively, on their sales of cocoa beans and cocoa semi-finished products and sales of chocolate and chocolate products under the Fairtrade label. The corresponding volumes are modest, representing less than 1% of the total cocoa market.

In 2003, producers sold 2,687.4 tonnes of cocoa beans and cocoa semi-finished products, in beans equivalent, under the Fairtrade label. Table 4.2 gives an overview of the share of total exports of Fairtrade cocoa by country of origin during the period 1994–2003. As country information on sales is related to a very limited number of co-operatives in some cases, the disclosure of volumes would have been against the FLO confidentiality policy. In 2003, more than 90% of the sales originated from two producers: Kuapa Kokoo Ltd. (Ghana) and Conacado Inc. (Dominican Republic). Consequently, the nine other producers concerned in 2003 sold less than 10% of the total, representing less than 200 tonnes. This highly concentrated market may be the result of lower costs associated with trading for larger cocoa producers. Since 2003, five new small cooperatives in Peru,

TABLE 4.2
Market Share of Fairtrade Cocoa Exported by Country Since 1994 (% of total)

Country	1994	1995	1996	1997	1998	1999	2000	2001	2002	2003
Ghana	66.0	70.0	81.5	78.4	55.0	68.0	37.0	15.2	27.7	45.5
Bolivia	20.0	26.9	12.2	10.8	20.0	5.2	25.7	24.7	7.0	3.0
Ecuador	13.0	3.1	3.7	4.6	0	5.2	11.4	7.4	9.0	2.0
Belize	0.0	0.0	2.6	3.2	4.0	1.8	2.4	8.9	0.3	0.0
Dom. Rep.	0.0	0.0	0.0	2.9	21.0	18.7	21.1	36.6	54.0	48.6
Cameroon	0.0	0.0	0.0	0.0	0.0	1.0	2.0	0.0	1.0	1.0
Costa Rica	0.0	0.0	0.0	0.0	0.0	0.0	0.0	7.4	0.09	0.0
Nicaragua	0.0	0.0	0.0	0.0	0.0	0.0	0.0	0.0	1.0	0.4
Total	100.0	100	100	100	100	100	100	100	100	100

Source: ICCO, *Consultative Board Report on the World Cocoa Economy. Fifth Meeting*, London, 2005. With permission.

and a large one in Côte d'Ivoire have received the Fairtrade certification. In Peru, this is mainly the result of an ongoing United Nations programme to convert cocoa producers to alternative crops.

Until 2003, cocoa and chocolate products were sold in the Fairtrade market in 16 countries, mainly in Europe. Table 4.3 provides sales information by country during the 1994–2005 period. Presently, only information on total sales of chocolate and chocolate products is available, as the National Initiatives do not provide detailed data on the cocoa content of these products. In 2003, 80% of the total sales of chocolate and chocolate products were realised in only five countries, the United Kingdom (35%), Italy (13%), Germany (13%), Switzerland (10%) and France (9%).

Since 2004, Fairtrade chocolate and chocolate products have also been sold in Japan, Australia, New Zealand and Mexico. The estimated market share of Fairtrade labeled cocoa in each country of destination is less than 1%, with the highest shares (0.9%) in Switzerland and Luxembourg in 2003. Further to these, in 2005, 83% of the total sales of Fairtrade cocoa were realised in only six countries including the United Kingdom (40%), Germany and France (13% each), Austria, Italy and Switzerland (6% each). However, as shown in Table 4.4, the estimated market share of Fairtrade labelled cocoa in each country is still less than 1%, with the highest shares in Luxembourg (1.55%), Austria (1.12%) and the United Kingdom (1.02%; ICCO 2005).

Fairtrade certified chocolate and chocolate products are composed of various ingredients. It is not always possible for all these ingredients to be sourced from a Fairtrade-certified producer organisation and consequently

TABLE 4.3
Fairtrade Cocoa Imported by Country Since 1994 (in Tonnes)

Country	1994	1995	1996	1997	1998	1999	2000	2001	2002	2003	2004	2005
Netherlands	123.9	147.6	187.8	177.4	172.1	95.0	92.5	102.7	105.9	147.0	177.0	176.0
Switzerland	83.2	171.1	179.6	124.0	141.4	151.8	180.9	213.2	253.8	275.0	331.0	322.0
Germany	0.0	0.0	311.1	330.8	328.7	290.0	321.1	394.2	339.4	343.0	603.0	746.0
UK	0.0	0.0	22.0	24.3	81.5	190.1	317.2	426.6	550.6	903.0	1,626.0	2,238.0
Austria	0.0	0.0	0.0	38.0	26.5	41.4	38.2	43.7	76.7	94.0	186.0	336.0
Luxembourg	0.0	0.0	0.0	13.7	15.6	12.9	17.0	19.9	17.1	21.0	29.0	34.0
Italy	0.0	0.0	0.0	0.0	45.6	108.3	151.5	194.7	162.8	346.0	296.0	329.0
Sweden	0.0	0.0	0.0	0.0	6.7	17.5	15.6	30.8	49.7	52.0	34.0	29.0
Denmark	0.0	0.0	0.0	0.0	0.0	14.4	15.7	21.3	13.0	13.0	12.0	38.0
Finland	0.0	0.0	0.0	0.0	0.0	0.0	2.9	4.9	6.5	9.0	11.0	15.0
Canada	0.0	0.0	0.0	0.0	0.0	0.0	0.0	0.7	42.7	54.0	118.0	231.0
Belgium	0.0	0.0	0.0	0.0	0.0	0.0	0.0	0.0	2.9	61.0	120.0	147.0
France	0.0	0.0	0.0	0.0	0.0	0.0	0.0	0.0	32.6	227.0	398.0	723.0
Norway	0.0	0.0	0.0	0.0	0.0	0.0	0.0	0.0	0.4	0.5	2.0	4.0
USA	0.0	0.0	0.0	0.0	0.0	0.0	0.0	0.0	2.1	92.2	249.0	251.0
Ireland	0.0	0.0	0.0	0.0	0.0	0.0	0.0	0.0	0.0	5.7	9.0	15.0
Total	207.1	318.7	700.5	708.2	818.1	921.4	1,152.6	1,452.7	1,656.2	2,643.4	4,2201.0	5,657.0

Source: ICCO, *Consultative Board Report on the World Cocoa Economy. Fifth Meeting*, London, 2005. ICCO, *Consultative Board Report on World Cocoa Economy. Twelfth Meeting*, Kuala Lumpur, 2007b.

TABLE 4.4
Fairtrade Market Share of Cocoa Consumption (in % of Apparent Cocoa Consumption)

Country	1996	1997	1998	1999	2000	2001	2002	2003	2004	2005
Luxembourg	0.00	0.69	0.78	0.65	0.83	0.95	0.81	0.90	1.32	1.55
Austria	0.00	0.16	0.12	0.13	0.11	0.14	0.27	0.34	0.56	1.12
United Kingdom	0.01	0.01	0.04	0.09	0.15	0.21	0.27	0.42	0.74	1.02
Switzerland	0.69	0.47	0.59	0.47	0.59	0.81	0.98	0.83	0.90	0.85
Netherlands	0.61	0.55	0.51	0.33	0.26	0.26	0.34	0.46	0.54	0.50
Canada	0.00	0.00	0.00	0.00	0.00	0.00	0.07	0.09	0.16	0.37
Italy	0.00	0.00	0.05	0.11	0.17	0.19	0.16	0.34	0.29	0.30
France	0.00	0.00	0.00	0.00	0.00	0.00	0.02	0.10	0.17	0.29
Belgium	0.00	0.00	0.00	0.00	0.00	0.00	0.01	0.11	0.20	0.28
Germany	0.13	0.13	0.11	0.10	0.12	0.13	0.12	0.12	0.20	0.27
Denmark	0.00	0.00	0.00	0.06	0.07	0.12	0.08	0.09	0.07	0.20
Sweden	0.00	0.00	0.03	0.10	0.10	0.18	0.27	0.30	0.17	0.15
Finland	0.00	0.00	0.00	0.00	0.04	0.05	0.06	0.08	0.10	0.13
Ireland	0.00	0.00	0.00	0.00	0.00	0.00	0.00	0.04	0.06	0.08
United States	0.00	0.00	0.00	0.00	0.00	0.00	0.00	0.01	0.03	0.03
Norway	0.00	0.00	0.00	0.00	0.00	0.00	0.00	0.00	0.01	0.02
Japan	0.00	0.00	0.00	0.00	0.00	0.00	0.00	0.00	0.00	0.00
Average	0.04%	0.04%	0.04%	0.05%	0.06%	0.07%	0.09%	0.14%	0.20%	0.27%
Yr-on-yr change	109%	–2.0%	11%	10%	18%	27%	17%	56%	47%	36%

Source: ICCO, *Consultative Board Report on the World Cocoa Economy. Fifth Meeting*, London, 2005. ICCO, *Consultative Board Report on World Cocoa Economy. Twelfth Meeting*, Kuala Lumpur, 2007b.

for a chocolate product to be 'fully Fairtrade'. FLO policy defines the conditions that allow a composite product to carry the Fairtrade label. It is stated that all the ingredients for which the certification exists must be sourced from Fairtrade certified producer organisations and 50% of all the ingredients, by dry weight, must be sourced from Fairtrade certified producer organisations. However, exceptions to this rule exist as many manufacturers find it difficult to apply this rule strictly. The implementation of these exemptions requires FLO approval.

4.4 MARKETING SYSTEMS AND ECONOMICS OF FAIRTRADE COCOA

World cocoa production has risen at an average annual growth rate of 2.7% during the 1996–2006 periods. Africa's share of world cocoa production increased from around 66% in the second part of the 1990s to a projected share of 71% in 2006/2007, with its production rising at an average annual rate of about 3%. Cocoa output in the Asia and Oceania region has grown at an average rate of 2% per annum and cocoa production in the Americas has declined at an average rate of 1% per annum. Following the liberalisation of the cocoa marketing systems in the 1990s, farm gate prices in most cocoa-producing countries have been largely determined by international prices. As a result, farm gate prices have shown greater fluctuations in most cocoa-producing countries between 1996 and 2006 reflecting, *inter alia*, changes in international cocoa prices, variations in the international value of the domestic currency, and specific local market structures and conditions, including taxation, competition, distance from port and quality. Although world market prices in real terms were 13% lower in 2005/2006 than in 1993/1994, real farm gate prices increased in some producing countries during that period, as national price inflation was more than offset by rises in nominal farm gate prices (ICCO 2007a,b).

Figure 4.2 provides the f.o.b. prices for Fairtrade cocoa beans that purchasers had to pay during the September 1998–February 2005 period and the corresponding monthly averages of the ICCO daily prices, which reflect the prices of cocoa beans in the London and New York futures markets. It is obvious that the incentive to sell under the Fairtrade market for producers and, conversely, the opportunity cost for Fairtrade purchasers is higher during periods of low market prices, as was the case in 1999–2001.

The 'Fairtrade price' or 'FLO price' represents the price received by the co-operatives. The use of the funds derived from the premium (US $150 per tonne) is decided by the General Assembly of the Co-operatives, which

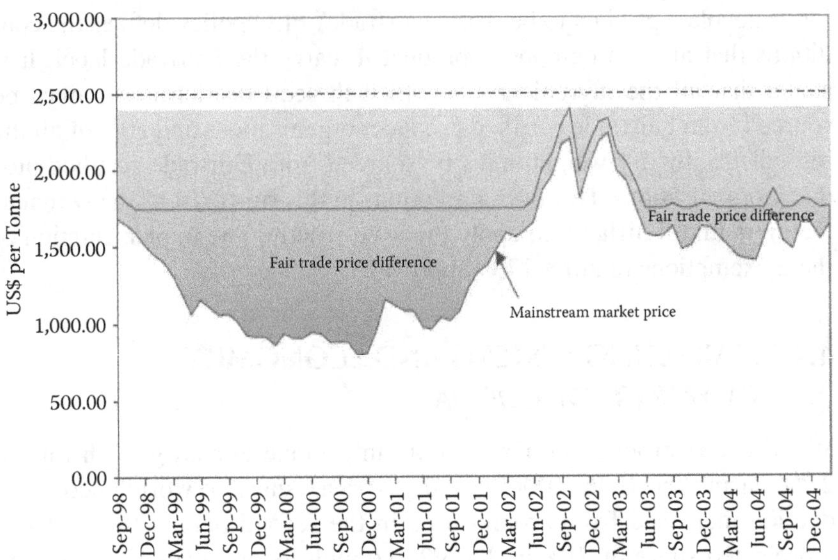

FIGURE 4.2 FLO f.o.b. prices versus f.o.b. market prices September 1998–February 2005. (Adapted from ICCO, *Consultative Board Report on the World Cocoa Economy. Fifth Meeting*, 2005. With permission.)

is required to act with total transparency. A proportion of the premium may be paid out to farmers but, in general, it is pooled in a social fund for the benefit of the community rather than passed on directly to the farmers (ICCO 2005).

Fairtrade cocoa price per tonne = Max {FLO floor price (US $1600); f.o.b. market price} + Fairtrade premium (US $150 per MT f.o.b)

FLO (2005) explained that the premium fee paid is used for either cocoa-related projects, such as farmer training and creation of nurseries for new planting materials, or for social projects, such as boreholes, schools and other community-based investments that directly provide benefits to the farmers and their communities.

In contrast to the premium of US $150 per tonne, there are no prescriptions under the Fairtrade arrangements on the use of the difference between the FLO minimum price of US $1,750 per tonne minus premium (US $150) and the mainstream price. For example, if the mainstream market price is US $1,200 per tonne, the difference with the FLO minimum price minus premium is US $400 per tonne. It is up to the individual co-operative to decide on the use of these funds. It is further noted that, in most cases, the farmers, at the moment of selling to their co-operatives, receive the same price for their cocoa as when the co-operative sells mainstream cocoa. In most cases, the co-operatives pay the same price to all farmers and they sell only part of their total trade volume under the Fairtrade arrangement. The co-operatives do not know which farmers have delivered to them the cocoa they sell under their Fairtrade arrangements. The situation naturally leaves open the possibility for a co-operative to distribute advantages of the Fairtrade arrangements among all their members (FLO 2005).

Consolidation in the cocoa trade and in the cocoa-processing industry which resulted in a highly concentrated industry structure, enabled cocoa processors and chocolate companies to reduce demand for physical stocks during the 1990s. In addition, developments in bulk transportation, information technology and communications contributed to greater efficiencies in stock management. The liberalisation of the marketing system in Côte d'Ivoire, the world's largest supplier of cocoa, was associated with the discontinuation of forward sales at the end of the last decade, reducing the availability of forward physical cover for manufacturers. As a result, cocoa processors and chocolate manufacturers are believed to have substantially reduced the amount of their forward purchases, greatly affecting the forward structure of the market, which consequently changed to become more of a spot market. Although the cocoa market has witnessed some changes in the behaviour of market participants since the 1990s, a period which witnessed high levels in the stocks-to-grindings ratio, world market prices continued, in general, to reflect adequately the degree of imbalance between supply and demand across time (ICCO 2007a,c).

The long-awaited turning point in world cocoa bean prices was delayed until the beginning of 2001 when the industry faced the prospect of a substantial decline in global stocks, sharply reducing the world stocks-to-grindings ratio. Market operators acknowledged that the world cocoa economy had indeed entered a phase of structural deficits. A period of low and falling prices had reduced farmers' incomes, and they were responding by reducing inputs, which was reflected in lower yields as husbandry standards declined. Cocoa farmers were unable to counter the increasing losses in yields from the spread of pests and diseases (black pod in most regions, especially Africa, witches' broom in the Americas, cocoa pod borer in Asia) because of insufficient resources (ICCO 2007a,c). By the end of the 1990s, productivity gains from maturing trees were also approaching peaks and expansion of cocoa cultivation through new plantings had apparently ceased. At the same time, demand had been reacting positively to the low prices, as cocoa consumption continued to recover in Eastern Europe and showed growth in new markets in Asia. After recording a 28-year low of US $800 per tonne in December 2000, the cocoa market changed direction in 2001, arresting the persistent decline in prices that had been a feature of the market since May 1998. However, the recovery was fragile, with prices remaining volatile, changing direction several times during the year and drifting downwards in the last two months of the 2000/2001 cocoa season. Nevertheless, international prices averaged US $980 per tonne in that year, representing an increase of 15 from the 1999/2000 season (ICCO 2007a).

A period of irregular but sustained price increases during the 2001/2002 cocoa year confirmed that the period of low market prices, a feature of most of the previous decade, had been reversed. The spectacular recovery in cocoa bean prices in 2001/2002 is illustrated by the fact that, for the year as a whole, cocoa prices averaged US $1,200 per tonne, representing an increase of almost 60% over the previous season. Such a dramatic annual increase in prices had been recorded on only two earlier occasions, once in 1972/1973 when cocoa prices also rose by around 60% and in 1976/1977, when they increased by almost 120%. The upward pattern in market prices seen in 2001/2002 was largely due to a deterioration of the fundamental supply–demand situation in the world cocoa market, but short-term technical and speculative factors also contributed to price increases by initiating a number of price rallies during that year. Cocoa futures prices were further boosted following an attempted coup on 19 September 2002 in Côte d'Ivoire.

Concerns over potential disruptions to the flow of cocoa at the beginning of the new crop year, originating from a continuing political and social crisis in the world's leading cocoa-producing country, pushed international prices to 16-year highs at US $2,400 per tonne in October 2002. The remaining prospect of a third successive production deficit that could reduce even more the stocks-to-grindings ratio in 2002/2003 and short-covering by trade, investment funds and speculators also contributed to an additional increase in cocoa prices in the first half of the 2002/2003 cocoa year. Nevertheless, the harvesting, transportation and commercialisation of cocoa proceeded normally in Côte d'Ivoire, despite the prevailing political and social unrest in the country. Moreover, the higher international cocoa bean prices were closely reflected in the rising trend in farm gate prices in Côte d'Ivoire which prompted higher standards of husbandry, as well as increased sales of pesticides and fertilisers. Thus, rising yields may have helped farmers to offset the impact of civil unrest in the country.

After a three-year period of constant increases, average international cocoa prices recorded a sizeable drop of 23% in the 2003/2004 season, at US $1,650 per tonne. The concerns over potential disruptions to harvesting, evacuation and export of cocoa from Côte d'Ivoire caused by the political and social unrest in the country gradually weakened during the 2003/2004 season. The other major underlying factors influencing the movement of prices during the 2003/2004 season were the weather conditions and the evolution of the consequent outlook for the crops in West Africa. Although most analysts expected a global production deficit at the beginning of the season, the production forecasts were progressively revised upwards towards a large surplus at the end of the season. However, the downward pattern broke sharply at the beginning of July 2004, reflecting concern among market participants about the weather in West Africa. At the end of August, futures prices reached their highest levels for the 2003/2004 season at US $2,350. At the beginning of September, larger mid crops in West Africa and improving rain conditions contributed subsequently to a fall in prices. At US $1,650 per tonne in the 2004/2005 cocoa year, the average international cocoa prices remained at the same level as during the previous season. This was mainly explained by the balanced supply and demand situation experienced during this campaign. Nonetheless, two rallies occurred during the season. In November 2004 and in February to March 2005, international prices increased to ~US $1,750. The first rally was due to the slow flow of cocoa to the ports because of renewed tensions in Côte d'Ivoire. This resulted in speculative buying in large volumes. Speculation and technically driven factors were the causes for the second rally which lasted almost six weeks (ICCO 2007a,c).

The most essential characteristic of Fairtrade is that producer organisations receive a higher price for their cocoa beans. The Fairtrade price represents the necessary condition for the producer organisations to have the financial ability to fulfil the above requirements, and to cover the certification fees. The differential in the price of cocoa beans between the conventional market and the Fairtrade market represents the consumers' willingness to pay for a certified product. The Fairtrade prices are calculated on the basis of world market prices, plus Fairtrade premiums. The Fairtrade premium for standard quality cocoa is US $150 per tonne. The minimum price for Fairtrade standard quality cocoa, including the premium, is US $1,750 per tonne. If the world market price of the standard qualities rises above US $1,600 per tonne, the Fairtrade price will be the world market price plus US $150 per tonne.

The cost of compliance includes the fees paid by the farmer organisations to the Fairtrade Organization and indirect costs to comply with the FLO requirements. The cost of certification used to be borne by the importers and not the producers. This made the FLO certification unique, by passing the whole cost of the certification to the buyer. However, since December 2004, both registered producer associations and traders have to support certification fees, mainly to provide additional resources to the newly created FLO-Cert Ltd. For traders, the costs are composed of an initial application fee (up to €2,000 and payable only once) and an annual certification fee (up to €3,000) dependent upon the total turnover of the trading company. To get the FLO certification, producer organisations have first to pay an initial application fee (up to €5,200 and payable only once). For the following years, the fee is composed of a fixed amount (€500 per year), and a variable amount depending on the value of cocoa sold under Fairtrade (0.45% of the f.o.b. value). This implies that a co-operative which sells 50 tonnes of cocoa in one season has to pay fees of US $20 per tonne. With a Fairtrade turnover of 500 tonnes the average fee is cut in half, amounting to US $10 per tonne.

The FLO-approved producer organisations must comply with a number of requirements. Only organisations of small farmers can be given the FLO certification. The FLO Fairtrade standards for cocoa require the following:

1. The Fairtrade activity is to promote the 'social development' of the organisations. To this end, the FLO-certified producer organisations have to consist mainly of farmers managing their own farms and the organisations must have a democratic and transparent structure.
2. The Fairtrade activity is supposed to enhance the 'economic development' of the organisation. To this end, the FLO-certified producer organisations must have the capacity to export their production to strengthen their business operations. Moreover, the Fairtrade premium is supposed to be managed democratically.
3. Under the heading of 'environmental development,' the FLO-registered producer organisations have to include the environment in farm management. More specifically, the use of certain listed pesticides is prohibited and the production of organic cocoa beans is encouraged.
4. Working conditions in FLO-registered producer associations have to follow the International Labour Organisation (ILO) conventions. The FLO 'standard on labour conditions' describes how child labour can be used, as well as the requirements in terms of freedom of association and collective bargaining. All farm workers have to work in a safe environment and under fair conditions of employment, especially regarding wages (Fairtrade Foundation 2004; FLO 2006).

4.5 SUPPLY-CHAIN MANAGEMENT OF FAIRTRADE COCOA

The certification system enables all companies to participate in Fairtrade by sourcing products from an open register of certified producer organisations and accredited exporters, importers and manufacturers, and thus encourages deeper engagement between producers and buyers. Along with a commitment to long-term equitable relationships, there is also a need for sharing of market information and assistance from buyers in developing the capacity of producer organisations to achieve the objective of empowering producer organisations; and businesses that deal exclusively in Fairtrade cocoa products will engage in these activities as part of their core operations.

The organisations engaged in Fairtrade cocoa trading can be divided into three groups:

1. The network and umbrella organisations of the Fairtrade movement, which consist of the following four organisations: the Fairtrade Labelling Organization (FLO), the International Fairtrade Association (IFAT), the European Fairtrade Association (EFTA) and the Network of European World Shops (NEWS). The Fairtrade Labelling Organization was established in 1997, and is the worldwide Fairtrade standard setting and certification organisation. Since 2004, it has been composed of two independent bodies, FLO-I for standard setting and FLO-Cert Ltd. for Fairtrade certification and auditing activities. The FLO has established common principles, procedures and specific certification requirements for Fairtrade and certifies mainly commodity products. This relates partially to the fact that non-commodity products are usually not subject to direct comparison of price and quality. The FLO deals with cocoa, as well as with coffee, bananas, tea, honey products, rice, fresh fruits, juices, sugar, sport balls, wine and flowers. It estimates its total retail sale value at US $500 million in 2004. The FLO membership consists of the 20 National Initiatives located across Europe, North America, Mexico and Australia/New Zealand, as listed in Table 4.1 and has recently introduced a common label to be applied across all products in all countries. The International Fairtrade Association (IFAT) was established in 1989 and is a worldwide membership organisation that brings together both producers and buyers. It is a federation to promote Fairtrade and a forum for exchanging information to help members increase benefits for producers. It consists of approximately 110 producer organisations and 50 buying organisations. The Network of European World Shops, established in 1994, acts as the umbrella body for the approximately 2,700 'world shops' that sell predominantly Fairtrade goods across Europe. The European Fairtrade Association (EFTA), established in 1990, is an association of 12 importing organisations in nine European countries.
2. The Southern producer organisations, which supply the products, are traditionally cooperatives or associations. Presently, 15 have the FLO certification to sell cocoa beans under the Fairtrade label. Although cocoa is mainly produced in Africa (71%), 12 FLO-registered cocoa-producer associations are located in the

Latin American and Caribbean region. FLO-registered traders only buy part of the cocoa beans produced by the participating co-operatives. The remainder is sold in the mainstream market. Table 4.5 provides an exhaustive list of the producer organisations with their main characteristics.
3. The Fairtrade importing organisations, known as Alternative Trading Organizations (ATOs) are traditionally non-governmental organisations (NGOs) in Northern countries. These are the buying organisations, which act as importers, wholesalers and retailers of the products purchased from the producer organisations. They focus on improving market access and strengthening producer organisations. In Europe, they sell their products through 'world shops', local groups, campaigns, wholesale and mail-order catalogues (EFTA 2006; FLO 2006). Table 4.6 provides an exhaustive list of the 47 FLO-registered cocoa traders.

Monitoring and auditing of Fairtrade-certified cocoa products are designed to provide an effective independent guarantee as efficiently as possible. Companies selling Fairtrade-certified cocoa products provide quarterly reports to verify their supply chain and trading terms for labelled products. These reports are audited annually by physical inspection. Intermediary traders and processors are accredited by FLO to supply primary and semi-finished cocoa products, and help to maintain an auditable supply chain (Fairtrade Federation 1999; Fairtrade Foundation 2006). FLO inspects producer organisations on behalf of all its members, therefore producers need only one certification to supply all Fairtrade cocoa markets. However, neither the Fairtrade Foundation nor FLO trades in Fairtrade products; their role is to certify products against Fairtrade standards, aimed at providing a robust and credible international certification system as efficiently as possible. To this end, the monitoring and audit process for Fairtrade cocoa labelling recognises a number of distinct actors:

- *Licensees:* Suppliers of finished cocoa products bearing the Fairtrade mark under a license agreement with the Fairtrade Foundation (in the United Kingdom) or another member of FLO International.
- *Manufacturers and Processors:* Traders of the primary Fairtrade cocoa and/or semi-finished cocoa products are accredited by FLO International to maintain an auditable supply chain.
- *Exporters and Importers:* Traders in the primary Fairtrade cocoa between countries and accredited by FLO International to maintain an auditable supply chain.
- *Fairtrade Producers:* The primary producing organisation, certified by FLO International against the Fairtrade producer standards.

Not all products go through all of these stages and some have more than one processing stage; for example, bananas have a very short supply chain, whereas that for cocoa and chocolate is more complex. When a single company fulfils more than one of these roles—for example, a coffee company that imports and roasts green beans and then markets the finished product—they will have a contractual and reporting relationship with both the Fairtrade Foundation and FLO International (Fairtrade Foundation 2006).

Figure 4.3 is a flow diagram showing the various stages of the supply chain management of Fairtrade cocoa. Every product that carries the Fairtrade mark requires a licensee to have overall responsibility for compliance with Fairtrade standards. The licensee should be the last supplier in the wholesale supply chain; in the case of proprietary brands this will usually be the brand owner, whereas for private label products the licence may be held by the brand owner or their immediate supplier. There can be only one licensee for any product and licensees sign just one contract with the Fairtrade Foundation that covers all their Fairtrade products sold in the United Kingdom. Under the terms of their contract with the Fairtrade Foundation, licensees must ensure they buy certified Fairtrade ingredients for use in their products that carry the Mark, and these must be supplied by a registered importer, manufacturer or processor. Similarly intermediary processors must buy from registered importers or be accredited as an importer in their own right.

The importer is responsible for ensuring that the primary Fairtrade certified product has been bought from a registered producer at the specified terms of trade. Both of these elements must apply for a product to be certified Fairtrade; neither products bought from registered producers at conventional market price, nor products bought from non-registered producers, even at the Fairtrade price, can be certified. All intermediary suppliers are required to denote products as 'Fairtrade Certified' only when they are sold to other accredited actors in the Fairtrade market in order to maintain the integrity of Fairtrade labelling. As intermediary suppliers are accredited by FLO International, they can buy from multiple producers and sell to multiple licensees under a single contract and reporting arrangement. Fairtrade producers are monitored and inspected by FLO International and so need only one certification to supply to any of the 20 Fairtrade cocoa markets (Fairtrade Foundation 2006).

4.6 CONCLUSION AND FUTURE PROSPECTS

The Fairtrade labelling, standards and certification system offers important economic and social support to farmers and their communities in the developing world, with potentials of fostering economic growth and

TABLE 4.5
FLO Certified Cocoa Producer Associations

Country	Organisation	Started in	Reported Total Sales, in 2003/2004 (in Tonnes)	Reported Total Sales Under Fairtrade, in 2003/2004 (in Tonnes)	Key Characteristics
Central America					
Belize	Toledo Cocoa Growers' Association (TCGA)	1993	30	30	All certified Fairtrade cocoa also organic
Costa Rica	APPTA	1997	N/A	N/A	Produces organic certified cocoa
Nicaragua	CACAONICA	1994	N/A	N/A	Organic certified
Caribbean					
Dominican Republic	Conacado Inc.	1994	8,500	1,200	Sold 5,000 tonnes of organic cocoa. Sales of Fairtrade cocoa are steady. Processes Fairtrade certified cocoa semi-finished products.
Haiti	FECCANO	2003	N/A	N/A	N/A
South America					
Bolivia	El Ceibo	2000	400	40	10% is Fairtrade certified and 90% is organic certified – Processes fairtrade certified cocoa semi-finished products.

Peru	COCLA	2004	0	0	Involved in coffee and tea – Expect to produce 140 tonnes of certified organic cocoa.
Peru	Cooperativa Agraria Cacaotera Acopagro Ltda	2005	380	0	Produced 150 tonnes of certified organic cocoa in 2004
Peru	Cooperativa Agraria Cafetalera El Quinacho	2003	N/A	N/A	Produces certified organic cocoa
Peru	Cooperativa Agraria Cafetalera Valle Rio Apurimac	2003	N/A	N/A	Produces certified organic cocoa
Peru	Cooperativa Agraria Industrial Naranjillo Ltda	2004	N/A	N/A	Produces certified organic cocoa
Peru	Cooperativa Agraria Cafetalera Pangoa	2005	34	0	Produced 24.7 tonnes of certified organic cocoa in 2004
			Africa		
Cameroon	MACEFCOOP	1997	N/A	N/A	Produces coffee
Ghana	Kuapa Kokoo Union	1993	62,901	1,800	Volumes of cocoa beans available to the Fairtrade depends on Cocobod policy.
Côte d'Ivoire	Kavokiva – Cooperative Agricole Kavokiva de Daloa	2004	15,000	0	—

Source: ICCO, *Consultative Board Report on the World Cocoa Economy. Fifth Meeting*, 2005.

TABLE 4.6
FLO Cocoa Registered Traders

Organisation	Country	FLO Status
Agglomeration Technology Ltd	UK	Cocoa Manufacturer
Alter Eco	France	Cocoa Importer Manufacturer Licensee
Anayate	Canada	National Member Cocoa Manufacturer Licensee
Barry Callebaut Belgium NV	Belgium	Cocoa Manufacturer
Barry Callebaut France	France	Cocoa Manufacturer
Barry Callebaut Sourcing AG	Switzerland	Cocoa Importer
CafeMa International	Switzerland	Cocoa Importer
Chocolat Bernrain	Switzerland	Cocoa Importer Manufacturer
Chocolaterie Pralibel N. V.	Belgium	Cocoa Manufacturer Licensee
Claro Fair-trade AG	Switzerland	Cocoa Importer Licensee
Commerce Equitable Oxfam Quebec Inc	Canada	National Member Cocoa Manufacturer Licensee
Commercio Alternativo	Italy	Cocoa Importer Licensee
Conapi S. c. r. l.	Italy	Cocoa Importer
Consorzio CTM- altromercato	Italy	Cocoa Importer
Daarnhouwer & Co. B.V	Netherlands	Cocoa Importer
Dutch Cocoa	Netherlands	Cocoa Grinder
Echange Equitable	France	Cocoa Importer Manufacturer
EcoTrade	USA	Cocoa Importer Licensee
Gepa, Fair Handelshaus	Germany	Cocoa Importer
Grand Foods Ltd.	UK	National Member Cocoa Manufacturer Licensee
Herza Schokolade GmbH & Co	Germany	Cocoa Manufacturer
Horsley, Hick & Flower Ltd	UK	Cocoa Manufacturer
ICAM S. p. a	Italy	Cocoa Importer Grinder, Manufacturer Licensee
Jesse Oldfield Ltd	UK	National Member Cocoa Grinder Licensee
Kerry Sweet Ingredients	UK	Cocoa Manufacturer
La Siembra Co-operative	Canada	Cocoa Importer Licensee
Maestrani	Switzerland	Cocoa Manufacturer licensee
Monbana	France	Cocoa Manufacturer
Nichols Foods Ltd.	UK	National Member Cocoa Manufacturer Licensee
Northern Tea Merchants	UK	National Member Cocoa Grinder Licensee
Nv ‘t Boerinneke- Marino	Belgium	Cocoa Manufacturer
Omanhene Chocolate	USA	Cocoa Importer Licensee

(Continued)

TABLE 4.6 (CONTINUED)
FLO Cocoa Registered Traders

Organisation	Country	FLO Status
OntarBio Organic Farmers' Co-operative Inc.	Canada	National Member Cocoa Manufacturer Licensee
Oxfam Fairtrade cbva	Belgium	Cocoa Importer Licensee
Petit'Grandeur	France	Cocoa Chocolate Manufacturer
Phoenix Foods Ltd.	UK	Cocoa Manufacturer
Progetti e Qualita P. s. c. a. r. l	Italy	Cocoa Importer Licensee
Pronatec AG	Switzerland	Cocoa Importer Grinder, Manufacturer Licensee
Rapunzel	Germany	Cocoa Importer Manufacturer
Satro quality drinks GmbH	Germany	National Member Cocoa Manufacturer Licensee
The Cool Hemp Company	Canada	National Member Cocoa Manufacturer Licensee
The House of Sarunds Limited	UK	National Member Cocoa Manufacturer Licensee
Trading Organic Agriculture B.V.	Netherlands	Cocoa Importer
Twin Trading Ltd	UK	Cocoa Importer
Urtekram A/S	Denmark	Cocoa Importer Licensee
Wakachiai Project	Japan	Cocoa Importer
Weinrich & Co. GmbH	Germany	Cocoa Manufacturer

Source: ICCO, *Consultative Board Report on the World Cocoa Economy. Fifth Meeting*, 2005.

development within the regions. The certification process has enhanced the sustainability of marketing systems in the Fairtrade cocoa market over the past decade, with seemingly continuous progressions in the future. This can be achieved by FLO International through continuous monitoring of the supply chains of Fairtrade cocoa; helping companies develop new Fairtrade-certified cocoa products; and working with others to raise consumer awareness and support for Fairtrade cocoa and its certified products. In the long term, with growing demand and supply for Fairtrade cocoa, these markets will mature and price premiums for Fairtrade cocoa are likely to decrease. In order to ensure sustainable further development of this new and growing market niche, farmers will need to improve their productivity and quality to maintain or increase their income. The various Fairtrade organisations should ensure strict producer standards and ensure that such organisations are democratic and accountable to their members, capable of ensuring compliance with social and economic criteria

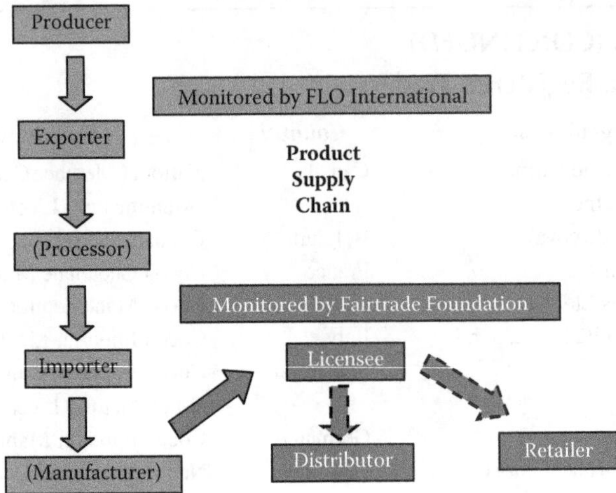

FIGURE 4.3 Flow diagram showing supply chain management of Fairtrade cocoa. (Adapted from Fairtrade Foundation, *Fairtrade Foundation Report*, 2006. With permission.)

and committed to a programme of ongoing improvement. Retailers have a vitally important role as 'gatekeepers' to the consumer market by making Fairtrade cocoa certified products available, informing their customers about Fairtrade cocoa, and protecting them from misleading claims. At the same time, the growing interest in Fairtrade cocoa offers retailers an opportunity to increase their sales of premium, added-value products and to boost customer loyalty by demonstrating a shared concern for a fairer world.

5 Organic Cocoa Industry

5.1 INTRODUCTION

Since the late 1990s, demand for certified organic food has emerged as a vibrant market within the global food trade. The organic food and drinks market was estimated to have reached in 2005 a value of US $18 billion in the United States and a value of US $5.4 billion in Germany and US $4.1 billion in the United Kingdom, the two major European markets for organic products. The recent strong growth was largely driven by increased consumer concern over food safety. The outbreak of 'food scares' in several countries around the world, such as bovine spongiform encephalopathy (BSE) and salmonella, together with the emergence of public awareness of the risks of food processing have raised concerns over food safety and production methods. In response to this strong growth, organic food production is increasing on all continents, with much of the increase occurring in third-world countries, where farmers are attracted by the higher prices for organic food products. For the same reason, many governments encourage farmers to convert to organic farming (ICCO 2006).

At present, there are many definitions in use in different regions of the world and these, even though complex, are not restricted to a 'chemical free' method. A more generalised definition has been formulated in the *Codex Alimentarius* (FAO/WHO 1999) as: 'Organic agriculture is a holistic production management system which promotes and enhances agro-ecosystem health, including biodiversity, biological cycles, and soil biological activity. It emphasises the use of management practices in preference to the use of off-farm inputs, taking into account that regional conditions require locally adapted systems (ICCO 2006).'

According to the International Federation of Organic Movements (IFOAM), organic agriculture is an agricultural system that promotes environmentally, socially and economically sound production of food, fibre, timber and so on. It relies on ecological processes, biodiversity and cycles adapted to local conditions, rather than the use of inputs with adverse effects. Organic agriculture combines tradition, innovation and science to benefit the shared environment and promote fair relationships and a good quality of life for all involved (International Federation of Organic Agriculture Movements; IFOAM 2006). Organic cocoa farming should sustain and enhance the health of soil, plant, animal, human and planet as one and indivisible. It should also be based on living ecological systems and cycles, work with them, emulate them and help sustain them. Organic cocoa farming should build on relationships that ensure fairness with regard to the common environment and life opportunities and be managed in a precautionary and responsible manner to protect the health and well-being of current and future generations and the environment.

Organic certified cocoa beans represent less than 1% of the worldwide cocoa crop (Barry Callebaut 2013) and are mainly produced by few countries, especially those producing very low quantities of the global bulk cocoa beans (Table 5.1). The majority of organic beans, about 75%, are produced in the Americas, with the Dominican Republic topping the list (Table 5.2). The category of organic chocolate confectionery was expected to grow 34.3% worldwide from 2008 to 2012 or from 30,300 tonnes to 40,700 tonnes, according to market intelligence provider Euro Monitor International. Despite the strong growth in recent years, organic cocoa and chocolate remains a niche market. Europe is the biggest market for imports of organic cocoa beans, and most organic products sold in the United States and in Canada are imported from Europe. Business and government leaders, economists and organic experts often have differing views about the value of actively increasing the amount of land under organic management. One of the challenges cited in debates about organic agriculture is increasing crop yield per hectare to match or exceed the yields produced under conventional farming practices. (Note: In view of the limitations of data availability, such estimates should be regarded as provisional and should be used with due caution.)

The general requirements for cocoa beans, cocoa and chocolate products to be labelled 'organic' are the following: cocoa beans must grow on land which has been free of prohibited substances for three years prior to harvest. Cocoa beans grown on land which is 'in transition' to organic during the first three years can be sold labelled as 'cocoa in conversion' and also cocoa production and processing methods are regulated (only organic fertilisers, improvement in soil fertility, biodiversity, etc.). For the production of chocolate, 95% of the ingredients (not counting added water and salt) must be organically produced. The processor must also be certified. However, special provisions allow labelling to state that a product is '100% Organic' if the product contains 100% organically produced ingredients; 'Made with Organic Ingredients' (or a similar statement), if the product contains at least 70% organic ingredients; and 'Has some organic ingredients' (or a similar statement), if the product contains less than 70% organic ingredients. Producers of cocoa, which are normally located in developing

TABLE 5.1
Organic Cocoa Producers

Country/Region	Co-Operatives/Organisations
	Africa (6)
Ghana	
Madagascar	Arco Ocean Indien/Millot – Remandraibe – Sagi
São Tomé	Cecab
Tanzania	Biolands/Kyela Co-op Union
Togo	
Uganda	ESCO
	Americas (14)
Belize	TCGA
Bolivia	El C
Brazil	
Colombia	
Costa Rica	APPTA
Cuba	
Dominican Republic	Five exporters, incl. Conacado and Yacao
Ecuador	
El Salvador	
Mexico	Asesoria Técnica en Cultivos Orgánicos
Nicaragua	La Campesina – Cacaonica
Panama	Cocabo, Servicio Múltiple de Cacao Bocatoreña
Peru	COCLA - Cooperativa Agraria Cacaotera Acoprago Ltda - Cooperativa Agraria Cafetalera El Quinacho - Cooperativa Agraria Cafetalera Valle Rio Apurimac - Cooperativa Agraria Industrial Naranjillo Ltda -
Venezuela	Aragua state cocoa farmers association
	Asia and Oceania (4)
Fiji	
India	
Sri Lanka	
Vanuatu	Malecoula

Sources: ICCO, *A study on the market for organic cocoa,* 2006; IFOAM, *The world of organic agriculture: Statistics & emerging trends*, 2006; SIPPO 2002. With permission.

countries, have to meet those standards. Changes in legislation in developed countries can affect the possibility for small producers to access the global market. Today, to import organic products to the EU market, an import permit is needed.

TABLE 5.2
Estimated Production and Exports of Organic Cocoa Beans

Country/Region	Date	Organic Production (in Tonnes)	Organic Exports (in Tonnes)
Africa (5)		**3,000**	**1,770**
Ghana	2005	n.a	n.a
Madagascar	2003	1,500	1,500
São Tomé		n.a	n.a
Tanzania and Uganda	2005	1,500	270
Togo		n.a	n.a
Americas (14)		**11,738**	**8,638**
Belize	2004/2005	33	33
Bolivia	2003/2004	400	400
Brazil	2005/2006	1100	50
Colombia		n.a	n.a
Costa Rica	2004/2005	300	300
Cuba		n.a	n.a
Dominican Republic	2004/2005	5,000	5,000
Ecuador		n.a	n.a
El Salvador	2005	30	30
Mexico	2005	2,500	600
Nicaragua	2004	98	98
Panama	2005	350	350
Peru	2005	1,850	1,700
Venezuela	2005	77	77
Asia and Oceania (4)		**762**	**762**
Fiji	2002	50	50
India	2005	12	12
Sri Lanka	2005	200	200
Vanuatu	2002	500	500
Total identified		**15,500**	**11,170**

Sources: ICCO, *A study on the market for organic cocoa*, 2006; IFOAM, *The world of organic agriculture: Statistics & emerging trends*, 2006; SIPPO, 2002. With permission.

5.2 BENEFITS OF ORGANIC COCOA FARMING

Organic cocoa farming has both environmental and economic benefits to the global cocoa industry. In general, organic cocoa farming has the following advantages over other conventional farming techniques:

1. Environmental advantages would include soil conservation, increased diversity of plants and animals, utilisation of local and renewable resources, reduced soil and groundwater pollution and can contribute towards specific habitat conservation.
2. There is an increase in demand for organic products because of the consumer perception that organic products are of high nutritional and health value, partly due to the restrictions on the use of fertilisers and pesticides, which reduces the likelihood of any harmful residues.
3. The existence of this niche market provides a means through which producers can be compensated for internalising external costs that would have been borne by society.
4. Lower production intensity in organic farming can help in limiting surpluses due to lower yields per unit area and reduced areas of intensive farming.
5. Organic farming offers opportunities for the diversification of farms and has the potential to contribute towards rural development.

5.3 CONSUMPTION PATTERNS OF ORGANIC COCOA

The world market for premium chocolate (including flavoured, single-origin, organic, ethically traded and high-cocoa chocolate) has grown significantly in recent years, and will continue to do so, even in times of economic recession, as consumers seek out affordable luxuries during bad times, having foregone larger indulgences (Cooper 2008). It was predicted that the global premium chocolate market would grow from US $7 billion in 2007 to US $12.9 billion (or US $3.6 billion in the United States alone) in 2011 (Moran 2008), driven by increasing consumer awareness of premium chocolate and a growing interest from the world's leading chocolate manufacturers in the premium chocolate segment.

Although the category is estimated to have expanded by 65% since 2002, premium chocolate accounts for less than 10% of the global chocolate market, and is believed to account for 12 and 18% of total chocolate sales in Europe and the United States, respectively. One in three consumers in these two regions is believed to have changed their consumption pattern in favour of premium chocolate products over the past few years. A survey by Barry Callebaut in 2007, explained that consumer awareness of organic and fair-trade chocolate is growing in both Europe and the United States. About 33% of consumers in Western Europe and the United States declared having purchased fair-trade chocolate, and 24% had tried organic chocolate. Purchasers are no longer confined to the higher income groups, and the segment is making inroads into the mainstream chocolate market (Cooper 2008).

5.4 CERTIFICATION AND MARKET FOR ORGANIC COCOA

Worldwide there are many systems and marks for certifying organic produce. Within the European Union, the logo bearing the words 'Organic Farming' or translations thereof (Figure 5.1) can be used on a voluntary basis by producers whose systems and products have been found to satisfy Council Regulation (EEC) No 2092/91. The 'Euro-leaf' logo (Figure 5.1 bottom right) became compulsory from 1 July 2009 for pre-packaged organic food produced in any of the 27 EU member states (ICCO 2010b).

Many of the major chocolate manufacturers now emphasise the need for traceability along supply chains and collaborate with various certification organisations, three of which are described below.

The Fairtrade Foundation (http://www.fairtrade.net) sets labor and economic as well as environmental and phytosanitary standards:

> Fairtrade Standards include requirements for environmentally sound agricultural practices. The focus areas are: minimised and safe use of agrochemicals, proper and safe management of waste, maintenance of soil fertility and water resources and no use of genetically modified organisms. Fairtrade Standards do not require organic certification as part of its standards. However, organic production is promoted and is rewarded by higher Fairtrade Minimum Prices for organically grown products.

FIGURE 5.1 Examples of different organic certification marks. (Adapted from ICCO, Report presented to the Executive Committee at the 142nd Meeting, 2010b. With permission.)

FIGURE 5.2 Examples of marks used by certification bodies involved with cocoa traceability and GAP. (Adapted from ICCO, Report presented to the Executive Committee at the 142nd Meeting, 2010b.)

The Rainforest Alliance (http://www.rainforest-alliance.org) works to conserve biodiversity and ensure sustainable livelihoods by transforming land-use practices, business practices and consumer behavior. Working with a network of environmental groups, farmers must comply with appropriate standards for protecting wildlife, wild lands, workers' rights and local communities in order to be awarded the certified seal (Figure 5.2).

UTZ CERTIFIED (http://www.utzcertified.org/index.php?pageID = 224) producers comply with the Code of Conduct covering good agricultural practices, and social and environmental criteria. Their compliance is checked yearly by an independent auditor. The developmental stage of its programme was to be completed by the end of 2009 with the first certified cocoa expected in early 2010. The initial focus is on Côte d'Ivoire, but the intention is to expand to other cocoa producing countries.

The labels of these organisations are shown in Figure 5.2.

Trade data regarding certified organic cocoa are extremely difficult to find. Three factors have been found to compound the absence of official statistics and these include: the extremely limited volumes produced and marketed, the various forms cocoa products take (beans, liquor, powder, cake, butter, paste and chocolate) and the disparity between quantities produced and traded due to stocks (Liu 2008). Data are not only incomplete and fragmented; even their reliability may be questioned. According to the International Cocoa Organization (ICCO), worldwide production and exports of organic cocoa stood at 15,500 tonnes and 11,170 tonnes, respectively, in 2005. However, Willer and Yussefi (2006) calculated a much larger figure exceeding 32,000 tonnes. Inasmuch as both sources do not include production volumes for important suppliers such as Colombia or Ecuador, Liu (2008) asserted that exported quantities ranged between 11,000 and 15,500 tonnes in 2005. Irrespective of the exact figures, it is certain that the organic cocoa market represents only a very small share of the total cocoa market – estimated at less than 1% of total production (www. icco.org) – and that is growing rapidly. According to a study by the Swiss Import Promotion Programme (SIPPO), the world market for organic chocolate grew by 10–15% per annum from 2000 to 2003 (Menter 2005), and Euro Monitor International estimated that global organic chocolate sales increased from US $171 million in 2002 to US $304 million in 2005.

According to industry sources, organic products still account for a small share of the total market, but this share is steadily increasing. These sources indicate that the total industrial demand for organic and Fairtrade beans amounted to around 25,000 to 30,000 tonnes in 2006, mostly in the European Union and the United States. The European Union takes by far the largest share of this demand, because part of American demand for organic cocoa is also sourced through European importers. The largest EU markets are Germany, the Netherlands and France, although non-EU member Switzerland is also of great importance. It can be estimated that this has increased substantially in the last years. Latin America is estimated to account for more than 70% of worldwide organic cocoa production, compared with its 13% share of the conventional cocoa market, which is dominated by African producers. The Dominican Republic is the world's largest organic cocoa supplier, with an annual production of around 5,000 tonnes (ICCO 2006). The National Confederation of Dominican Cacao Producers, which groups 182 producer associations, is the world's leading single supplier with a 60% share in worldwide exports (Velasquez-Manoff 2009).

Organic cocoa is subject to strong price fluctuations, which are mainly due to the small volumes traded, the lack of consistency in quality and the irregularity of supplies. Because of the volatility of prices, as well as the difficulties of estimating premiums along the supply chain (Liu 2008), price premiums for organic cocoa fluctuate considerably according to the source of information used. ICCO (2006) indicates a premium of US $100–300 per tonne, whereas Liu (2008) indicates premiums of up to US $1,600 in 2006. Other sources indicate an organic premium of between 10 and 40% (New Agriculturalist 2007), or 10 and 50%, over non-organic cocoa. Consumers accept a higher price for organic chocolate, especially in the United Kingdom and the United States, where organic chocolate commands a price that is about three times higher than that of conventional chocolate. Organic chocolate is considered a premium product, and its demand is relatively price insensitive (Rano 2008). According to Koekoek (2003), there might be an opportunity for African cocoa to distinguish itself on the organic market, as conventional African cocoa is generally valued higher on world markets than Dominican cocoa.

5.4.1 European Market for Organic Cocoa and Cocoa Products

Europe is by far the major market for organic cocoa beans, as well as for processing and manufacturing activities to obtain certified cocoa and chocolate products. Imports of organic cocoa into Europe were estimated at 14,000 tonnes of cocoa bean equivalents in 2003, of which an estimated 2,000 tonnes were re-exported to the United States (CBI 2007). According to Bakker and Bunte (2009), Europe currently processes around 15,000 to 20,000 tonnes of organic chocolate annually. Organic chocolate and other cocoa products are estimated to hold a market share of 0.8% by volume (Koekoek 2003). This market share is growing quickly, fuelled by an increase in ethical consumerism and a trend towards premium chocolate. The UK organic chocolate market is the largest in the European Union, with sales reaching €27.3 million in 2005 (Boal 2006), up 30% annually since 2002. This compares to an annual growth of just 1.5% for the overall UK chocolate market over the same period (Benkouider 2005). According to industry sources, the current organic chocolate market accounts for around 1.4% of the total UK chocolate market. This share is estimated to grow to around 2.4% over the next five years (Halliday 2008).

As European processors are seeing increased demand for organic cocoa products from their buyers (mostly chocolate makers), demand for organic beans is increasing. The organic market can be an important niche market for smaller producers of cocoa beans and butter. Several industry sources (large processors) indicated that they are currently examining potential suppliers of organic cocoa, and interest from specialised importers also remains strong.

5.4.2 Market for Organic Cocoa and Cocoa Products in the United States

According to industry sources, the US organic chocolate market was worth US $70.8 million in 2007, up 49% from 2006 and leading the growth in the organic snack food category. Organic chocolate sales remain only a fraction of the overall US chocolate market, worth US $6 billion (The Green Guide s.d., 2004); however, there has been a sharp rise in demand in recent years, driven by consumers' growing environmental awareness. Although the US market remains smaller, in absolute terms, than the European market, it is growing as fast as, or faster than, the European market (Menter 2005). According to the Nature Conservancy, organic chocolate sales have increased by around 70% annually since 2002, and industry players agree that the market potential remains high (Liu 2008). However, access to sufficient supplies of organic chocolate of an acceptable quality remains a major challenge. Dagoba, founded in 2001, was the first company to market organic chocolate bars in the United States. The company, which was acquired by Hershey in 2006, sources organic cocoa from Costa Rica, Ecuador, the Dominican Republic, Madagascar and Peru. Following the success of Dagoba, Green & Black's was imported into the United States in 2003. Other leading organic chocolate companies include Organic Commodity Products and Endangered Species Chocolate Company.

5.4.3 Organic Cocoa Market in Japan

The market for organic chocolate is still in its early stages in Japan and it is very difficult for a Japanese consumer to find this product on the shelves of organic stores. According to a study conducted by IFOAM Japan, one of the reasons is that most domestic confectioners and processors do not make organic chocolate products. This is partly because the Organic JAS certification system does not include milk, thus preventing processors from making organic chocolate products containing milk. Nonetheless, a few manufacturers make bitter chocolate domestically (e.g., *Nisshin Kako*). So far, Japan imports organic cocoa not as beans but as cocoa processed products. However, the market may grow, especially if certain conditions are met, such as amendments regarding the Organic JAS certification for milk, and also if Africa, from which most of the conventional cocoa imported in Japan is sourced, increases its volume of organic cocoa beans produced or if Japan sources more of its imports from the Americas.

5.5 COSTS AND BENEFITS ASSOCIATED WITH ORGANIC COCOA EXPORTS AT PRODUCER LEVEL

Organic cocoa commands a higher price than conventional cocoa, which should cover both the cost of fulfilling organic cocoa production requirements and certification fees. The differential in cocoa bean price between conventional and organic markets represents the consumers' willingness to pay for an organic certified product. A significant share (around 10%) of organic certified cocoa beans in the market is, in addition, Fairtrade certified. For reference, FLO-certified organic cocoa beans receive a fixed price premium of US $200 per tonne. For 'non-Fairtrade' certified cocoa beans, there is no fixed price premium, as it is subject to market fluctuations, usually ranging from US $100 to US $300. For instance, according to trade statistics provided by the *Centro de Exportación e Inversión de la República Dominicana* (CEI-RD), the organic price premium averaged US $100 in 2003/2004 and US $275 in 2004/2005. However, some origins with small volumes can fetch much higher premiums. For several exporters, a premium of US $200 seems to be the minimum at which they can continue with organic exports.

The cost of compliance to organic standards includes the fee to be paid by the farmer organisation/exporter to the certification body and indirect costs to comply with organic requirements. The total fee can be composed of an initial application fee and an annual certification fee (on a fixed basis or in proportion to sales). According to The Organic Standard (TOS), the average certification fees at the farm level (for all types of products) were 3% of farm turnover. This may underestimate costs for cocoa farmers, as they are subject to additional costs due to the international aspect and the usually small size of their business. In addition to certification fees, the farmer organisations have to bear additional administrative costs, additional labour costs (as organic production is more labour-intensive than conventional production) and opportunity costs associated to the loss in yields after discarding synthetic inputs and converting their operations from conventional to organic production. Before restoration to a full organic system, pest suppression and fertility problems are common. The degree of yield loss varies and depends on the inherent biological attributes of the farm, farmer expertise and the extent to which synthetic inputs were used under the previous management regime.

5.6 DEMAND FOR ORGANIC COCOA

Promotion and marketing strategies of retailers and supermarkets have stimulated the demand for organic chocolate products. Food-retailing chains also promote organic foods as a tool to improve their public image. Concerns about growth-stimulating substances, GM food, dioxin-contaminated food and livestock epidemics (such as BSE) have given further impetus to organic food demand in general, as consumers increasingly question the safety of conventional foods. Several governments have responded with declarations of targets for the expansion of organic production. Many consumers perceive organic products as safer and of higher quality than conventional ones. According to the market research company Global Insight, the consumers' growing interest in organic chocolate was evident during the April 2006 Easter period when sales of organic products surged.

5.7 SUPPLY OF ORGANIC COCOA AND MARKET SIZE

Future growth of the organic cocoa market will depend more on supply constraints than on developments in demand, at least in the medium term. Many market analysts state that the tendency so far has been for the rate of demand growth to outstrip the rate of supply growth. 'To meet this growth target will be impossible unless more farmers are persuaded to convert to organic', declared the marketing director of one of the major players in the market. However, the general level of price premiums for organic cocoa does not support this tenet.

The supply side faces many challenges to meet the growing demand. At the producer level, issues of supply consistency and quality must be addressed. Producing countries face many constraints, such as the high costs of certification by foreign organisations and a lack of knowledge about organic channels. Trade channels have to allow for increased volumes of organic cocoa, for instance through the entry of bigger players in the market (e.g., ED&F Man through Corigins). An emerging development of the processor and manufacturer industry in North America would boost the availability of organic chocolate to American consumers.

6 Traditional and Modern Cocoa Cultivation Practices

6.1 ENVIRONMENTAL REQUIREMENTS FOR COCOA CULTIVATION

Certain environmental and edaphic conditions must be at their optimum for successful growth of the cocoa plant. These factors include temperature, rainfall, humidity, soil types, soil pH and soil nutrition. Cocoa cultivation requires an appropriate climate that is mostly found within the area bounded by the Tropics of Cancer and Capricorn. The majority of the world's cocoa is grown at small or large plantations within 10° north and south of the equator (Figure 3.1), and best suited for sea level up to a maximum of about 1,000 metres, although most of the world's cocoa grows at an altitude of less than 300 metres.

6.1.1 Temperature

The ideal range of temperatures for cocoa is minimums of 18–21°C and maximums of 30–32°C, thus temperatures generally within 18–32°C (65–90°F). Commercial cocoa production is limited to where the average minimum in the coldest months is greater than about 13°C. If the absolute minimum temperature falls below 10°C for several consecutive nights, the yield is likely to be reduced. Defoliation and dieback occurs between 4–8°C.

6.1.2 Rainfall

Environmental factors such as rainfall and temperature have significant effects on flowering and subsequent pod setting. Variations in the yield of cocoa trees from year to year are affected more by rainfall than by any other climatic factor. Trees are very sensitive to a soil water deficiency. Rainfall should be plentiful and well distributed through the year. Cultivation requires rainfall well distributed across the year, with a range between 1,000–4,000 mm (40–160 in) per year, but preferably between

1,500 and 2,500 mm (60 and 100 in). Dry spells where rainfall is less than 100 mm per month should not exceed three months.

6.1.3 Soils and Nutrition

Cocoa is grown on a wide range of soil types. The trees require soils containing coarse particles which allow free space for root development, and a reasonable quantity of nutrients to a depth of 1.5 m to allow the development of a good root system. Below that level it is desirable not to have impermeable material so that excess water can drain away. Cocoa can withstand waterlogging for short periods but excess water should not linger. The cocoa tree is sensitive to a lack of water so the soil must have both water retention properties and good drainage.

The chemical properties of the topsoil are mostly important as cocoa has a large number of roots that absorb nutrients. Cocoa can grow in soils with a pH in the range of 5.0–7.5, and thus can therefore cope with both acid and alkaline soil, but excessive acidity (pH 4.0 and below) or alkalinity (pH 8.0 and above) must be avoided. Cocoa is tolerant of acid soils provided the nutrient content is high enough. The soil should also have a high content of organic matter, 3.5% in the top 15 centimetres of soil. Soils for cocoa must have certain anionic and cationic balances. Exchangeable bases in the soil should amount to at least 35% of the total cation exchange capacity (CEC), otherwise nutritional problems are likely. The optimum total nitrogen/total phosphorus ratio should be around 1.5 (ICCO 2005). About 200 kg N, 25 kg P, 300 kg K and 140 kg Ca are needed per ha to grow the trees prior to pod production. For each 1,000 kg of dry beans harvested, about 20 kg N, 4 kg P and 10 kg K are removed; if the pod husks are also removed from the field, the amount of K removed increases to about 50 kg.

6.2 TRADITIONAL COCOA CULTIVATION PRACTICES

During cultivation, cocoa prefers high humidity, typically ranging between 70–80% during the day and 90–100% at night. Cocoa trees are usually planted to achieve a final density of 600–1,200 trees/acre (1,500–3,000 trees/hectare) and inter-cropped with food crops. Due to the fragility of the cocoa trees during the early stages of growth, they are mostly protected from strong winds using food crops, for instance, plantain or banana trees are used as wind shields on plantations in Ghana and Côte d'Ivoire; or tree plants (Leuceana or Gliricidia, or coconut or cashew) as used in Indonesia, Malaysia and Vietnam.

The trees grow well on most soils but preferably well-aerated soils with good drainage and a pH of neutral to slightly acidic (5–7.5), and pests and diseases carefully controlled. Cocoa trees used to grow to a height of ~10 metres tall at maturity, preferably under the shade of other trees. However, modern breeding methods have led to the development of trees to a standard of ~3 metres tall to allow for easy harvesting.

6.2.1 Growth and Propagation

Seedlings are generally used as planting materials for cocoa. They are raised in nurseries where shade, wind protection, nutrition and irrigation are provided as shown in Figure 6.1. The seeds are obtained from ripe pods and, if the fresh beans are planted immediately, at least 90% should germinate within two weeks. Cocoa seedlings have a single main stem that grows vertically to a height of one to two metres. A typical cocoa nursery is as shown in Figure 6.1.

The bud then forms three to five branches (the jorquette) that grow out at an angle as fan branches. Further upright suckers (chupons) emerge below the jorquette and grow up through the fan branches forming more jorquettes and further whorls of fan branch growth. In this way the tree becomes higher, forming several layers of jorquettes, each successively weakening and eventually fading out.

6.3 MODERN COCOA CULTIVATION PRACTICES USING VEGETATIVE PROPAGATION

In modern times, a combination of both traditional seed system and vegetative propagation system are used for growing cocoa plantations

FIGURE 6.1 Traditional nurseries for raising cocoa seedlings for transplanting.

FIGURE 6.2 Batch budding techniques used in multiplication of planting materials. (*Continued*)

in almost all of the cocoa-growing regions of the world. Vegetative propagation is used where selected characteristics of cocoa are desired, including higher yields, disease resistance, drought resistance, early maturing, special flavour profiles or unique growth characteristics. The trees raised using a vegetative propagation system are much more uniform in growth and performance than those raised from the traditional seed system.

The modern vegetative propagation system is developed using various techniques including rooted cuttings, budding and grafting. The processes used for the propagation of these different techniques are outlined in Figures 6.2 to 6.4.

(e)

(f)

FIGURE 6.2 (CONTINUED) Batch budding techniques used in multiplication of planting materials.

Cocoa presents special problems for *in vitro* propagation and reliable economic methods for mass tissue culture propagation are generally now being developed in many cocoa-growing countries around the world.

6.4 ESTABLISHMENT AND SHADE

Young cocoa plants may be field planted after three to six months. Protection from strong sun and wind ensures that cocoa producers interplant their cocoa with other trees. These shade trees include *Gliricidia sepium*, Leuceana and other commodity trees such as banana, plantain, coconut, rubber or oil palms (Johns 1999; Afoakwa 2010). Typical young cocoa plantations shaded by various shade trees are as shown in Figures 6.5

(a)

(b)

(c)

FIGURE 6.3 Top grafting techniques used in multiplication of planting materials. (*Continued*)

(d)

(e)

FIGURE 6.3 (CONTINUED) Top grafting techniques used in multiplication of planting materials.

to 6.9. The shade provided by these tree systems has substantial effects on the growth and productivity of the cocoa tree throughout its development into a mature tree. Some degree of shade control is needed through pruning and thinning to achieve the desired level of shade and maximise growth and production.

The effect of shade on cocoa is very complex. Shade influences the microclimate of the cocoa block through its effect on the amount of solar radiation received by the cocoa trees, the wind and the relative humidity, and through its effect on the metabolic rate of the cocoa trees

(a) (b)

FIGURE 6.4 Side grafting techniques used in multiplication of planting materials. (*Continued*)

it indirectly influences the nutrient status of the soil. The micro-climate, in turn, influences the incidence of pests and diseases. Establishment without shade is rarely successful so the shade must be well established prior to field planting.

The absence of shade places significant ecological stress on the cocoa trees, which become susceptible to pests attack. Shade strategies include retaining remnant forest; planting temporary and permanent shade species and interplanting with species that also provide a commercial return. Cocoa can be planted under thinned forest canopy, naturally regenerating, or the canopy of artificially planted trees (Greenberg 1998; N'Goran 1998).

Shade removal is possible after three to four years. Removing shade from cocoa has resulted in significant increases in yield with a positive interaction between increased light and applied nutrients (Cunningham and Arnold 1962). Planting density depends on factors such as tree vigour, light interception and the farming system. It may range from 800–3,000 trees/ha with about 1,200 trees/ha being common in Malaysia under permanent shade.

6.5 FLOWERING AND POD DEVELOPMENT

The cocoa flowers arise in groups directly from old wood of the main stem or older branches at points which were originally leaf axils (Figure 6.10). Each flower has five prominent pink sepals, five smaller yellowish petals, each of which forms a pouch, an outer whorl of five staminodes, and an inner whorl of five double stamens, each stamen bearing up to four anthers.

(c)

(d)

FIGURE 6.4 (CONTINUED) Side grafting techniques used in multiplication of planting materials.

The staminodes are about as tall to twice as tall as the upright style and form a 'fence' around the style. The stamens are curled so that the anthers develop inside the petal pouches. The ovary consists of five united carpels each having 4 to 12 locules and one style that has several linear stigmatic lobes.

The emergence of the bud through the bark of the tree marks the beginning of the cocoa bean development. This takes about 30 days from its histological beginnings to its culmination on the bark surface and within hours of its emergence, the bud matures, the sepals split and the flower continues to mature during the first night following the budding (Figure 6.11). On the next morning after budding, the flower is fully opened and the anthers release their pollens. If not pollinated and fertilised on this day by

FIGURE 6.5 Young cocoa plantation inter-cropped with plantain trees.

FIGURE 6.6 Cocoa plantation inter-planted with banana trees.

insects, the flowers continue to abscission on the following day. It is interesting to note that a single healthy cocoa tree produces about 20,000 to 100,000 flowers yearly but only 1 to 5% of these get pollinated and develop into young pods (Figure 6.12).

Once successfully pollinated and fertilised, the various stages of embryo and ovule growth continue, the pods reaching maximum size after about 75 days following pollination. The pods then mature for another 65 days, making a total of about 140 days after pollination (Figure 6.13). The fruit are then allowed to ripen for about 10 days (Figure 6.14) and the pods harvested. The matured cocoa fruit measure between 100 and 350 mm

FIGURE 6.7 Cocoa plantation inter-planted with leuceana trees.

FIGURE 6.8 Cocoa plantation inter-planted with coconut trees.

(4 and 14 in) long and have a wet weight of ~200 g to ~1 kg. Maximum crop yields are usually achieved 3–5 years later and most cocoa trees will produce commercially acceptable yields until 25–30 years old (Wood and Lass 1985).

A key determinant of properly ripened cocoa fruit is the external appearance. There are considerable variations in the shape, colour and surface texture on the pods, depending on genotype. The ripening is

FIGURE 6.9 Cocoa plantation inter-planted with gliricidia trees.

FIGURE 6.10 Budding and flowering of cocoa from bark of old tree.

visible as changes in the colours of the external pod walls occur and the nature of colour changes is dictated by the genotype of cocoa involved. However, cocoa fruit ripening is generally thought to be from green or purple to varied shades of red, orange or yellow, depending on genotype. The composition of the internal content, comprising the bean and pulp, is extensively discussed in the next section, with emphasis on the bean composition and their influence on chocolate flavour precursor formation and development.

FIGURE 6.11 Matured flower with opened sepals from bark of cocoa tree.

FIGURE 6.12 Cocoa pod development.

6.6 HARVESTING OF COCOA PODS

Harvesting is the start of the post-harvest process that determines the quality of the cocoa beans to be sold to the cocoa and chocolate industry (David, 2005). The development of the pod takes five to six months from fertilisation of the flower to full maturity (Sukha, 2003). Only ripe and undiseased pods are harvested for optimal processing and attainment of high-quality beans for chocolate and other cocoa base products. The external appearance of the pod is used as the key indicator for determining the extent of ripening. The nature of the colour change in the external pod wall is

FIGURE 6.13 Matured unripe cocoa pods.

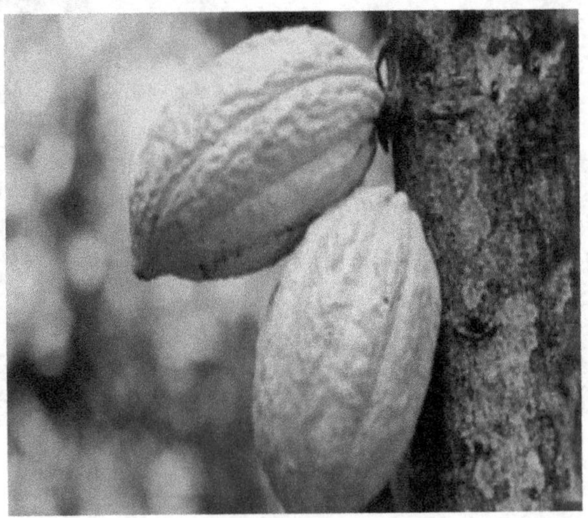

FIGURE 6.14 Matured and ripened cocoa pods.

dependent on the genotype or variety of the cocoa plant (Afoakwa, 2010). For example, the *Trinitario* variety when unripe is purple but when fully ripe looks yellowish with slight purple grooves whereas a mixed hybrid changes from green to yellowish and likewise *Amelonado* and *Amazonica* varieties (Mikkelsen 2010). Cocoa harvest is not limited to one discrete period but spread over several months and in some regions there may be pods available for harvest throughout the year. Typically, there are one or two peak harvest periods influenced by flowering in response to rainfall.

The main season in Ghana for cocoa harvest is from September to January and the minor season is from May to August (Mikkelsen 2010). However, temperature and the crop already on the tree will also influence flowering so that the yearly-cropping pattern can vary in areas with a relatively uniform climate.

Harvesting of cocoa fruits involves the removal of pods from the trees and the extraction of the beans and pulp from the interior of the pod. Although the ripening process occurs in a 7–10-day period, the pods can safely be left on the trees for up to two weeks before harvesting. Thus, a 3-week window exists during which the cocoa may be considered fit to harvest. There are two concerns that dictate how quickly the harvest is completed: potential for pod diseases and the possibility of bean germination in the pod if delayed for too long. These tend to have a negative influence on the quality characteristics of the cocoa beans.

Harvesting of cocoa pods can be done by hand or using an assisted tool. Pods from lower trees can be harvested by plucking the pods from the trees by hand as shown in Figure 6.15. On the other hand, a knife or cutlass is normally used to remove the pod from the tree, but there exists a special long-handled tool (made of hooked knives on long poles) for removing pods which are higher up the tree (Figure 6.16). Cocoa pods are normally harvested every 2 to 4 weeks over a period of several months, as ripening does not occur at the same time.

After removing the pods from the trees, they may be gathered into heaps (Figure 6.17) and opened immediately or allowed to sit for a few days before opening, a technique known as pod storage which has been reported to have significant beneficial effects on the flavour quality of

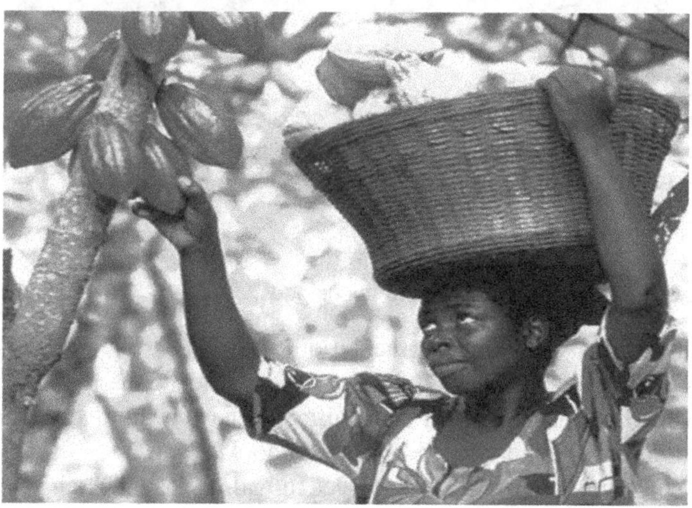

FIGURE 6.15 Harvesting of cocoa pods by hand.

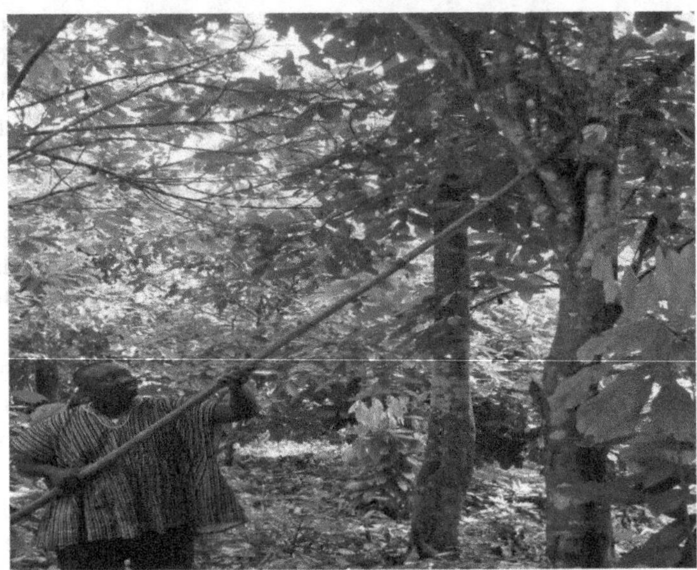

FIGURE 6.16 Harvesting of cocoa pods by a hook and pole tool.

FIGURE 6.17 Heaping of harvested cocoa pods.

the bean during subsequent fermentation and processing. Much of this depends on the geographic and historical practices encountered in the various growing regions. Details of its significance have been investigated and are discussed in Chapters 15 and 16 of this book.

Over-ripe pods are more likely to get fungal diseases which affect the beans and should then not be used. Pods that have fungal diseases should be picked from the trees and placed on the ground away from the cocoa trees or can be buried. This will help stop infection of other pods. Mice and rats will also eat ripe pods. If pods are left too long on trees, the beans will start to germinate (as stated above) and this is undesirable for the general quality and flavour of the cocoa after fermenting and drying.

6.7 POD BREAKING

The harvested cocoa pods are broken or cut open to extract the beans. This is done using either a club or mallet or a cutlass to break the pod open, after which the beans are scooped out (Figure 6.18). The latter technique, although a considerable skill can be acquired by its application, suffers from the disadvantage that the beans can be easily damaged. In Ghana, where the cutlass is favoured as the means of pod breaking, the technique is to strike the pod once on its longer dimension. A second blow is given on the opposite side and the two halves of the husk parted by twisting the cutlass. The beans, together with the placenta, are then scooped out with the point of the tool (Rohan, 1963; Afoakwa, 2010). The practice of cutting with a cutlass or machete requires considerable skill as the beans can easily be damaged during the process and subsequent penetration by mould and stored product pests, rendering them defective.

Surveys by Rohan (1963), Carr, Davies, and Dougan (1979), Carr and Davies (1980), Tomlins et al. (1993) and Baker et al. (1994) found that many farmers store their harvested unopened pods for a few days to up to two weeks prior to splitting. In some cases this was to allow a small

FIGURE 6.18 Opening of heaped cocoa pods with wooden clubs for fermentation.

producer time enough to gather sufficient pods for fermentation over several weeks of harvest. A more commonly cited reason for pod storage was that it improved initiation of fermentation and gave better quality beans (Biehl et al., 1989; Tomlins et al., 1993; Baker et al., 1994). Research has confirmed that fermentation may be affected by ripeness (Wood and Lass, 1985; Ardhana and Fleet, 2003; Schwan and Wheals, 2004) and pod storage (Carr et al., 1979; Biehl, 1989; Afoakwa et al., 2011a,b; 2012; 2013a,b). Pod storage in excess of five days has been found to affect flavours and free fatty acid levels in cocoa (Afoakwa et al., 2011a,b). Therefore, a lack of control at the harvest, storage and splitting stages may be clearly seen to contribute to the variable quality of cocoa beans produced worldwide.

Quality control operations start at pod breaking when experienced pod breakers reject all defective beans (germinated, flat and diseased). During pod breaking field baskets are emptied on a bed covered with banana leaves. When the heap gets to the desired size they are fashioned into cone-shapes and covered with banana leaves and held in place by pieces of wood (Duncan, 1984). There are 30–40 beans or seeds inside the pod attached to a central placenta. The beans are oval in shape and enveloped in a sweet white mucilaginous pulp. After breaking the pod, the beans are then separated by hand and the placenta is removed. A seed coat or testa separates the seed cotyledons from the pulp. Beans taken directly from the pod to controlled drying conditions develop virtually no chocolate flavour after processing and fresh beans are free from compounds necessary for the development of chocolate flavour. The process of fermentation is therefore necessary for the formation of constituents or flavour precursors that undergo further development during the roasting process.

6.8 COCOA (GOLDEN) POD

The cocoa pod is ovoid or ellipsoidal in shape, 15–30 cm long and 8–10 cm wide, surrounded by a strong 10–15 mm thick husk (Mossu, 1992; Afoakwa, 2010). The pod at its early stage of growth and development is called a cherelle (Hebbar, Bittenbender, and O'Doherty 2011). The cocoa fruit attains a full size after 5 to 6 months of fertilisation and takes about a month more for ripening. A typical ripened cocoa pod is as shown in Figure 6.19. The pod contains approximately 30 to 40 cocoa beans attached to a central placenta embedded in a whitish acidic mucilaginous pulp called aril (Dezaan Cocoa Manual, 2009; Fowler, 2009; Afoakwa, 2010; Crozier et al., 2011; Guehi et al., 2010; Hebbar et al., 2011).

This pulp consists of 80–90% water; 6–13% fermentable sugars (Mikkelsen, 2010) such as glucose, fructose and sucrose; 0.5–1% citric acid and small amounts of aspartic acid, asparagines, and glutamic acid with a pH of 3.0–3.5 (Mossu, 1992; Ardhana and Fleet, 2003; Guehi et

FIGURE 6.19 Typical ripe cocoa pod (the golden pod).

al., 2010; Nazaruddin et al., 2006b; Afoakwa et al., 2013a). The pulps surrounding the seeds are used in the production of jam, wine and gin.

Figures 6.20 and 6.21 show opened cocoa pods with their constituent cross-sectional and longitudinal bean arrangements, respectively. The mature in-pod cocoa bean is made up of three components: pulp, testa and cotyledons.

6.9 WEED CONTROL

Weed control is mainly an issue during establishment or early stages of growth; traditionally young cocoa is weeded by manual slashing along the tree rows or around young plants. More recently, herbicides are also being

FIGURE 6.20 Longitudinal view of bean arrangement in matured cocoa pod.

FIGURE 6.21 Cross-sectional view of bean arrangement in matured cocoa pod.

used. When cocoa is mature and a complete canopy is formed, heavy shading and leaf mulch inhibit weed growth so that only occasional attention to removing woody weeds is required. Weeds will be an issue wherever the canopy allows light to penetrate. Adopting appropriate weed control measures will keep the ground around the cocoa tree and the shade tree free from weeds. In weed control, two different techniques can be distinguished: manual/mechanical and chemical control. Manual/mechanical control involves the use of grass knives or mechanical slashers. Chemical control involves the use of spraying machines to apply herbicides to the weeds that need to be controlled.

6.10 PRUNING

Pruning is the removal of unwanted branches from a cocoa tree. It is an important operation and can affect yield for months, even years, as well as affecting the shape and structure of the tree for the rest of its life. Insects and diseases multiply more on unpruned cocoa trees with dense canopies than on trees that have been opened up by pruning and display well-aired canopies. Pruning can also stimulate flowering and pod production. Pruning can be carried out properly by using good tools such as a bow saw, a secateur, a chupon knife and a long-handle pruner. Cocoa propagated from seed is pruned to develop the preferred structure shown in Figure 6.22. Pruning is mainly used to limit tree height. The first jorquette should be formed at 1.5–2 m. Further chupons are continually removed, preventing subsequent jorquettes and restricting further vertical growth. Some pruning of fan branches may be required to maintain evenness in the structure and remove low-hanging branches. The end

(a)

(b)

FIGURE 6.22 Pruned cocoa trees.

result is the formation of a tree with the canopy at a convenient height for management. Vegetative propagated plants have a different structure and will require different management. There is little evidence of the value of pruning strategies to promote high yields. Mechanical pruning (hedging) is not practiced.

7 Pests and Diseases of Cocoa

7.1 INTRODUCTION

The cocoa tree is susceptible to a number of diseases and pests that affect the yield of pods from the trees. Many diseases and insect pests are known to attack the cocoa tree and the pods leading to economic loss. Most of these diseases are caused by fungi and viruses. Five major diseases, namely witch's broom (WB), black pod (*Phytophthora* pod rot [PP]), moniliasis pod rot (MO), cocoa swollen shoot virus disease (CSSVD) and vascular streak die back (VSD), affect the crop, causing about 40% yield loss per year. More than 1,500 insect pests have been found to attack the cocoa plant in different cocoa-growing regions of the world but a small number are of economic importance. These are the red borer, tea mosquito bug, mealy bug, gray weevil, cockchafer beetle, rat, striped squirrel and a host of others (Nair 2010).

7.2 MAJOR COCOA DISEASES

Across the major cocoa-producing countries in the world, three major plant diseases and three major groups of insect pests are reported to afflict the cocoa industry. The diseases are cocoa swollen shoot virus disease, the black pod disease and the witches broom. The major pests of cocoa are the mirids, *Distantiella theobroma* and *Sahlbergella singularis*. The economic importance of these diseases and pests are derived from their adverse effects on per hectare yields and consequent losses to the industry's aggregate cocoa output (Asante 1995). Due to their versatility in infesting other pods, it is recommended that all diseased pods be harvested with the healthy ones and then separated for destruction. The cocoa pod diseases and pests are as described.

7.2.1 Cocoa Swollen Shoot Virus Disease (CSSVD)

This is a viral disease affecting cocoa and is spread by small whitish insects known as mealy bugs. The pods assume a roundish shape and also diminish in size (Figure 7.1), causing a drastic reduction in yield

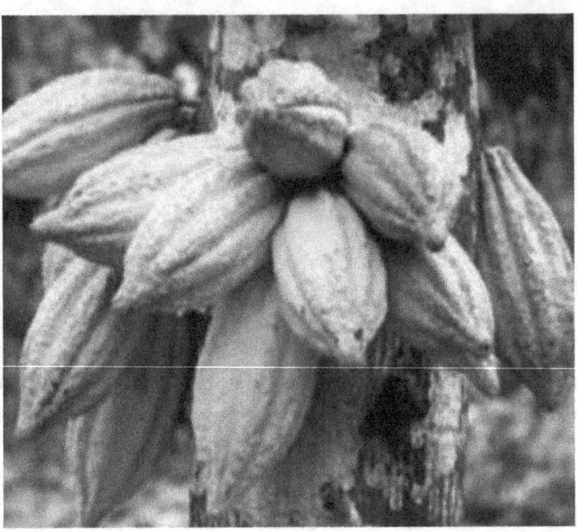

FIGURE 7.1 Swollen shoot disease (results in small pods).

from infested trees. It is reported to be the most serious disease in Ghana, Nigeria, Togo, Côte d'Ivoire and Sri Lanka (Nair 2010). The economic importance of the CSSVD lies in its debilitating and destructive effect on the cocoa tree, sometimes within as short a period as three years (Ollennu, Owusu, and Thresh 1989). *Amelonado* cocoa is generally found to be more susceptible to African CSSVD than *Upper Amazon* and *Trinitario* types (Nair 2010). Control measures involve cutting down infected trees and adjoining trees and burning them completely. There is, however, no evidence that this disease has any adverse effect on the quality of fruits of the cocoa tree or on the quality of the products after fermentation. This is because a full investigation on these has not been conducted and it would be dangerous to assume that no evidence exists.

7.2.2 Black Pod Disease (BP)

The black pod disease, also known as pod rot, is a fungal disease caused by three *Phythophthora* species: *P. palmivora*, *P. megakarya* and *P. capsici* (ICCO 2011). *Phytophthora palmivora* occurs in the center of origin of cocoa and causes 44% of global crop loss and *Phytophthora megakarya* is restricted to Cameroon, Nigeria, Togo, and Ghana, causing about 10% crop loss. However, *Phytophthora megakarya* is the more aggressive and destructive. *P. capsici* is widespread in Central and South America, causing significant losses in favourable environments (ICCO 2011).

This disease is characterised by browning, blackening and rotting of cocoa pods and beans (Figure 7.2). The pods can be attacked at any stage

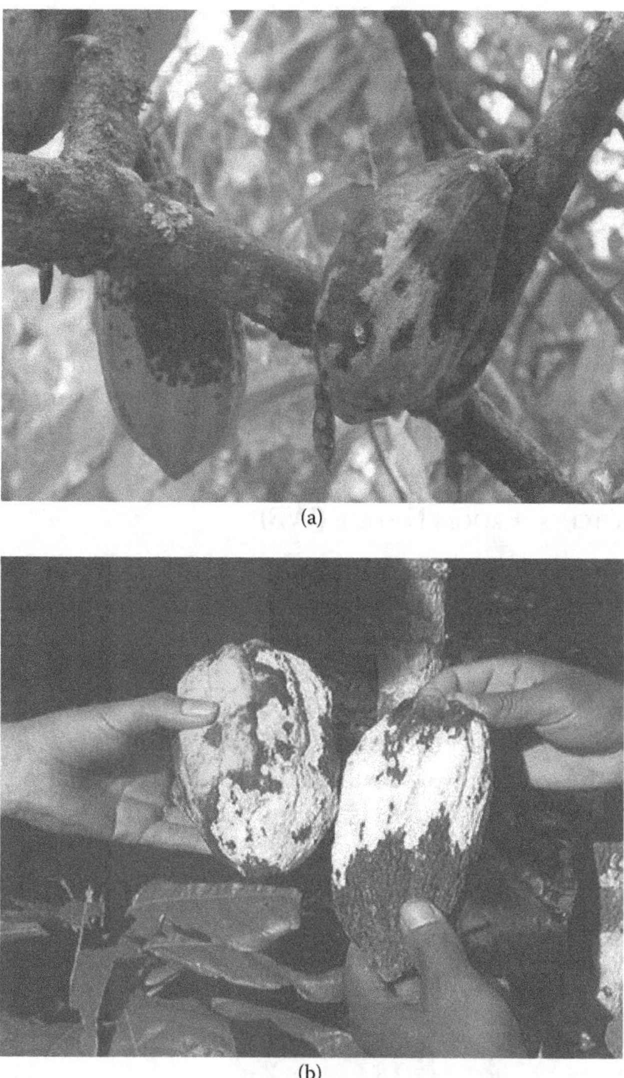

(a)

(b)

FIGURE 7.2 Black pod disease (a) on infested cocoa tree; (b) harvested infested cocoa pod.

of development, and the initial symptoms are small, hard, dark spots on any part of the pod (ICCO 2011). Internal tissues, including the beans, are colonised and shrivel to form a mummified pod. The fungi attack every portion of the cocoa tree and are controlled by good cultural practices by the removal of infected pods and by spraying with approved fungicides. Their rate of infestation could be reduced by reducing the humidity and by increasing aeration on the cocoa farm. Chemical control involves coating the pod surface with recommended fungicides which stops the germination of fungal spores (Opoku et al. 2007a). Spraying of cocoa farms against black pod begins in the rainy season at 3–4 weekly intervals or at any time a farmer spots 1–2 infected pods (CRIG 2008). Current research efforts are directed towards the possible use of phosphonic acid, a fully systematic fungicide for the control of *P. megakarya* (Opoku and Owusu 1995) and the breeding of resistant planting materials (Baah and Anchirinah 2011).

The pods harvested from infected trees may be used with the healthy pods if the fungal attack has not penetrated the pod walls, leaving the beans unaffected. If, however, harvesting is delayed and the attack is severe, there is some evidence (Awua 2002) that the free sugars of the pulp are utilised by the fungus, giving rise to a dry pulp similar to that of an unripe pod. If such pods occur in quantity, fermentation is impaired and a product of poor quality results.

7.2.3 Witches' Broom Disease (WB)

This disease is caused by the fungus *Marasmius perniciosus* and is indigenous to South America. It has, however, spread to surrounding cocoa-growing countries and has caused considerable damage to cocoa trees in Brazil and Trinidad and Tobago. The fungus attacks only actively growing tissue (shoots, flowers and pods) causing cocoa trees to produce branches with no fruit and ineffective leaves (ICCO 2011). It is characterised by abnormal tufted vegetative growth on the trees at the expense of pod formation (Figure 7.3). The pods show distortion and present green patches that give the appearance of uneven ripening. Unless the cocoa pod is almost ripe when attacked, the infection destroys the diseased pods and renders them useless.

FIGURE 7.3 Witches' broom disease.

Cultural practices such as phytosanitary pruning is the only effective means of controlling witches' broom (ICCO 2011). Complete removal of all infected material is advocated, but it is an impossible task because hidden inoculum sources always remain. The infected trees are also controlled by spraying with fungicides. However, this disease is absent in the West African cocoa-growing region (Afoakwa 2010).

7.3 COCOA PESTS: POD BORERS (CAPSIDS, COCOA THRIPS AND MEALY BUGS)

Several insect pests attack and feed on young shoots and pods of the cocoa tree. However, only a small number are of economic importance. Among the major pests infesting cocoa, the significant ones are the red borer, tea mosquito bug, mealy bug, gray weevil, cockchafer beetle, rat, striped squirrel, capsids and a host of others. Red borer (*Zeuzera coffeae*) infests mainly the young cocoa plant. Their larvae bore into thick shoots and into the main stem below the first jorquette along the center and cut a traverse tunnel before pupation (Nair 2010). Figures 7.4 to 7.6 show images of typical capsids, mealy bugs and thrips, respectively.

Capsids and moths feed on young shoots and pods of the cocoa tree. The two main species responsible for crop losses are *Sahlbergella singularis* and *Distantiela theobroma* (Baah and Anchirinah 2011). These insects are capable of reducing yields of healthy farms to less than 25% of their potential in one year. Seedlings may completely fail to become established due to the presence of capsids. Even when seedlings are not killed outright, capsids delay cocoa coming into bearing for several

FIGURE 7.4 Capsid.

FIGURE 7.5 Mealy bug.

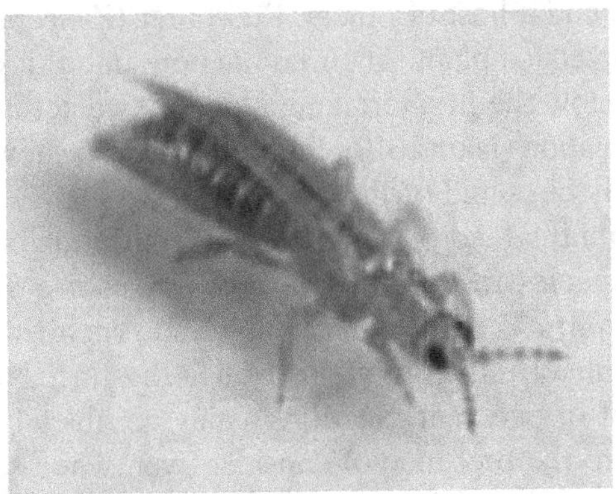

FIGURE 7.6 Thrips.

years. They damage the young soft tissues of the trees by piercing the young shoots with their mouth parts, injecting poisonous saliva and then sucking out the fluid food from the wound, causing the death of the young trees. These infections can be controlled by the application of the recommended insecticides and by leaving a reasonable amount of shade between the young trees. None of these insect pests have been reported to have any direct influence on the quality of manufactured chocolate products. However, it is feared that large-scale insecticide spraying exercises used in their control may result in taints in the prepared products. These control techniques may also increase the pesticide levels in the fermented and dried cocoa beans, and may pose problems of high unacceptable pesticide doses on the international markets. It is therefore recommended that cocoa with these infections be controlled under supervision by agricultural extension officers.

7.4 COCOA CROP PROTECTION

In controlling cocoa diseases, all trees should receive individual attention, as a single infected plant is likely to act as a source of infection for all the other trees on the farm. If left unattended, one sick tree will eventually lead to all the others also contracting the disease. There are four methods used to prevent diseases from developing or controlling them if they do become established. These methods are: regulatory, cultural, biological and chemical.

In regulatory control, measures are taken, usually by law, to prevent material contaminated with a pathogen from being transported from one area that already has a particular disease to another area which does not yet have the disease. Cultural control is a broad approach that involves preventing the pathogen from coming into contact with and infecting the cocoa trees or eradicating the pathogen or significantly reducing its numbers in an individual plant or within an area. Biological control involves a range of measures that include directly introducing other micro-organisms that are enemies of the pathogen. Chemical control usually seeks to remove the pathogen from the disease location. Chemicals that are toxic to the pathogen are applied to the cocoa or shade trees, either to prevent pathogen inoculum from establishing in a host, or to cure an infection that is already in progress. Some of these chemicals (e.g., Lindane, etc.) have been found to contain harmful or toxic constituents and have thus been banned from their application on cocoa. Routine application of broad-spectrum insecticides to prevent pests from establishing themselves should therefore not be carried out. Thus, cocoa farmers are advised to contact the appropriate agricultural extension officers within their area for support in providing them with the approved chemicals and the appropriate application procedures.

8 Post-Harvest Treatments and Technologies of Cocoa

8.1 INTRODUCTION

Raw cocoa beans have an astringent unpleasant taste and have to be taken through the post-harvest treatments of fermentation, drying and roasting to attain the characteristic cocoa taste, color and flavour. The major post-harvest treatment of cocoa involves harvesting, gathering of the pods, pod storage as a means of pulp pre-conditioning, opening or pod breaking, fermentation of the beans, drying of fermented beans, sorting, packaging and storage. Fermentation and drying are the principal operations for the curing of cocoa beans (Garcia-Alamilla et al. 2007). The purpose of the treatments is therefore to develop the right flavour precursors and colour in the beans and also preserve them before distribution to manufacturing areas. Chocolate flavour precursors such as organic acids, peptides, sugars and fatty acids are formed during fermentation and are modified through drying and roasting (Faborode, Favier, and Ajayi 1995; Knight 2000; Galvez et al. 2007; Afoakwa et al. 2008).

8.2 TECHNIQUES FOR IMPROVING COCOA BEAN QUALITY

Current market trends have heightened the overall demand for cocoa beans. At the same time, greater attention is being paid to the quality of the cocoa beans being produced worldwide. Over the last two decades, much of the research activities have been centred on cocoa bean pre-fermentation and fermentation treatments as well as drying, and these processes have been aimed at solving certain quality or flavour problems. This chapter outlines the progress that has been made in improving cocoa quality, focussing on the role of pre-fermentation treatments, fermentation technologies and to a lesser extent, drying.

8.2.1 Pre-Fermentation Treatments of Cocoa Pods

The impact of post-harvest treatment on fresh cocoa pods and the effects of these treatments on fermentation and final bean quality have been investigated. Three basic processes have been evaluated for the treatment of fresh cocoa beans prior to fermentation: pod storage, depulping (mechanical or enzymatic) and bean spreading (Rohan 1963; Wood and Lass 1985; Schwan and Wheals 2004). All of these treatments were developed or investigated in attempts to reduce the problem of acidity in dried fermented cocoa beans. Over-acidity in processed cocoa beans has been linked to the production of high levels of lactic and acetic acid during fermentation. By removing a portion of the pulp, or reducing the fermentable sugar content of the beans, it has been shown that less acid is produced during fermentation, leading to less acid beans (Duncan et al. 1989; Sanagi et al. 1997). Removal of up to 20% of the cocoa pulp from fresh Brazilian cocoa beans significantly improved the flavour quality of the beans produced (Schwan and Wheals 2004).

8.2.2 Pulp Pre-Conditioning and Cocoa Bean Quality

Pulp pre-conditioning entails changing the properties of the pulp of the cocoa beans prior to fermentation (Meyer et al. 1989). Because the pulp is the substrate metabolised during fermentation, changes in the pulp affect the production of acids by lactic acid bacteria, yeasts and acetic acid bacteria (Ostovar and Keeney 1973; Meyer et al. 1989, Biehl 1984; Nazaruddin et al. 2006a). The main objective of pulp pre-conditioning is to reduce the formation of acids throughout the course of subsequent fermentation without enhancing the degradation of acids at the end of fermentation.

Methods of pulp pre-conditioning include bean spreading, bean pressing, depulping and post-harvest pod storage (Meyer et al. 1989; Biehl 1984; Nazaruddin et al. 2006a). In recent times, pulp pre-conditioning in the form of post-harvest storage of the beans in the harvested ripe cocoa pod has become a major area of concern to researchers. Previous research has shown that pulp pre-conditioning by post-harvest storage of cocoa pods led to the reduction of nib acidification during subsequent fermentation, reduction of the acid note and an increase in cocoa flavour in the resulting raw cocoa (Biehl 1984; Meyer et al. 1989; Nazaruddin et al. 2006a). The extent of the pre-conditioning in terms of duration and how it affects the various biochemical constituents of the flavour precursors in different cocoa varieties grown across the different cocoa regions of the world is yet to be understood.

8.2.3 Pod Storage and Cocoa Bean Quality

Post-harvest pod storage as a method of pulp pre-conditioning involves keeping the harvested pod (fruit) for a period of time before breaking the pod. In West Africa, the pods are usually left on the ground to be gathered together for breaking by a team of workers after a few days of harvesting activity. Ghanaian farmers have unknowingly adopted this technique of pod storage simply by their practice of using family labour to collect the harvested pods into piles before organising friends and neighbours to help break open the pods (Duncan 1984). Researches on pod-stored beans demonstrated that fermentations of beans from stored pods are more rapid and result in higher brown bean counts (Afoakwa et al. 2011a). Also, in most Asian countries pod storage is conducted to reduce pulp volume and acidity (Biehl 1984; Biehl et al. 1989; Meyer et al. 1989).

During pod storage, the beans within the pod lose moisture. This allows more air to penetrate the cocoa once the pods are broken and start to ferment. More air causes the fermentation to happen more rapidly and temperature rises are faster than if pods are broken when they are freshly harvested. The faster fermentation and temperature rise results in an improved quality cocoa (Biehl 1984; Biehl et al. 1989; Meyer et al. 1989; Nazaruddin et al. 2006b). In other research findings involving pre-conditioning of cocoa before fermentation, polyphenol components were found to reduce considerably using the pod-storage technique (Nazaruddin et al. 2006b). It is best to store pods in a cool and dry position, under cover from rain, as this will tend to reduce the possibility of fungal contamination. Pods which become infected with fungi should be discarded. Pods are ready to ferment when enough moisture has been lost from the pod to allow a good fermentation. Pod storage in excess of six days has been found to affect flavours in cocoa (Clapperton et al. 1994; Nazaruddin et al. 2006b). Comprehensive studies aimed at clarifying changes in the key biochemical constituents in pulp pre-conditioned (pod storage) and fermented cocoa beans under different pod-storage durations is still scarce. Thus, the last two chapters in this book contain findings from research conducted to determine the effect of different pod-storage durations (extended up to 21 days and reduced up to 10 days) on the biochemical constituents (key flavour precursors), physico-chemical properties, polyphenolic constituents and fermentative qualities of the pulp pre-conditioned and fermented cocoa beans.

8.2.4 Mechanical Depulping

Methods for mechanically depulping fresh cocoa beans include pressing (Rohan 1963; Wood and Lass 1985), centrifuging (Schwan, Rose, and Board 1995) or simply spreading beans onto a flat surface for several hours

FIGURE 8.1 Bean pressing equipment.

prior to fermentation, causing a significant increase in the sweating produced in the first 24 h of fermentation. In addition to reducing acidity, benefits of depulping include shorter fermentations and increased efficiency and the ability to use the excess pulp in the manufacture of jams, marmalade, pulp juice, wine or cocoa soft drinks (Buamah, Dzogbefia, and Oldham 1997; Schwan and Wheals 2004; Dias et al. 2007). Figures 8.1 and 8.2 are, respectively, typical bean pressing and mechanical depulping equipment used.

8.3 COCOA BEAN FERMENTATION

The importance of cocoa bean fermentation in contributing to chocolate quality has been recognised for over 100 years, and numerous studies have been conducted from different countries to determine the chemical and biochemical changes as well as the microbial species associated with this process (Schwan et al. 1995; Schwan and Wheals 2004; Afoakwa et al. 2008; Afoakwa 2010). Cocoa fermentation aids the degradation

FIGURE 8.2 Mechanical depulping equipment.

of the bulk of the pulp surrounding the cocoa beans so that the beans can easily be dried. The process induces biochemical transformation within the beans that leads to formation of the color, aroma and flavour precursors of chocolate. Without fermentation, cocoa beans are excessively bitter and astringent and, when processed, do not develop the flavour that is characteristic of chocolate (Schwan and Wheals 2004; Ardhana and Fleet 2003; Beckett 2009; Afoakwa 2010). The character and strength of chocolate flavour are governed primarily by the genetic constitution of the cocoa variety, and the fermentation process releases and develops this flavour potential (Faborode, Favier, and Ajayi 1995; Afoakwa et al. 2009).

Fermentation usually lasts between three to eight days depending on weather conditions and time during the cocoa season (Thompson, Miller, and Lopez 2001; Sukha 2003; Afoakwa 2010). The process usually takes longer at the start and peak of the cocoa crop but shortens towards the end of the crop when there is less mucilage available for fermentation. Initial turning during fermentation is done after 48 hours and an additional turning 48 hours thereafter to facilitate adequate aeration of the fermenting mass and to ensure that beans from the top and bottom are thoroughly mixed together (Baker, Tomlins, and Gray 1994; Sukha 2003; Afoakwa and Paterson 2010).

During fermentation volatile acidity formed reaches approximately 2% of the dry basis. This acidity is the result of some components such as acetic, propionic, butyric, isobutyric and isovaleric acids but 90% of these components are acetic acid, which has an important role in the catalysis of enzymatic reactions for producing components of desirable sensorial characteristics (Barel 1997). Cocoa beans that are not fermented do not develop the aroma characteristics and are more bitter and astringent (Faborode et al. 1995; Afoakwa et al. 2008). After fermentation, a drying process is adopted with the purpose of reducing moisture content (from 55–60% to 8%) and volatile acidity content for stopping undesirable reactions and the oxidation of polyphenols (Jinap, Siti, and Norsiati 1994; Faborode et al. 1995; Fowler 1999; Afoakwa and Paterson 2010).

Cocoa aroma is crucially dependent on genotype, harvesting time, pre-fermentation treatment, fermentation, drying and roasting (Schwan and Wheals 1993, 2004; Afoakwa et al. 2008). The fresh beans have the odour and taste of vinegar but appropriate amounts and ratio of precursors are essential for optimal flavour volatiles production during roasting. Subcellular changes in the cotyledons release key enzymes effecting reactions between substrates pre-existing in unfermented beans (Hansen, del Olmo, and Burri 1998, Schwan and Wheals 2004; Afoakwa et al. 2008).

8.4 FERMENTATION TECHNIQUES

Many different fermentation techniques have been developed for the curing of cocoa beans. These techniques vary considerably from continent to continent, region to region, country to country, and in some cases different farmers from the same producing region or country practice different fermentation techniques depending on the quantity of beans, available materials, climatic conditions and the type of system available. Five fermentation techniques are mostly employed throughout the world for the curing of cocoa beans. These include

1. Heap fermentation
2. Box fermentation
3. Basket fermentation
4. Tray fermentation
5. Curing on drying platforms

8.4.1 Heap Fermentation

Fermentation in heaps is a popular method among smallholding farmers in Ghana and many other African cocoa-producing countries. It is also practised to a lesser extent in the Amazon region of Brazil. Judging from the final quality attributes of Ghanaian cocoa, fermentation in heaps can produce good-quality products. Varying quantities of cocoa beans from

FIGURE 8.3 Fermentation platform made of plantain leaves in preparation for heap fermentation.

FIGURE 8.4 Heaped cocoa beans for fermentation.

25 to 1,000 kg are heaped in the field on plantain leaves (Figures 8.3 and 8.4) and covered with the same material (Figure 8.5). The beans are mixed (turned) periodically to ensure even fermentation and to decrease the potential for mould growth. This is often done daily or every other day by forming another heap. Mixing is laborious and small heaps may not be turned at all. The duration of fermentation is from 4 to 7 days.

FIGURE 8.5 Heaped cocoa beans undergoing heap fermentation.

8.4.2 Box Fermentation

This is the most common type of fermentation conducted around the world and is also used in West Africa. Fermentation in boxes is considered to be an improvement over the other methods. It requires a fixed volume of cocoa and is the method of choice on large estates. It consists of wooden or concrete boxes in which the wet cocoa beans are fermented.

Box fermentations can be used to ferment different quantities of cocoa from 25 kg up to about 1,000 kg. Figure 8.6 shows the construction of

FIGURE 8.6 Arrangement in box fermentation system.

boxes for the box fermentation system. In this system, boxes are made from wood 15-cm wide and 2.5-cm thick. If timber of this thickness is not available, the boxes can be made of plywood but should be insulated with polystyrene on the outside to hold in the heat. It requires a fixed volume of cocoa and is the method of choice on large estates. The sweat box may be a single unit or one of a number of compartments within a large box created by subdividing the space into smaller units with either fixed or movable internal partitions.

A board weighted with stones is placed on top to help retain heat and prevent the surface from drying. The boxes are always raised above ground level and have holes to facilitate drainage and aeration and are usually placed indoors. The boxes may be arranged in rows or in tiers, that is, with one slightly raised above the other. In the single boxes, the beans are mixed by hand whereas in the tier arrangement the beans are mixed by simply transferring beans from one box to the next in line below (Figure 8.6).

The beans are fermented in the boxes for six days and are mixed after two and four days and may be covered with banana or plantain leaves, or jute sacks.. During fermentation the temperature should be monitored, as is the case in all fermentation methods. In box fermentations with quantities of beans 100 kg or less, turning can be done by hand. In the case of larger quantities of beans (more than 250 kg), partitioned boxes are used. In this case, turning of beans is performed by shovelling them from one side of the partition to the other. Jute bags conserve heat better than banana leaves, so jute bags or a combination of banana leaves and jute bags is recommended as a covering for the boxes during fermentation. As a general rule, the closer to 50°C that fermentations reach, the better the quality of the dried cocoa. For all types of fermentation, beans should be turned on various days. Turning means that the beans should be mixed around with a shovel or hands to help get air into the fermenting cocoa and to help make the fermentation even.

8.4.3 Basket Fermentation

Fermentation in baskets is practiced principally by small-scale producers in Nigeria, the Amazon region, the Philippines and some parts of Ghana. Woven baskets are used in this method. They are filled with wet beans, and then covered with palm or banana leaves (Figure 8.7). The sweatings drain through the sides and bottoms of the basket and mixing is effected by transferring the beans from one basket to another. It is sometimes advised that the baskets be lined with palm leaves, but it is doubtful if this necessary. The size of the basket varies, and in Ghana

FIGURE 8.7 Basket fermentation of cocoa beans.

the range is from about 10 to 150 kilograms wet bean capacity. Basket fermentation is often used when cocoa fermented in heaps is vulnerable to larceny.

8.4.4 Tray Fermentation

Tray fermentation has evolved from studies carried out on heap fermentation and involves the fermentation of cocoa beans in a stack of shallow trays with perforated bases (Figure 8.8). The trays are usually 0.9×0.6×13 cm deep with battens fixed across the bottom of the trays and matting rests on the battens to hold the beans in place. The effective depth of the tray is 10 cm and can hold about 45 kg of wet beans. The beans are packed into the trays and may be stacked up to about 8 trays. The bottom tray is usually left empty to allow easy drainage of the sweatings and to improve aeration. After the first day, the stack of trays is covered with plantain/banana leaves or jute sacks to retain heat. Tray fermentation is faster than in the other methods and is usually completed within four days and is thought to produce better-quality fermented cocoa beans.

The beans are not mixed during tray fermentation which is therefore less labour demanding, yielding a higher output because of the shorter fermentation period. The appearance of the fermented beans is, however, less uniform than in the box fermentation.

FIGURE 8.8 Tray fermentation system showing stacked trays with wet cocoa beans.

8.4.5 Curing on Drying Platforms

This technique is confined almost entirely to Ecuador, where the pods are broken and the fresh beans heaped onto the drying platforms (Figure 8.9). It is customary for the beans to be spread during the day and heaped each night and, during the main crop (starts towards the end of the wet season), a remarkably fine cocoa is obtained. In the minor crop, however, the product is of much lower quality. An alternative method is employed in some parts of Ecuador, and is known as the Tendal method of fermentation. A long trench is filled with stones and sand and covered with bamboo matting.

FIGURE 8.9 Curing of cocoa beans on drying platform.

Wet cocoa is heaped in clumps on this base to a depth of approximately 40 cm. The cocoa is covered with leaves from Bihau, Ecuador, and the whole covered with a tin roof. Treatment lasts 24 to 36 hours, with mixing after 24 hours. Fermenting cocoa on drying platforms is convenient, but unless properly managed, the process tends to produce under-fermented cocoa with the added danger of undesirable mould growth and its consequences of off-flavour development.

9 Changes during Fermentation of Cocoa Beans

9.1 PHYSICAL STRUCTURE OF UNFERMENTED COCOA BEANS

The cocoa bean consists essentially of a shell (testa), which represents 10–14% dry weight of the cocoa bean, and the kernel or cotyledon (86–90%), which confers characteristic flavours and aromas of chocolate (Osman, Nazaruddin, and Lee 2004; Afoakwa 2010). The minimum average bean size is 1 g (Nair 2010). A transverse section of a cocoa bean showing two cotyledons (nibs) and a small germ or embryo, all enclosed in a leathery seed coat or testa is shown in Figure 9.1, The cotyledons contain about one-third water and one-third fat (cocoa butter) and the remainder consists of phenolic compounds, starch, sugar, theobromine, non-volatile acids and many other components in small concentrations. The cotyledons of the *Criollo* varieties are normally white in colour, those of the *Forastero* dark purple, whereas *Trinitario* have white to deep purple cotyledons (Afoakwa and Paterson 2010). The cotyledons have two main functions, as the storage organs containing nutrients for the development of the seedling and as the premier leaves of the plant when the seed germinates (Nielsen 2006; Nair 2010; Afoakwa 2010).

The testa acts as a semi-permeable barrier to the flow of substances between the seed and pulp. It has been demonstrated that the testa is freely permeable to water, ethanol, acetic and lactic acids and some volatile organic compounds (Wood and Lass 1985; Biehl and Voigt 1996). It also acts to contain the substances that are released from the cocoa seed cells when they are lysed during fermentation (Roelofsen 1958; Biehl et al. 1985). It also affects the mass transfer rates during the bean drying process. Several researchers have found that certain methods of drying may make the testa more impermeable to water and acetic acid and limit the diffusion of these substances (Faborode, Favier, and Ajayi 1995; Augier et al. 1998; Fowler 2009).

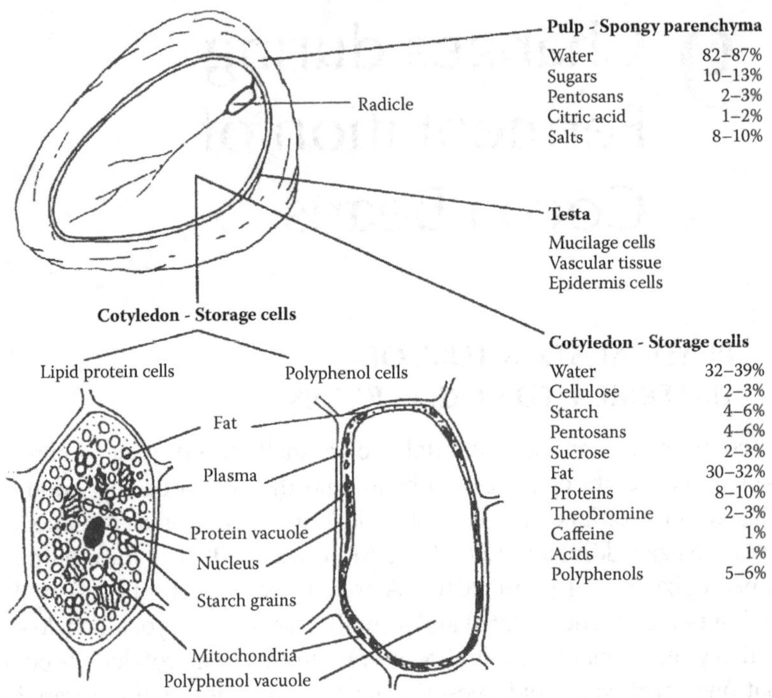

FIGURE 9.1 Anatomy of the cocoa seed. (Adapted from Lopez and Dimick, in *Biotechnology: A Comprehensive Treatise*, Vol. 9, Weinheim: VCH, 1995. With permission.)

9.2 COCOA PULP: FERMENTATION SUBSTRATE

Cocoa pulp is a whitish, sugary and acidic mucilaginous soft tissue that covers the cocoa beans (Figure 9.2). It is the fermentation substrate and its composition is therefore a critical factor on the outcome of fermentation (Beckett 2009; Thompson, Miller, and Lopez 2001; Nielsen 2006; Afoakwa 2010). It consists of 82–87% water, 10–15% sugar, 2–3% pentosans, 1–3% citric acid and 1–1.5% pectin making it a rich medium for microbial growth (Amoa-Awua 2006, Afoakwa et al. 2008). This composition clearly makes the pulp an ideal medium for a medley of micro-organisms to proliferate. Proteins, amino acids, vitamins (predominantly vitamin C) and minerals are also present but are in the minority. Research conducted on Ghanaian *Forastero* cocoa by Nielsen (2006) noted that the main sugars of the fresh pulp were glucose (5.4–6.6%) and fructose (6.3–7.4%) with only small amounts of sucrose (less than 0.3%) present. The glucose and fructose (monosaccharides) to sucrose ratio is a function of the maturity of the cocoa fruit, thus it changes with the degree of maturity with unripe pods containing a higher proportion of sucrose and ripe pods containing mainly fructose and glucose. This clearly suggests that pulp pre-conditioning (pod storage)

FIGURE 9.2 Fresh cocoa beans surrounded by white mucilaginous cocoa pulp.

plays a central role in the permutation of pulp composition (Afoakwa et al. 2012). Berbert (1979) determined the sugar content of freshly harvested (ripe) pods and those stored for six days after harvest.

More glucose and fructose were present in samples six days after harvest than in fresh ones. A slight increase in total sugar concentration was also observed in pods stored for six days in a comparative analysis of pulp from beans collected in the Côte d'Ivoire, Nigeria and Malaysia. The pulp is not only viscous due to its sugar content but also due to a relatively high content of pectin and other polysaccharides 1–2% (Nielsen 2006; Afoakwa 2010). Pettipher (1986) found differences in the amounts of water, citrate, hemicelluloses, lignin and pectin. Pectin content (ca. 1% on a fresh weight basis) was found to vary between 37.5 and 66.1 kg dry weight pulp. With a relatively low pH of the fresh pulp of about 3.94–4.12, which might be due to the citric acid content (approximately 0.6–0.7%), and absent or only low amounts (less than 0.2%) of acetic acid, lactic acid and ethanol were detected in the fresh pulp (Nielsen 2006).

9.3 CHEMICAL COMPOSITION OF UNFERMENTED COCOA BEANS

9.3.1 Fat

The cocoa bean has a high fat content typically known as cocoa butter or cocoa fat. Cocoa butter or fat constitutes about half the weight of the cocoa bean and its quantity and quality are critical to chocolate manufacture.

Ranging from 45–60%, the majority of the lipid profile exists as triacylglycerides (95%) with a minute amount of approximately 5% existing as mono- and di-glycerides, glycolipids, sterols and phospholipids (Belitz, Grosch, and Schieberle 2009). From previous research it has been established that the general fat content ranges from 50.40–53.35% and 52.27–55.21%, respectively, for the pulp pre-conditioned fermented and unfermented beans (Afoakwa et al. 2011a).

The fatty acid composition of the triglycerides is typically 26.5% palmitic, 35.4% stearic, 34.7% oleic and 3.4% linoleic. This composition may vary depending on the average ambient temperatures at which the beans are grown. This fatty acid composition, in turn, affects the melting point of the cocoa butter in the finished beans, a property critical in chocolate manufacture (Lehrian and Keeney 1980). The lipids found in cocoa butter are not absorbed well by the human body and minimally affect serum cholesterol levels (Kris-Etherton et al. 1993).

9.3.2 Proteins

Proteins are a significant component of cocoa seeds, both in quantity (1.5–2.0%) and their role in the development of flavour precursors (Wright et al. 1982). A large number of proteins have been isolated from cocoa beans and they are usually classified according to their solubility when fractionated and their electrophoretic mobility. Cotyledons contain as storage proteins single albumin and globulin species (Biehl, Wewetzer, and Passern 1982a). The globulin, with two polypeptides of 47 and 31 kDa (Pettipher 1990; Spencer and Hodge 1992; Afoakwa et al. 2008) is degraded in fermentation, whereas the albumin fraction (21 kDaltons) is not. In addition to polyphenols, the unfermented cocoa bean contains lipid–protein cells, which have cytoplasms tightly packed with multiple small protein and lipid vacuoles and other components such as starch granules, all of which play roles in defining cocoa flavour and aroma characters.

The compositions of fermented and air-dried cocoa beans include moisture, fat, caffeine, theobromine, proteins, polyhydroxyphenol, sugars, starch, pentosans, cellulose, carboxylic acids and traces of minerals (Reineccius et al. 1972; Beckett 2009; Afoakwa 2010). The non-protein nitrogen is found as amino acids of which about 0.3% is in amide form, and 0.02% as ammonia, which is formed during fermentation of the beans. Cotyledon protein degradation into peptides and free amino acids appears central to the flavour formation (Afoakwa et al. 2008).

9.3.3 Carbohydrates

Starch is the predominant carbohydrate in cocoa. It is present in nibs but not in shells, a fact useful in the microscopic examination of cocoa powders in methods based on the occurrence of starch as a characteristic constituent (Belitz and Grosch 2004). Components of the dietary fiber are, amongst others, pentosans, galactans, mucins containing galacturonic acid, and cellulose. Soluble carbohydrates present include stachyose, raffinose and sucrose (0.08–1.5%), glucose and fructose. Sucrose hydrolysis, which occurs during fermentation of the beans, provides the reducing sugar pool important for aroma formation during the roasting process. Mesoinositol, phytin, verbascotetrose and some other sugars are found in cocoa nibs (Schieberle 2000; Belitz and Grosch 2004).

Prior to fermentation, the cotyledons may contain up to 2% sucrose. During fermentation, nearly all the sucrose is hydrolysed by native cocoa seed invertase into the reducing sugars glucose and fructose (Hansen, del Olmo, and Burri 1998). Reineccius et al. (1972) found that the fresh unfermented *Trinidad* cocoa beans contained 1.5–8.0% sucrose and trace amounts of penitol, fructose, sorbose, mannitol and inositol. Total sugars comprised sucrose and total reducing sugars, whereas the total reducing sugars comprised reducing sugars such as fructose, glucose, mannitol and inositol. According to Berbert (1979), the sucrose concentration of the unfermented beans generally comprised about 90% of the total sugars (2.4–8.0%), whereas both fructose and glucose made up about 6% (0.090 and 0.070%, respectively); other sugars were found to be less than 4%. Afoakwa et al. (2011b) noted that during fermentation, both the non-reducing and total sugars decreased significantly to 2.03 and 5.02 mg/g (89 and 75%) reduction, respectively. The decreases are an indication of the production of reducing sugars; however, the rate of decrease in both the total and non-reducing sugars slowed down towards the end of day six (6) of fermentation. The reducing sugars, fructose and glucose form about 0.9 and 0.7 mg/g, respectively, and others (including mannitol and inositol) at less than 0.50 mg/g. Differences have been attributed to method and time of harvesting, type and origin of cocoa beans (Afoakwa and Paterson 2010). See Table 9.1.

9.3.4 Organic Acids

Citric acid is the predominant acid present in fresh cocoa seed (0.3–2.0%, or 3–20 mg/g wet basis; 5–25 mg/g dry basis). Smaller amounts (0.05–0.5%; 0.5-5 m/g wet basis; 5–25 mg/g dry basis) of oxalic, succinic and malic acids may also be present (Lopez and Quesnel 1973; Bucheli et al. 2001; Ardhana and Fleet 2003). Of the organic acids (1.2–1.6%), citric,

TABLE 9.1
Bean Composition of Unfermented West African (*Forastero*) Cocoa

Constituents	Dried Beans (%)	Fat-Free Materials (%)
Cotyledons	89.60	—
Shell	9.63	—
Germ	0.77	—
Fat	53.05	—
Water	3.65	—
Ash (Total)	2.63	6.07
Nitrogen		
Total nitrogen	2.28	5.27
Protein nitrogen	1.50	3.46
Theobromine	1.71	3.95
Caffeine	0.085	0.196
Carbohydrates		
Glucose	0.30	0.69
Sucrose	1.58	3.86
Starch	6.10	14.09
Pectins	2.25	5.20
Fibre	2.09	4.83
Pentosans	1.27	2.93
Mucilage and gums	0.38	0.88
Polyphenols	7.54	17.43
Acids		
Acetic (free)	0.014	0.032
Oxalic	0.29	0.67

Source: E.O. Afoakwa, *Chocolate Science and Technology*, Wiley-Blackwell, 2010. With permission.

acetic, succinic and malic acid contribute to the taste of cocoa and they are formed during fermentation (Belitz and Grosch 2004; Schwan and Wheals 2004). The amount of acetic acid released by the pulp and partly retained by the bean cotyledons depends on the duration of fermentation and on the drying method used (Biehl et al. 1982b; Barel 1997; Belitz and Grosch 2004; Schwan and Wheals 2004).

9.3.5 Polyphenols

Polyphenolic cells (14–20% dry bean weight) contain a single large vacuole filled with polyphenols and alkaloids including caffeine (0.1–0.2%), theobromine (2.5–3.2%), and theophylline (Osman, Nazaruddin, and Lee 2004). The pigmented polyphenols, when undisturbed, confer a deep purple colour to fresh *Forastero* cotyledons. The major polyphenolic compounds contained in cocoa seeds are catechins (3.0–6.0%), leucocyanidins (2.5%) and tannins (2.0–3.5%). The polyphenols have bitter, astringent flavours and their antioxidant properties help protect the seed from damage and disease (Kim and Keeney 1984; Kyi et al. 2005). Nazaruddin et al. (2001) reported that the total polyphenols ranged from 45–52 mg/g in cocoa liquor, 34–60 in beans, and 20–62 in powder: (–)-epicatechin contents were 2.53, 4.61 and 3.81 mg/g, respectively. Polyphenol reactions with sugar and amino acids contribute flavour and colour to the cocoa bean whereas the alkaloids contribute to the bitterness (Lehrian and Patterson 1983; Afoakwa and Paterson 2010). Theobromine (3, 7-dimethylxanthine), which occurs at 1.2% in cocoa, provides a stimulating effect, which is less than that of caffeine in coffee (Belitz and Grosch 2004). About 400 ml of cocoa contain 0.1 g of theobromine and 0.01 g of caffeine (Craig and Nguyen 1984).

The nib cotyledons consist of two types of parenchyma cells. More than 90% of the cells are small and contain protoplasm, starch granules, aleurone grains and fat globules. The larger cells are scattered among them and contain all the phenolic compounds and purines. Polyphenol and alkaloids are central to bean flavour character (Kim and Keeney 1983; Belitz and Grosch 2004; Schwan and Wheal, 2004). These polyphenol storage cells (pigment cells) make up 11–13% of the tissue and contain anthocyanins and, depending on their composition, are white to dark purple. Three groups of phenols are present in cocoa: catechins (about 37%), anthocyanins (about 4%) and leucoanthocyanins (about 58%).The main catechin is (–)-epicatechin, as well as (+)-catechin, (+)-gallocatechin and (–)-epigallocatechin. The anthocyanin fraction consists mostly of cyanidin-3-arabinoside and cyanidin-3-galactoside, (Kim and Keeney 1984; Schwan and Wheals 2004; Afoakwa et al. 2008).

9.3.6 Enzymes

Fresh unfermented cocoa beans contain various enzymes: α-amylase, β-fructosidase, β-glucosidase, β-galactosidase, pectinesterase, polygalacturonase, proteinase, alkaline and acid phosphatases, lipase, catalase, peroxidase and polyphenol oxidase (Afoakwa et al. 2008). These enzymes exhibit different stabilities during fermentation and may be inactivated by heat, acids, polyphenols, proteases and processing. During fermentation, aminopeptidase, cotyledon invertase, pulp invertase and polyphenol oxidase are significantly inactivated, carboxypeptidase is partly inactivated, whereas endoprotease and glycosidases remain active (Hansen et al. 1998; Afoakwa et al. 2008).

Hansen et al. (2000) noted that differences in enzyme activities can be partly explained by pod variation and genotype but in general, activities present in unfermented beans seem not to be a limiting factor for optimal flavour precursor formation in fermentation. Significant fermentation effects may relate to factors such as storage protein sequence and accessibility, destruction of cell compartmentalisation, enzyme mobilisation and pulp and testa changes (Afoakwa et al. 2008).

9.4 CHANGES IN CHEMICAL AND BIOCHEMICAL COMPOSITION

The development of chocolate flavour begins with the chemical and biochemical changes occurring within the bean during fermentation and drying. Fermentation is essential for development of appropriate flavours from precursors. From previous research, it is imperative to note that pulp pre-conditioning (pod storage, mechanical and enzymatic depulping and bean spreading), mode of fermentation and drying provide the necessary conditions for complex biochemical reactions to occur (Lopez and Dimick 1995; Thompson et al. 2001; Nazaruddin et al. 2006a; Nielsen et al. 2007a; Afoakwa et al. 2012a,b). After pod harvest, beans and adhering pulp are transferred to heaps, boxes, or baskets for fermentations lasting from five to six days for *Forastero* beans but for *Criollo* only one to three days. Typical heap fermentation of beans covered with plantain leaves in a West Africa cocoa farm is as shown in Figure 8.5. In the first day, the adhering pulp liquefies and drains off, with steady rises in temperature. Under anaerobic conditions micro-organisms produce acetic acid and ethanol that inhibit germination and contribute to structural changes such as removal of the compartmentalisation of enzymes and substrates with movements of cytoplasmic components through the cocoa cotyledon generally between 24–48 h of bean fermentation. By the third day, the bean mass will have heated typically to around 45°C, remaining at 45–50°C until fermentation is complete.

Mucilaginous pulp of beans undergoes ethanolic, acetic and lactic fermentations with consequent acid and heat stopping germination, with notable swelling and key changes in cell membranes facilitating enzyme and substrate movements. Differences in pH, titratable acidity, acetic and lactic acid concentrations, fermentation index and cut test scores for cocoa beans from different origins are reported (Jinap and Dimick 1990; Luna et al. 2002; Misnawi et al. 2003).

During fermentation, the rate of diffusion of organic acids into the cotyledons, timing of initial entry, duration of the period of optimum pH and final pH are crucial for optimum flavour formation (Biehl and Voigt 1999).

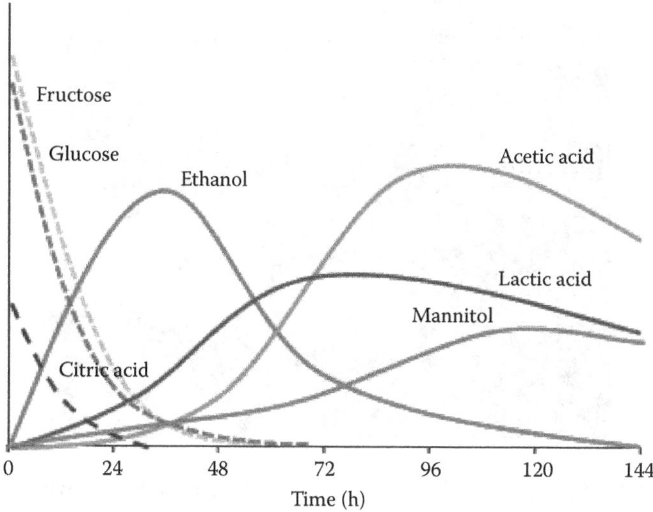

FIGURE 9.3 Changes in volatile acids, sugars and alcohol during fermentation of cocoa.

Beans of higher pH (5.5–5.8) are considered unfermented—with low fermentation index and cut test score—and those of lower pH (4.75–5.19), well fermented. Fermentation techniques can reduce acid notes and maximise chocolate flavours. Ziegleder (1991) compared natural acid (pH 5.5–6.5) and alkaline (pH 8) cocoa extracts obtained by direct extraction; the former possessed a more intense and chocolate aroma than the latter, attributed to high contents of aromatic acids and sugar degradation products with persistent sweet aromatic and caramel notes. Cocoa beans of lower (4.75–5.19) and higher pH (5.50–5.80) were scored lower for chocolate flavour and higher for off-flavour notes, respectively, and chocolate from intermediate pH (5.20–5.49) beans was scored more highly for chocolate.

Sucrose and proteinaceous constituents are partially hydrolysed, oxidised phenolic compounds and glucose are converted into alcohols, oxidised to acetic and lactic acids during fermentation (Figure 9.3). Beans subsequently undergo an anaerobic hydrolytic phase, followed by aerobic condensation. Timing, sequence of events and degree of hydrolysis and oxidation vary between fermentations. Concentration of flavour precursors is dependent on enzymatic mechanisms. Colour changes also occur with hydrolysis of phenolic components by glycosidases accompanied by bleaching, influencing final flavour character.

Nitrogenous flavour precursors formed during anaerobic phases are dominated by the amino acids and peptides available for non-oxidative carbonyl–amino condensation reactions promoted in elevated temperature phases such as fermentation, drying, roasting and grinding. Although

FIGURE 9.4 Cocoa beans before fermentation surrounded by white mucilaginous pulp.

FIGURE 9.5 Cocoa beans after fermentation.

degraded to flavour precursors, residual protein is also diminished by phenol–protein interactions. During aerobic phases, oxygen-mediated reactions occur, such as oxidation of protein–polyphenol complexes formed anaerobically. Such processes reduce astringency and bitterness: oxidised polyphenols influence subsequent degradation reactions.

Figures 9.4 and 9.5 show typical appearances of unfermented and fermented cocoa beans, respectively. The fermentation method determines the final quality of products produced especially for flavour. Previous studies on post-harvest pod storage and bean spreading had shown marked improvement in chocolate flavour and reductions in sourness, bitterness and astringency. In commercial production, similar effects were obtained through combinations of pod storage, pressing and air-blasting. Variations in such factors as pod storage and duration affect the pH, titratable acidity and temperature achieved during fermentation, influencing enzyme activities and flavour development.

Important flavour-active components produced during fermentation include: ethyl-2-methylbutanoate, tetramethylpyrazine and certain pyrazines. Bitter notes are evoked by theobromine and caffeine, together with diketopiperizines formed from roasting through thermal decompositions of proteins. Other flavour precursor compounds derived from amino acids released during fermentations include 3-methylbutanal, phenylactaldehyde, 2-methyl-3-(methyldithio)furan, 2-ethyl-3,5-dimethyl- and 2,3-diethyl-5-methylpyrazine. Immature and unfermented beans develop little chocolate flavour when roasted and excessive fermentation yields unwanted hammy and putrid flavours.

9.5 MICROBIAL SUCCESSION DURING COCOA BEAN FERMENTATION

During fermentation, microbial activity on the cocoa pulp generates heat, and produces ethanol, acetic and lactic acids that kill the bean. Until the pods are split, the beans are microbiologically sterile. Once the pod is split, the beans and pulp are exposed to numerous sources of micro-organisms, including the farmer's hands and implements, the pod exterior and largely insect activity on the farms. The immediate effect of this exposure is the initiation of the microbiological attack of the sugar-rich acidic pulp. At the initial stages of the fermentation process also known as the anaerobic hydrolytic phase, the pulp condition is anaerobic and anaerobic yeast flourish.

The yeasts quickly generate an alcoholic fermentation and the sugars in the pulp are converted to alcohol and carbon dioxide. The citric acid is used in the metabolism of the yeasts. This initiates a slow rise in the pH of the pulp material. The yeasts dominate the first 24–36 hours of the fermentation process, after which the rising pH creates a self-limiting factor on further proliferation. In addition, enzymes released by the yeasts attack the pectin constituents of the cell walls of the pulp mass. The subsequent release of the fluid cell contents runs off the fermenting pulp as what is referred to as 'sweatings'. Examples of yeasts isolated during cocoa fermentation include *Saccharomyces cerevisiae, Kluyveromyces marxianus, Saccharomyces exiguous, Candida castelli, Candida saitoana, Candida guilliermondii, Schizosaccharomyces pombe, Pichia farinosa* and

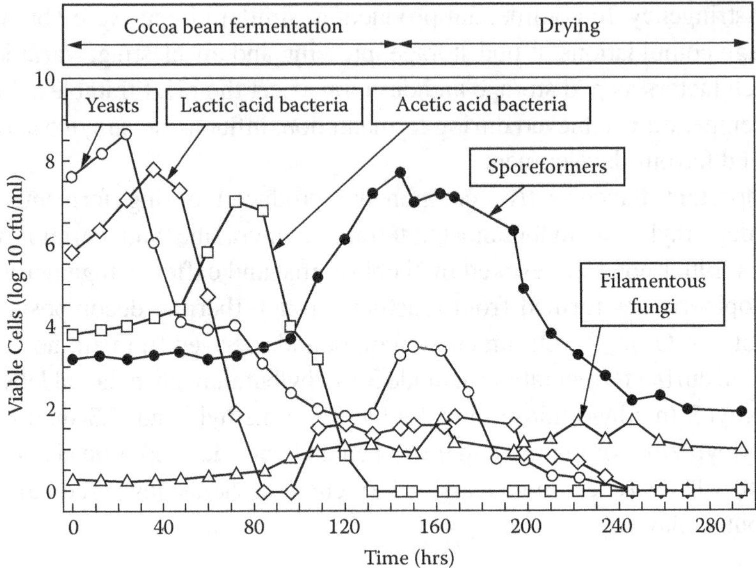

FIGURE 9.6 Changes in microbial activities during cocoa fermentation. (Reprinted from Schwan and Wheals, *Critical Reviews in Food Science and Nutrition* 44:205–221, 2004. With permission.)

Torulopsis spp. (Schwan and Wheals 2004). Figure 9.6 shows the changes in microbial activities during the fermentation period.

The continuous breakdown of the pulp and its liquefaction result in the formation of voids between the cells in the pulp. The loss of fluids through the sweating process increases the rate of acid depletion as it is carried away in the run-off. These voids increase in size and allow air to percolate through the pulpy mass. The combination of this change from anaerobic to aerobic conditions in the substrate, the rise in pH as the citric acid is consumed and loss through sweating and increasing alcohol content being generated by the fermentation of the sugars leads to the eventual inhibition of yeast activity. This signals the end of the anaerobic phase of the process.

The second phase, known as the oxidative condensation phase, occurs under aerobic conditions and is initially dominated by lactic acid bacteria. Lactic acid bacteria increase in numbers when part of the pulp and sweatings have largely drained away, and the yeast population is declining. Yeast metabolism favours the growth of acidoduric lactic acid bacteria. Of the lactic acid bacteria isolated from cocoa fermentations *Acetobacter lovaniensis, A. rancens, A. xylinum, Gluconobacter oxydans, Lactobacillus fermentum, Lb. plantarum, Leuconostoc mesenteroides* and *Lactococcus* (*Streptococcus*) *lactis* were the most abundant species in the first 24 h of fermentation (Schwan and Wheals 2004). As the microbial activity increases, the temperature of the bean mass also begins to increase until it reaches about 45°C (113°F). The conditions at this temperature are more favourable for the promotion of the growth of acetic acid forming bacteria replacing lactic acid formers as the dominant microflora. Figure 9.6 shows the changes in microbial activities during both fermentation and drying periods.

After the decline in the populations of yeasts and lactic acid bacteria, the fermenting mass becomes more aerated. This creates conditions suitable for the development of acetic acid bacteria. These bacteria are responsible for the oxidation of ethanol to acetic acid and further oxidation of the latter to carbon dioxide and water. The acidulation of cocoa beans and the high temperature in the fermenting mass, which causes diffusion and hydrolysis of proteins in the cotyledons, has been attributed to the metabolism of these organisms. Thus, the acetic acid bacteria play a key role in the formation of the precursors of chocolate flavour. In general, the members of genus *Acetobacter* have been found to be more frequent than those of *Gluconobacter.* Species of *Acetobacter aceti* and *Acetobacter pasteurianus* have been isolated in most cocoa beans. The acetic acid formers go on to become about 80 to 90% of the microbial population and their activities (heat and the acidity) eventually lead to the death of the seeds. This results in the breakdown of cellular components and a variety of reactions are initiated.

The increased aeration, increased pH value (3.5–5.0) of cocoa pulp, and a rise in temperature to about 45°C in the cocoa mass in the later stages of fermentation are associated with the development of aerobic spore-forming bacteria of the genus *Bacillus.* Many *Bacillus* spp. are thermotolerant and others grow well at elevated temperatures. *B. stearothemophilus*, *B. coagulans* and *B. circulans* were isolated from cocoa beans that had been subjected to drying and roasting (150°C) temperatures. Aerobic spore-forming bacteria produce a variety of chemical compounds under fermentative conditions. These contribute to the acidity and perhaps at times to the off-flavours of fermented cocoa beans. Indeed it has been suggested that C3–C5 free fatty acids found during the aerobic phase of fermentation and considered to be responsible for off-flavours of chocolate are produced by *B. subtilis*, *B. cereus* and *B. megaterium.* Other substances such as acetic and lactic acids, and 2,3-butanediol, all of which are deleterious to the flavour of chocolate, are also produced by *Bacillus* spp. Pulp fermentation products penetrate slowly into beans causing swelling and stimulating enzymic reactions that yield flavour precursors and, on roasting, characteristic flavour and aroma notes. Fresh beans with low contents of flavour precursors will have limited commercial usage and activities in fermentation will be unable to rectify this shortfall. Appropriate amounts and ratio of precursors are essential for optimal flavour volatiles production in roasting.

9.6 CHANGES IN COCOA PULP DURING FERMENTATION

During fermentation, micro-organisms grow in the cocoa pulp, produce a range of metabolites and cause changes to the pH, temperature and oxygen levels. The main chemical changes in the cocoa pulp, caused by microbial growth and metabolism, are

1. The degradation of pectin by microbial pectinases, causing liquefaction and drainage of some (10–50%) of the pulp (Outtara et al. 2008).
2. Decreases in the concentration of glucose, fructose and sucrose as they are consumed by the micro-organisms. Most researchers report ≥70% reduction in the level of pulp sugars by 36–48 h of fermentation (Schwan, Rose, and Board 1995; Galvez et al. 2007).
3. A decrease in concentration of citric acid (30–50% reduction by the end of fermentation; Holm, Aston, and Douglas 1993; Jinap 1994; Thompson et al. 2001).
4. Increases in the concentration of ethanol, lactic, acetic and to a lesser extent, other organic acids as they are produced by micro-organisms (Schwan and Wheals 2004). Ethanol and acetic acid tend to reach peak levels during the middle of fermentation (24–60 h), and then decrease again. In contrast, maximum levels of lactic acid are usually reached at the end of fermentation (Carr, Davies, and Dougan 1979; Ardhana and Fleet 2003; Nielsen et al. 2007a).
5. The production of small quantities of various other substances by micro-organisms. This includes polysaccharides, enzymes, proteins, fatty acids and volatile organic compounds (Thompson et al. 2001; Afoakwa et al. 2008)
6. An increase in the pH of the pulp, caused by the consumption of citric acid (final pH 4.5–6.5; Nielsen et al. 2007b).
7. An increase in temperature, due to the exothermic nature of many of the metabolic processes (maximum temperature 45–52°C).

9.7 CHANGES IN ENZYMATIC ACTIVITIES

Subcellular changes in the cotyledons release key enzymes effecting reactions between substrates pre-existing in unfermented beans (Hansen et al. 1998). Enzymes exhibit different stabilities during fermentation and may be inactivated by heat, acids, polyphenols and proteases. Aminopeptidase, cotyledon invertase, pulp invertase and polyphenol oxidase are significantly inactivated and carboxypeptidase partly inactivated, whereas endoprotease and glycosidases remain active during fermentation. During the anaerobic phase, the complex pigment components are attacked by glycosidases and are converted by hydrolysis to sugars and cyanidins. As well, sucrose is converted to glucose and fructose by invertase, proteins are converted to peptides and amino acids by proteinase and polyphenols are converted to quinones by polyphenols oxidase. During these processes, the colour of the cotyledons slowly changes and in the case of *Forastero* varieties, the deep purple tissue is converted to a red-brown colour. As the anaerobic phase nears its termination, the products of the enzymatic actions remain to be further converted in subsequent reactions.

In the aerobic phase, cyanidin and protein–phenolic complexes undergo oxidative reactions which are eventually expressed as the final spread of brown colour across the cotyledon surfaces as the red-purple pigments react. Quinone, generated by the actions of the polyphenols-oxidase, now reacts with hydrogen-bearing compounds. These in turn, form complexes with amines, amino acids and sulphur-bearing compounds, leading to the lessening of astringency and bitterness during subsequent roasting of the nibs. Clearly, the changes that occur within the bean during fermentation are very complicated and the hydrolytic and subsequent oxidative reactions generate numerous biochemical complexes which serve as flavour precursors during the roasting process. The genetic make-up of the bean is also certainly crucial to this process. Differences in enzyme activities can be partly explained by pod variation and genotype but in general, activities present in unfermented beans seem not a limiting factor for optimal flavour precursor formation in fermentation. Significant fermentation effects may relate to factors such as storage protein sequence and accessibility, destruction of cell compartmentalisation, enzyme mobilisation and pulp and testa changes.

A processing sequence is required to produce cocoa beans with good flavour. Pulp sugar fermentation should yield high levels of acids, particularly acetic acid. As seed pH decreases, cell structure is disrupted which triggers mobilisation or activation of the primary aspartic proteinase activity with massive degradation of cellular protein. Fermentation proteinase and peptidase activities seem critical for good flavour quality. Significant differences in enzyme activities exist between cocoa genotypes but simple and general relationships have not been established between genotype flavour potential and key enzyme activities in unfermented beans. Therefore, how enzymatic processes are regulated, and substrates and products that relate to desirable flavours, and limiting factors for the enzymatic contribution to fermentation processes still remain unclear.

9.7.1 Hydrolytic Enzyme Reactions

During the anaerobic phase of fermentation hydrolytic enzymes, namely invertase, glycosidases and proteases, have the highest activity (Lopez and Dimick, 1995; Thompson et al. 2001; Afoakwa 2010). The activity of invertase yields reducing monosaccharides (glucose and fructose) from sucrose that natively cannot partake in non-enzymatic browning reactions that occur during roasting to contribute to chocolate flavour. These reducing sugars represent more than 95% of the total reducing monosaccharides in cocoa beans (Forsyth and Quesnel 1957; Hansen et al. 2000).

In addition to the minuscule amounts of amino acids existing in the unfermented bean, proteases (endo- and exoproteases) account for the hydrolysis of proteins to amino acids and peptides, and their activity is dependent on pH and temperature. Proteins in cocoa are vicilin-like, globular in nature and are the main target of the proteases. According to Kirchhoff, Biehl, and Crone (1989) the types and ratio of free amino acids and peptides sequences are unique to cocoa. These amino acids and peptides participate in non-enzymatic browning reactions by forming complexes with reducing sugars during roasting to form important chocolate flavour precursors and also colour formation (Kirchhoff et al. 1989; Afoakwa et al. 2011a)

Glycosidase is a unique enzyme that hydrolyses anthocyanins to cyanidins and sugars (galactose and arabinose) and has more of an impact on colour development and some minor flavour components (Biehl et al. 1985; Lopez and Dimick 1995; Thompson et al. 2001; Afoakwa 2010). Anthocyanins are located in specialised vacuoles within the cotyledon and are responsible for the characteristic deep purple colour of the unfermented bean. These compounds are highly affected by the pH of the medium. The colours range from purple in a neutral state, violet in weak alkaline solutions and pink in acidic conditions (Konczak and Zhang 2004).

9.7.2 Oxidative Enzyme Reactions

Polyphenol oxidase (PPO) is the major oxidase occurring in the aerobic phase of fermentation but continuing well into the drying of cocoa beans and is responsible for some flavour modifications (Thompson et al. 2001). Oxygen facilitates the activity of PPO, however rising temperatures and insufficient moisture become inhibiting factors for the polyphenol oxidase enzyme during drying. Catechins, of which epicatechin makes up more than 90%, and leucocyanidins are the major classes of polyphenols subject to oxidation in cocoa beans. Oxidation of epicatechin during the aerobic phase of fermentation and drying is largely responsible for the characteristic brown colour of fermented cocoa beans. The dihydroxy configuration of polyphenols is oxidised to form quinones which in turn can polymerise with other polyphenols or complex with amino acids and proteins to yield characteristic coloured compounds and high-molecular-weight insoluble material (Thompson et al. 2001) that result in the reduction in astringency and bitterness.

10 Drying Techniques, Storage Practices and Trading Systems

10.1 DRYING OF COCOA BEANS

After fermentation, the moisture content of the beans needs to be reduced from 60% to 7%, an appropriate moisture content for secure storage of cocoa for a couple of months prior to marketing and processing. Above 8% moisture there is the danger of moulds developing within the beans, whereas below 5% the beans are very brittle. Drying also allows some of the chemical changes which occur during fermentation to continue and improve flavour development. This helps to reduce bitterness and astringency and also the development of the chocolate-brown colour of well-fermented cocoa beans. The drying process relies on air movement to remove water; this environment favours aerobic micro-organisms which proliferate at rates that decrease with moisture loss. Various methods of drying the fermented cocoa are used and may roughly be classified into natural or sun-drying and artificial drying.

Drying of cocoa is an important step in cocoa processing as some of the reactions which produce good flavoured cocoa are still proceeding during the drying process. Ideally, cocoa should be dried over a five-to-seven-day period. This allows acids in the cocoa to evaporate off and produce a low acid, high cocoa-flavoured product. Slow drying can lead to mould contamination and this will lead to low quality and low pricing by buyers. In many countries, the major harvest occurs in the wet season and sun drying of cocoa cannot be achieved. In these countries, diesel and wood fired dryers are used. This leads to rapid drying resulting in high acid and low chocolate-flavoured cocoa.

Tropical countries are characterised by relatively high ambient temperature, relative humidity and rainfall. In these countries agricultural products like cocoa are harvested all the year round and the beans must be dried immediately to reduce mass losses and prevent spoilage. These losses might occur as a result of microbial activities, especially mould. Drying of cocoa beans is done to retain chocolate flavour and for safe storage after fermentation from the moisture content of 60% to 7% dry basis (Cunha 1990).

10.2 DRYING METHODS

Drying can be achieved naturally by making use of solar energy. This involves spreading the cocoa beans on the concrete floor or on a raised platform under the sun. These beans are stirred manually to provide even drying of the beans. However, when the condition is not conducive, artificial drying is employed. An artificial drying system consists of mainly a motor, fan and heating element. The fan drives the heated or unheated drying air into the bed. When heat is added to the drying air, the rate of drying increases, depending on the selected drying temperature and air velocity (Jayas and Sohkansanj 1989).

10.2.1 Natural or Sun Drying

Sun drying of cocoa beans is inexpensive and best for optimal quality. It is the preferred method because it allows a slow migration of moisture throughout the beans, which transports flavour precursors that had been formed during fermentation (Sukha 2003). The main harvest of the majority of cocoa-producing countries coincides with the dry season; therefore cocoa beans are usually dried in the sun. This technique is employed by smallholders laying the wet beans on raised platforms covered with bamboo mats (Figure 10.1), or on raised platforms made of wooden boxes

FIGURE 10.1 Drying of cocoa beans on bamboo mats.

FIGURE 10.2 Drying of cocoa beans on raised platform.

(Figure 10.2) or, less satisfactorily (for hygienic considerations), on bamboo mats or concrete platforms on the ground (Figure 10.3).

The beans are constantly mixed and turned to promote uniform drying, to break aggregates that may form, and to prevent mould growth. The duration of drying depends on the weather and under sunny conditions the beans dry in about a week, but it may be extended to two weeks under cloudy or rainy conditions. Under favourable and sunny conditions, cocoa

FIGURE 10.3 Sun-drying of cocoa beans on mats on the ground and raised platforms.

beans at two to three beans thick can still be dried without any significant loss in quality, but quality loss will be expected under adverse weather conditions (Cunha, 1990). Low-quality beans are not recommended to be made into finished products owing to the presence of off-flavours and microbiological contaminations (Knight 2000; Thompson, Miller, and Lopez 2001).

10.2.2 Artificial Drying

Artificial drying of cocoa is brought about by the frequent raining, coupled with frequent turning which can be tiresome when the drying is done manually. However, engineers are leery of the problem of over-drying and quick-drying of cocoa beans by heated dryers. Over-drying reduces the dry matter and causes an increase in energy cost (Arinze et al. 1996). Knowledge of the drying rate and drying constant in relationship to the drying temperature and air velocity is very important to the scientists and engineers who are involved in the design of the dryers and other post-harvest machines involved in cocoa processing. Various studies on the drying of biomaterials indicated that the drying constant and drying rate are important factors in predicting the drying time of biomaterials. In regions such as Cameroon, Costa Rica, Malaysia, Fernando Pó, Panama and Brazil, their harvesting coincides with frequent rainfall hence artificial drying is necessary and desirable (Knight 2000; Thompson et al. 2001). A favourable aspect of artificial drying is its great economy of time and space. These problems can be overcome by the use of a solar dryer

FIGURE 10.4 Drying of cocoa beans on constructed solar dryers.

(Figure 10.4). These dryers use rocks painted black to collect heat from the sun and vent the hot air through the drying bed. They also have a roof which can be lowered when it rains and at night and this saves having to move the cocoa around; therefore less time and labour is required for the drying of cocoa (Knight 2000; Thompson et. al. 2001).

It is recommended that the drying beds not be loaded at more than 50 kg wet beans per sq. metre. Loading at amounts higher than that could result in mould contamination of the cocoa. The simplest forms of artificial dryers are convection dryers or Samoan dryers which consist of a simple flue in a plenum chamber and a permeable drying platform above. Air inlets must be provided in order to allow the convection current to flow without allowing smoke to taint the beans. These dryers are simple to construct and have been used in Western Samoa, Cameroon, Brazil (the Secador dryer) and the Solomon Islands (Jayas and Sohkansanj 1989).

Other artificial dryers are platform dryers using heat exchangers, where the hot air is kept separate from the products of combustion which pass to the atmosphere, or direct fired heaters, where the products of combustion mix with the hot air and are blown through the beans. These dryers can use oil or solid fuels as a source of power. The addition of a fan forces the hot air through the beans and creates a forced draught dryer. Another type of dryer uses conduction. Drying platforms built of slate or cement are heated at one end by a fire or heat source. Small versions of these using oil drums with flues embedded in cement were used in Cameroon at one time and were known as Cameroon Dryers. Heat distribution is not uniform with this type of dryer.

In other countries, diesel and wood fired dryers are used. This leads to rapid drying and a high acid, low chocolate flavoured cocoa will be produced. Platforms, trays, and rotary dryers of various designs, coupled to a furnace, are used, but in every instance, the initial drying must be slow and with frequent mixing to obtain uniform removal of water, which results in volatilisation of acids and sufficient time for oxidative biochemical reactions to occur. For all drying methods beans should be turned about twice each day so that they dry evenly. During turning, clumps of cocoa or beans that stick together should be separated. Flat beans and beans with fungal or insect contamination should be removed during drying (Knight 2000; Thompson et al. 2001; Afoakwa 2010).

10.3 CLEANING AND BAGGING OF DRIED COCOA BEANS

After drying and polishing, the beans are cleaned of any extraneous matter and packed in food-safe jute bags. New food safety requirements dictate that 'food safe' bags whose fibres have been treated with vegetable oil must

FIGURE 10.5 Packaging of cocoa beans in jute sacks.

be used to store cocoa beans. These bags are only used once and must be clean, sound, sufficiently strong and properly sewn (Figure 10.5).

10.4 STORAGE AND TRANSPORT OF COCOA BEANS

Traditional marketing and manufacturing procedures require that the fermented and dried cocoa beans be placed in open weave jute sacks in quantities of 60–100 kg and stored in stacks in warehouses for periods of 3 to 12 months prior to shipment. During longer storage periods, the quality of the cocoa beans can seriously be affected if proper care is not taken. Uniformly dried beans with a moisture content of 7 to 8%, when stored at a relative humidity of 65 to 70% will generally maintain that moisture, resist mould growth and insect infestation and not require repeated fumigation. During the initial storage period slow oxidation and loss of volatile acids may improve flavour but prolonged periods will eventually lead to staling of the beans. The cocoa quality can change during storage depending on temperature, relative humidity and ventilation conditions.

10.5 TRADING AND SHIPPING OF COCOA BEANS

When discussing cocoa trading, a clear distinction has to be made between the actual or physical markets and the futures or terminal markets. Nearly all cocoa coming from origin countries is sold through the physical market. The physical market involves the type of business that most people normally think of when talking about trading in commodities. The structure and length of the cocoa marketing channels differ from region to region within the same producing country as well as across producing countries. At one extreme of the spectrum, the marketing channel between cocoa farmers and exporters encompasses at least two middlemen: small traders and wholesalers. Small traders buy cocoa beans directly from farmers, visiting them one by one. In a second stage, small buyers sell the beans to wholesalers, who in turn will re-sell them to exporters. At the other extreme of the spectrum, cocoa beans are sold directly to exporters by farmers' cooperatives or even directly exported by the co-operative.

Once cocoa beans reach the port of export, they are stocked in warehouses, while being graded and subsequently loaded into one-tonne bags (Figure 10.6) and loaded onto cargo or shipping vessels. Warehouses should have cement and non-flammable floors without cracks and crevices in which insects can hide. Ideally, the floor level of the warehouse should be higher than the surrounding land to prevent flooding and to allow water to flow away.

In some producing countries, cocoa beans are processed in the conditioning plants, most of them located in port warehouses because of the high moisture level of the beans and a high variance in their quality. Conditioning – done either by hand or mechanically – is also used to blend poor quality with good quality beans.

Cocoa grading differs across producing and consuming countries. However, over the years, the physical market has developed standard practices set out by the main international cocoa trade associations: the

FIGURE 10.6 Packaging of cocoa beans in tonnes for the international market.

Federation of Cocoa Commerce Ltd (FCC) and the Cocoa Merchants' Association of America, Inc. (CMAA). For example, the FCC distinguishes two grades: good fermented cocoa beans and fair fermented cocoa beans. Samples of good fermented cocoa beans must have less than 5% mould, less than 5% slate and less than 1.5% foreign matter. A sample of fair fermented cocoa beans must have less than 10% mould, less than 10% slate and less than 1.5% foreign matter. These tests are carried out through the so-called cut test. Such a test involves counting off a given number or weight of cocoa beans, cutting them lengthwise through the middle, and then examining them. Separate counts are made of the number of beans which are mouldy, slaty, insect damaged, germinated or flat.

Once cocoa beans have been graded and loaded into cargo vessels, they are shipped either in new jute bags or in bulk. In recent years, shipment of cocoa beans in bulk has been growing in popularity because it can be up to one-third cheaper than conventional shipment in jute bags. Loose cocoa beans are loaded either in shipping containers or directly into the hold of the ship, the so-called 'mega-bulk' method. The latter mode is often adopted by larger cocoa processors. In general, cocoa futures contracts are not used to secure the supply of cocoa beans, but rather to offset the risk of adverse price movements. A cocoa futures contract is a commitment to make or to take delivery of a specific quantity and quality of cocoa beans at a predetermined place and time in the future. All contract terms are standardised and set in advance. As a result, cocoa futures contracts are interchangeable, except for delivery time.

10.6 FUTURES MARKETS FOR COCOA

There are only two places where cocoa futures contracts can be exchanged: the London International Financial Futures Exchange (LIFFE) and the New York Board of Trade (NYBOT). These organised exchanges provide the facility and trading platforms that bring buyers and sellers together. Moreover, they set and enforce rules to ensure that trading takes place in an open and competitive environment. For this reason, all bids and offers must be made through the exchange's clearinghouse, either through the exchange's electronic order-entry trading system, as in LIFFE, or in a designated trading pit by open outcry, as in NYBOT. As a result, the exchange's clearinghouse is acting as the buyer to all sellers and the seller to all buyers.

To enter into a transaction with the exchange's clearinghouse, a broker must deposit a specified amount of money to guarantee his or her commitment to the terms of the contract. This money is called the 'initial margin', and is a small proportion (i.e., 2–10%) of the total value of the contract. Once a contract is open, the position is 'marked to the market' daily. If the futures position loses value (i.e., if the market moves against it, e.g., the trader is long and the market goes down), the amount of money in the margin account will decline accordingly. For example, if the price of cocoa declines by one dollar per tonne or $10 per contract (i.e., a cocoa futures contract calls for delivery of a lot size of 10 tonnes of cocoa beans), this amount is subtracted from the accounts of all buyers and added to the accounts of all sellers. If the amount of money in the margin account falls below the specified maintenance margin (which is set at a level less than or equal to the initial margin), the futures trader will be required to post additional variation margin to bring the account up to the initial margin level. On the other hand, if the futures position is profitable, the profits will be added to the margin account. It is worth noting that, although the initial margin is small, a trader with a large and consistently losing position may have to tie up significant volumes of cash to maintain the margin.

Futures market participants fall into two general categories: commercial (i.e., hedgers) and non-commercial traders (i.e., speculators). Commercial traders are market participants who try to avoid or reduce a possible loss in the cash market by making counterbalancing transactions in the futures market. On the other hand, non-commercial traders do not produce or use a commodity, but risk their own capital by trading futures in that commodity in the hope of making a profit on price changes.

11 Ochratoxin A (OTA), Pesticides and Heavy Metals Contamination in Cocoa

11.1 OCHRATOXIN A (OTA) IN COCOA

11.1.1 Introduction

Ochratoxin A (OTA) is a polyketide secondary metabolite produced by many *Aspergillus* and *Penicillium* species (Abarca et al. 1994; Dalcero et al. 2002; Varga et al. 2003). It is a mycotoxin which results from a group of secondary metabolites produced by some fungal species and they can cause diseases or death when ingested by humans or animals (de Magalhães et al. 2011). This mycotoxin consists of a polyketide derived from a dihydroiso–coumarin moiety linked through the 12-carboxyl group to phenylalanine, via an amide linkage (Figure 11.1). It is a nephrotoxin which also displays hepatotoxic, teratogenic and immunosuppressive properties, and has been classified by The International Agency for Research on Cancer as a possible human carcinogen (group 2B; Kuiper-Goodman and Scott 1989; Petzinger and Ziegler 2000; Codex Alimentarius Commission 2013). Mycotoxin biosynthesis is related to environmental conditions such as temperature, humidity and rainfall during the cultivation, harvesting, post-harvesting and storage periods of agricultural products (de Magalhães et al. 2011). Ochratoxin A is mainly produced by *Aspergillus carbonarius*, *A. niger* and *A. ochraceus* in tropical zones, and by *Penicillium verrucosum* and *P. nordicum* in temperate zones (Pitt et al. 2000; Abrunhosa et al. 2001; O'Callaghan, Caddick, and Dobson 2003).

Over the past few decades, Ochratoxin A has received more attention because, in addition to being suspected of causing cancer of the urinary tract and damage to kidneys (Pittet 2001), it has also been found in several foods such as cocoa, coffee, cereals, wines, grapes for wine production, red paprika, dry fruits, beer and poultry feed processing (Almela et al. 2007; Burdaspal and Legarda 1998; Chiodini, Scherpenisse, and Bergwerff 2006), and detected in lower amounts in meat, eggs and milk

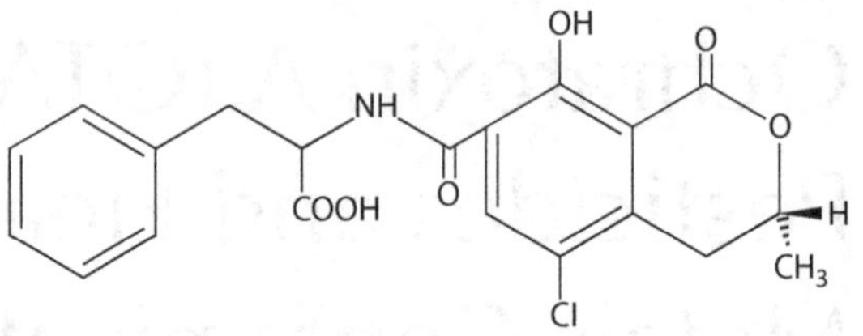

FIGURE 11.1 Chemical structure of Ochratoxin A. (Adapted from Ringot et al., *Chemico-Biological Interactions* 159:18–46, 2006. With permission.)

as a product of the carryover of animals fed with contaminated feedstuffs (Brera et al. 2011). Typical target organs of OTA are the kidney and liver; OTA residues have also been found in significant amounts in human milk, during the breastfeeding period, and in blood as an excretion product (Brera et al. 2002; Micco et al. 1991; Miraglia et al. 1995; Palli et al. 1999).

11.1.2 Chemical Data and Biosynthesis of OTA

Ochratoxin A was first isolated from *Aspergillus ochraceus* Wilh. in a laboratory screening for toxigenic fungi (van der Merwe et al., 1965). Its configuration was determined using optical rotator dispersion spectroscopy (Steyn and Holzapfel 1967; Steyn 1971). The empirical formula is C20H18O6NCl and the molecular weight is 403.82. The IUPAC-developed formula of OTA is l-phenylalanine- N-[(5-chloro-3,4-dihydro-8-hydroxy-3-methyl-1-oxo-1H-2-benzopyran-7-yl)carbonyl]-(R)-isocoumarin. The chemical abstract specification (CAS) of OTA is 303-47-9. It is a white crystalline compound, highly soluble in polar organic solvents, slightly soluble in water and soluble in aqueous sodium hydrogen carbonate. The melting points are 90 and 171°C, when recrystallised from benzene (containing 1 mol benzene/mol) or xylene, respectively (Betina 1989). OTA exhibits UV adsorption: λMeOH max (nm; ε) = 333 (6,400) [5]. The fluorescence emission maximum is at 467 nm in 96% ethanol and 428 nm in absolute ethanol. The infrared spectrum in chloroform includes peaks at 3,380, 1,723, 1,678 and 1,655 cm–1 (Kuipper-Goodman and Scott 1989). OTA has weak acidic properties. The pKa values are in the ranges 4.2–4.4 and 7.0–7.3, respectively, for the carboxyl group of the phenylalanine moiety and the phenolic hydroxyl group of the isocoumarin part (Galtier 1991; Valenta 1998).

OTA production is dependent on different factors such as temperature, water activity (*a*w) and medium composition, which affect the physiology of fungal producers. The biosynthetic pathway for OTA has not yet been completely established. However, labelling experiments using both 14C- and 13C-labelled precursors showed that the phenylalanine moiety originates from the shikimate pathway and the dihydroisocoumarin moiety from the pentaketide pathway (Figure 11.2). The key chain-building step of this reaction scheme is a decarboxylative condensation analogous to the chain elongation step of classical fatty acid biosynthesis (Birch and Donovan 1953; Kao, Katz, and Khosla 1994). The first step in the synthesis of the isocoumarin polyketide consists in the condensation of one acetate unit (acetyl-CoA) to four malonate units. Recent data showed that this step requires the activity of a polyketide synthase (O'Callaghan et al. 2003). Moreover, the gene encoding polyketide synthase appears to be very different between *Penicillium* and *Aspergillus* species (O'Callaghan et al. 2003; Geisen et al. 2004). In *A. ochraceus,* the gene of polyketide synthase is expressed only under OTA permissive conditions and only during the early stages of the mycotoxin synthesis (O'Callaghan et al. 2003).

11.1.3 OTA Contamination in Cocoa and Cocoa Products

Ochratoxin A is the major mycotoxin occurring in cocoa (Mounjouenpou et al., 2008). The presence of OTA in cocoa is mainly related to cocoa bean shells and fat-free cocoa solids (cocoa powder; Amézqueta et al. 2005; Gilmour and Lindblom 2008). Studies have shown that in cocoa, ochratoxin A is mainly produced by *Aspergillus carbonarius* and *Aspergillus niger* (Copetti et al. 2010; Mounjouenpou et al. 2008; Sanchez-Hervas et al. 2008). However, the presence of ochratoxigenic isolates of *Aspergillus melleus*, *Aspergillus westerdijkiae* and *Aspergillus ochraceus* have also been reported (Copetti et al. 2010).

Cocoa beans are principally produced in West Africa (~73–75% of world production) but significant production also occurs in Asia and Oceania and in Central and South America (ICCO, 2013a). Temperate and humid climate conditions of the above-mentioned countries, together with agricultural practices, favor *Aspergillus* and *Penicillium* growth and OTA biosynthesis. OTA is produced when favorable conditions of water activity, nutrition and temperature required for growth of fungi and OTA biosynthesis are present (Codex Alimentarius Commission 2013). OTA occurrence has been reported in cocoa powder from the Côte d'Ivoire, Guinea, Nigeria and Cameroon (Bonvehi 2004) and in cocoa powder sold in Italian shops (Tafuri, Ferracane, and Ritieni 2004). Both toxigenic fungi and OTA can be present at all stages of the farm processing, namely fermentation, drying, storage and transport (FAO/WHO/UNEP 1999). Inasmuch as the cocoa beans are extracted from a fruit, contamination by micro-organisms may occur

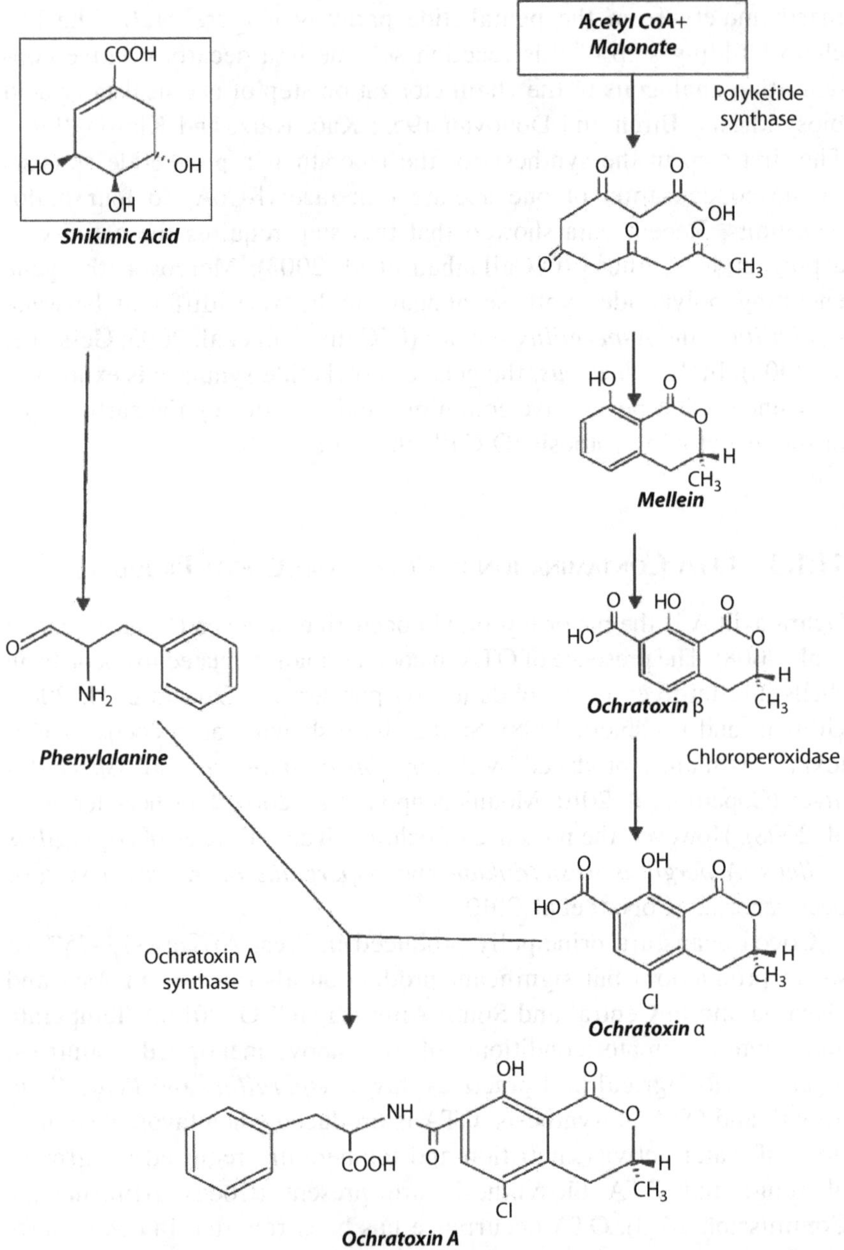

FIGURE 11.2 Biosynthesis of Ochratoxin A. (Adapted from Ringot et al., *Chemico-Biological Interactions* 159:18–46, 2006. With permission.)

and the development of OTA-producing fungi could begin when conditions become appropriate for growth. Generally the fermentation and drying processes could create this favorable condition when these processes are not properly done. In recent studies it was evidenced that, even though fungi capable of producing OTA were not found before fermentation, a rapid increase of OTA was registered in the later processing steps (Copetti et al. 2010). In addition, during the post-harvest phases OTA, being mainly a storage mycotoxin, can increase if storage and processing conditions are not properly controlled. Another co-factor responsible for mould development and mycotoxin production is due to the distance between the cultivation sites and the cocoa transforming sites, with resulting long delivery times. It is important to emphasise that the next manufacturing steps that involve removing shells, roasting (or vice versa), liquoring and refining, only the stage of shell removal can significantly reduce OTA levels. As these steps are performed at the industry level, industry should establish food-safety specific programs to reduce the OTA level in the processed cocoa products meant for human consumption.

High contamination frequencies have been found in cocoa samples and by-products. Burdaspal and Legarda (2003) showed that OTA was found in 99.7% of chocolate and cocoa powder samples. Contamination of 81.3% was also described in cocoa by-products by Miraglia and Brera (2002). Tafuri et al. (2004) found OTA contamination of between 0.22 and 0.77 $\mu g\ kg^{-1}$ in 10 samples of cocoa powder found on the Italian market. A study involving 46 cocoa samples of different origins found that 63% of samples were contaminated by OTA, with an average contamination of 1.71 $\mu g\ kg^{-1}$ (Amézqueta et al. 2004). A maximum content of 100 $\mu g\ kg^{-1}$ was obtained with cocoa contaminated artificially (Hurst and Martin 1998). Work by Amézqueta et al. (2005) revealed that shelling by hand helped to reduce contamination levels in cocoa beans by more than 95%. However, as cocoa is mostly processed on a large scale by various processing companies around the world, it would be impossible to shell the beans by hand during processing as a means of reducing the OTA concentrations.

According to the Working Document of the Expert Committee on 'Agricultural Contaminants' of the European Commission Scientific Committee on Food, the propositions for OTA regulation in cocoa and cocoa products are: (a) 2 mg/kg for raw material for the transformation in food products (cocoa beans, peeled beans, cocoa cake, nibs and cocoa powder) and (b) 1 mg/kg for consumer goods (chocolate powder, chocolate, chocolate beverages). However, with the maximum admissible value of OTA established in 2 mg/kg, about 40% of the cocoa which arrives in Europe may be rejected at the ports (Bonvehi 2004). JECFA established a provisional tolerable weekly intake (PTWI) of 100 ng/kg bodyweight for OTA (Codex Alimentarius Commission 2013). At present, it is noticed that the external market shows itself more and more rigorous concerning the mycotoxin levels in foods. Despite the fact that cocoa beans are not ingested raw, and most of OTA (80%) are in the shells, cocoa processing does not always completely eliminate the toxin (Amézqueta et al. 2005; Tafuri et al. 2004). Thus, beans and cocoa derivatives can only be exported if there are well-established quality criteria, including the analysis of OTA occurrence.

11.1.4 Toxicokinetics of OTA

Both toxicokinetic (the changes of concentrations of a compound in the organism over time) and toxicodynamic (the dynamic interactions of a compound with biological targets and their downstream biological effects) factors determine the toxicity of OTA. Upon absorption from the gastrointestinal tract, OTA binds to serum proteins. Considerable variations in serum half-lives across species are known to be dependent on the affinity and degree of protein binding. Reabsorption of OTA from the intestine back to the circulation, as a consequence of biliary recycling, favours the systemic redistribution of OTA towards the different tissues. In addition, reabsorption of OTA occurs in the kidney proximal and distal tubules. Accumulation occurs in blood, liver and kidney. The liver and kidney are also the major organs of OTA biotransformation.

The metabolism of OTA has not been elucidated in detail, and at present data regarding OTA biotransformation are controversial. In all species, both faecal and urinary excretions play important roles in plasma clearance of the toxin. In addition, mammalian milk excretion appears to be relatively effective (Tafuri et al. 2004; Amézqueta et al. 2005).

11.2 PESTICIDE RESIDUES IN COCOA

11.2.1 Introduction

Cocoa, *Theobroma cacao L.*, is a major cash crop cultivated in the tropical regions of West Africa, the Caribbean, South America and Asia. Like all living organisms, the cocoa plant can also be attacked by a wide range of pests and diseases. When this happens expected production targets are not met, and the economies of the producer nations are adversely affected. Preventive and curative measures are therefore necessary in the cocoa industry to maintain and even increase output (Akrofi and Baah 2007). Although non-chemical means of managing pests and diseases in the industry are widely recommended for health and other reasons, the use of some amounts of chemicals in the form of fertilisers, insecticides and fungicides is unavoidable in the effective management of cocoa farms (Moy and Wessel 2000; Opoku et al. 2007b; Adjinah and Opoku 2010).

11.2.2 Pesticides Used during Production and Post-Harvest Handling of Cocoa

A pesticide is any substance or mixture of substances intended for preventing, destroying or controlling any pest, including vectors of human or animal disease, unwanted species of plants or animals causing harm during or otherwise interfering with the production, processing, storage, transport or marketing of food, agricultural commodities, wood and wood products (WHO/FAO 2009). These substances are classified into several groups. The main pesticide groups by target pest include

1. Fungicides: For crop diseases such as black pod in cocoa
2. Herbicides: Kill weeds
3. Insecticides: Control insect pests, but they may also be
    a. Acaricides: Controlling mites
    b. Nematicides: Controlling nematodes (eelworms)
4. Rodenticides: Kill rats and mice (they are often much less effective against squirrels)

Other pesticide types that are not used on cocoa include molluscicides and bacteriacides (ICCO 2008). Among these groups of pesticides, the insecticides, specifically organochlorines have enjoyed much research publicity worldwide. Perhaps this may be due to their peculiar characteristics: low cost, their versatility against various pests, bioaccumulative nature and potential toxic effects to wildlife and humans (Bempah et al. 2011). Also their tendency for long-range transport and trans-boundary dispersion and their capacity to bioaccumulate in the food chain have also been known (Laws 2000).

Pesticides may be a necessary evil. They protect crops including cocoa from pest attack, and thus maximise productivity in agricultural businesses. However, if pesticide residues are found to be above their maximum residue limits, they pose adverse health effects to humans. 'Pesticide residues' means residues, including active substances, metabolites or breakdown or reaction products of active substances currently or formerly used in plant protection products.

11.2.3 Pesticide Safety in Cocoa and Cocoa Products

Safety aspects are of course by far the greatest concerns for the general public and thus regulators, but pesticides can be important tools for farmers and cannot simply be wished away. Consumers do not always appreciate the high levels of disease and insect pressure that occur in tropical countries, and solving pest control problems for growers remains a crucial part of the 'package' (ICCO 2012b).

There are at least four aspects to pesticide safety: acute (short-term) risks to farmers and other spray operators, impact of pesticides on the environment, residues remaining on food (and animal feed) and real and perceived concerns about longer term effects of pesticides (including combinations of substances). Exporting nations tend to lose foreign exchange if produce are found to contain high residue levels. There are repeated cases of excessive levels of pesticide residues being found in agricultural produce and the safety of these products has become an issue of concern. Recently, changes in regulations in the European Union, North America and Japan have called for a reflection on crop protection practices in cocoa and other commodity crops (ICCO 2007d).

The quality of cocoa imported into the European Union and elsewhere is now assessed based on traces of pesticides and other substances that have been used in the supply chain. The cocoa bean has a high content of butter or fat which absorbs the active ingredients in insecticides. Secondly, the accumulation of any chemicals in the cocoa fat may change the taste of the beans and eventually that of the chocolate made from them. This is known as tainting. It is therefore the task of entomologists to ensure that recommended chemicals do not leave any residues, and that the dosage is the minimum that would give the optimum control under the agricultural conditions in the country.

11.2.4 Regulations of Pesticide Use in Cocoa

The Food and Agriculture Organisation (FAO) of the United Nations and other international bodies have consistently encouraged national pesticide registration schemes, which have now been implemented in most countries. However, it is not always easy to implement regulations (especially those that are technical in nature) in remote rural areas, and products may also pass through 'porous national borders'. The farmer therefore may be faced with a bewildering array of products, with little advice provided on their appropriate use (ICCO 2010c).

In all countries the primary role of registration is to protect human health. The FAO code of conduct on the importation of chemicals is based on the principle of *prior informed consent* (see below), where importing countries have a right to know about pesticides that have been banned or restricted in other countries. It is the responsibility of governments to provide appropriate guidance on the use of hazardous compounds, ranging from easily comprehensible labelling to outright banning of the most toxic products.

The acceptable levels of active ingredients in foods are determined by the committee on pesticide residue of FAO/WHO, known as the Codex Alimentarius Commission, CAC. Created in 1963 the CAC implements the Joint FAO/WHO Food Standards Programme which is aimed at protecting the health of consumers and ensuring fair trade practices in the international food trade (Moy and Wessel 2000). The commission has set maximum levels of residue poisons in commodities going through the international market, including cocoa. If for any reason the residual levels in any commodity exceed the Codex levels, that particular commodity could be rejected by the importing country.

Pesticide residues on crops are monitored with reference to maximum residue limits (MRL) and are based on analysis of quantity of a given AI (active ingredient) remaining on food product samples. MRLs are defined as the maximum concentration of pesticide residue (expressed as milligrammes of residue per kilogramme of food/animal feeding stuff) likely to occur in or on food and feeding stuffs after the use of pesticides according to good agricultural practice (GAP), that is, when the pesticide has been applied in line with the product label recommendations and in keeping with local environmental and other conditions. For active substances for which no MRL is included in the Regulation, a default MRL of 0.01 mg/kg will apply.

The MRL for a given crop/AI combination is usually determined by measurement during a number (on the order of 10) of field trials where the crop has been treated according to GAP and an appropriate preharvest interval has elapsed. For many pesticides, however, this is set at the limit of determination (LOD), because only major crops have been evaluated and understanding of ADI is incomplete (i.e., producers or public bodies have not submitted MRL data, often because these were not required in the past). LOD can be considered a measure of presence/absence, but true residues may not be quantifiable at very low levels. For this reason the limit of quantification (LOQ) is often quoted in preference (and as a 'rule of thumb' is usually approximately twice the LOD).

It was further suggested to all producing countries to carry out a pesticide audit, prioritising the issues. As an example, it was thought that many residues originating from the treatment of cocoa beans in storage constituted the highest risk group, followed by insecticides applied in the field, fungicides and herbicides. The expert on pesticide matters was subsequently requested to produce a manual on the safe use of pesticides in cocoa growing to provide the necessary guidance to the relevant stakeholders.

Testing for residues is carried out following internationally agreed and validated methods (and good laboratory practice [GLP] standards apply in some countries). Procedures include extraction and 'clean-up' from samples, followed by analysis using various instruments, depending on the residue being analysed. Analysis techniques include: gas chromatography (GC), gas–liquid chromatography (GLC), gel permeation chromatography (GPC), high-pressure liquid chromatography (HPLC) and various mass spectrometry techniques, so such laboratories are expensive to set up and maintain.

11.2.5 Pesticide Control and Management in Cocoa

Pesticides have been divided into 'positive' and 'negative' lists, with reference to the new EU legislation (ICCO, 2010c). This is to support the control and management of pesticides over a wider geographical scope across the various producing countries. For this purpose, pesticides have therefore been divided into four categories, A to D. Detailed information outlining the different categories of pesticides used to control pesticide use during cocoa production, storage and transport provided in the ICCO (2010c) document on pesticide use in cocoa, a guide for training administrative and research staff, is as outlined below.

A. *Lists of Strategic/Recorded Pesticides for Use in Cocoa:* These pesticides have shown demonstrable efficacy in at least one regional cocoa-growing country and have either been given EU MRL or tMRL. The status of the active ingredients and acceptable concentrations for their use should be checked with the appropriate national agency within the cocoa-producing countries.

1. *Black Pod Diseases:* Active ingredients:
   - Benalaxyl
   - Copper hydroxide
   - Copper oxide
   - Copper oxychloride
   - Fosetyl aluminium
   - Metalaxyl-M (mefenoxam)
2. *Insects:* Active ingredients:
   - Acetamiprid
   - Beta-cyfluthrin
   - Cypermethrin (a isomer - ß)
   - Deltamethrin
   - Dimethoate
   - Imidacloprid
   - Lambda-cyhalothrin

Thiamethoxam

Termite control

Fipronil

3. *Weeds:* Active ingredients:

2,4-D dimethylamine salt

Glyphosate trimesium

Glyphosate isopropylamine

4. *Stored produce:* Active ingredients:

Aluminium phosphide

Magnesium phosphide 0.01 (as hydrogen phosphide)

Pyrethrins (pyrethrum) for fogging

Pyrethroids (treating sacks, etc.)

B. *Compounds to Be Used with Great Caution (limited time span, etc.):* These active substances have permitted MRLs in some markets. Many of these are temporary (tMRL) but not others or are likely to be phased out within 2–3 years. They have shown demonstrable efficacy in at least one regional cocoa-growing country.

1. *Black Pod Diseases:* Active ingredients:

Metalaxyl (unresolved)

2. *Insects:* Active ingredients:

Bifenthrin

Diazinon

Chlorpyrifos (ethyl)

Fenitrothion

Fenvalerate

Fenobucarb (bpmc)

Isoprocarb (mipc)

Malathion

Pirimiphos methyl

3. *Weeds:* Active ingredients:

Picloram

4. *Stored Produce*: Active ingredients:

Bioresmethrin

Methyl bromide

C. *Lists of Experimental Control Agents for Possible Future Inclusion in Category A:* These are subject to current or recent field testing and may well conform to criteria in category A when it is established that they conform to the established criteria.

1. *Black Pod Diseases:* Active ingredients:

Dimethomorph

Iprovalicarb

Mandipropamid

2. *Insects:* Active ingredients:
   a. Mirids (including *Helopeltis* spp.)
      Emamectin benzoate
      Thiacloprid
      Clothianidin
      Novaluron
      Teflubenzuron
      Spiromesifen
      Spirotetramat
   b. Cocoa pod borer
      Emamectin benzoate
      IGRs: novaluron, teflubenzuron,
      Methoxyfenozide
3. *Weeds:* Active ingredients:
   Safer contact herbicides required
4. *Stored produce:* Active ingredients:
   Sulfuryl fluoride

D. *Pesticides That Must Not Be Used for Cocoa:* These have been recorded as used on cocoa (e.g., by the ECA/CAOBISCO project), but have been rejected by major importing countries (usually for toxicological/ecotoxicological reasons) and have no residue tolerances in major markets. Active ingredients:
   Insecticides:
      Acephate
      Amitraz
      Aldrin
      Azinphos-methyl
      Cabaryl
      Carbofuran
      Carbosulfan
      Cartap
      Cyhalothrin (unresolved)
      Cyhexatin (acaricide)
      Dichlorvos (ddvp)
      Dieldrin
      Dioxacarb
      Endosulfan
      Lindane, gamma
      Methyl-parathion (= parathion-methyl)
      Methomyl
      Monocrotophos
      Profenfos
      Promecarb

- Propoxur
- Terbufos

Herbicides:

- Ametryn
- Atrazine
- Diuron
- Fomesafen
- Msma (methyl arsenic acid)

Fungicides:

- Benomyl
- Captafol
- Hexaconazole
- Pyrifenox
- Triadimefon
- Tridemorph
- Zineb

Stored produce:

- Allethrin (esbiothrin)
- Fenitrothion
- Isoprocarb (mipc)
- Permethrin
- Resmethrin
- Tetramethrin

Cocoa growers are strongly advised to stop using any products containing any AI listed under D. Where they have been used in the past for cocoa pests, there should now be recommended satisfactory substitutes for them; if this is not the case please contact the Secretariat of the International Cocoa Organization for advice (ICCO, 2010c). These lists may not be exhaustive. They have been based on ICCO records and the findings of the ECA/CABI/CAOBISCO project (Global Research on Cocoa, June 2008).

11.3 HEAVY METALS IN COCOA

11.3.1 Introduction

The term heavy metal refers to any metallic chemical element that has a relatively high density and is toxic or poisonous at low concentrations. Examples of heavy metals that are harmful to humans include mercury, cadmium, lead, and arsenic. Chronic exposure to these metals can have serious health consequences. In most cases, humans are exposed to heavy metals through inhalation of air pollutants, consumption of contaminated drinking water, exposure to contaminated soils or industrial waste, or consumption of contaminated food (http://www.psr.org/environment-and-health/confronting-toxics/heavy-metals/). Food sources such as vegetables, grains, fruits, fish and shellfish can become contaminated by accumulating metals from surrounding soil and water. Heavy metal exposure causes serious health effects, including reduced growth and development, cancer, organ damage, nervous system damage, and in extreme cases, death. Exposure to some metals, such as mercury and lead, may also cause development of autoimmunity, in which a person's immune system attacks its own cells. This can lead to joint diseases such as rheumatoid arthritis, and diseases of the kidneys, circulatory system, and nervous system.

Metals are particularly toxic to the sensitive, rapidly developing systems of fetuses, infants, and young children. Some metals, such as lead and mercury, easily cross the placenta and damage the fetal brain. Childhood exposure to some metals can result in learning difficulties, memory impairment, damage to the nervous system, and behavioral problems such as aggressiveness and hyperactivity. At higher doses, heavy metals can cause irreversible brain damage. Children may receive higher doses of metals from food than adults, because they consume more food for their body weight than adults.

Heavy metals are common trace constituents in the earth crust that have densities above 5 g/cm3. The most frequently reported heavy metals with potential hazards in soils are cadmium, chromium, lead, and zinc and copper (Alloway 1995). The concentration of these toxic elements in soils may increase from various sources including anthropogenic pollution, weathering of natural high background rocks and metal deposits (Senesi et al. 1999). Most heavy-metal contamination stems from high-temperature combustion sources, such as coal-fired power plants and solid waste incinerators. Local metal sources may include metal-plating industries and other metal industries. The use of leaded gasoline has led to global lead pollution even in the most pristine environments, from arctic ice fields to alpine glaciers.

11.3.2 Primary Sources of Heavy Metals

The primary anthropogenic sources of heavy metals are point sources such as mines, foundries, smelters, and coal-burning power plants, as well as diffuse sources such as combustion by-products and vehicle emissions. Humans also affect the natural geological and biological redistribution of heavy metals by altering the chemical form of heavy metals released to the environment. Such alterations often affect a heavy metal's toxicity by allowing it to bio-accumulate in plants and animals, bio-concentrate in the food chain or attack specific organs of the body (http://www.osha-slc.gov/SLTC/metalsheavy). Heavy metals are not biodegradable and hence persist in the environment (MacFarlane and Burchett 2001). Plants show several response patterns in heavy metals uptake (Kabata-Pendias and Pendias, 1997). Although most are sensitive to very low concentrations, others have developed resistance, and there are a small number of plants that are hyperaccumulators of toxic metals (Chapin, 1983; Mingorance, Valdes, and Oliva 2007).

11.3.3 Effect of Heavy Metals on the Environment

Pollution of the natural environment by heavy metals is a worldwide problem as these metals are indestructible and most of them have toxic effects on living organisms (Dalman, Demirak, and Balci 2006). Heavy metals are of high ecological significance inasmuch as they are not removed from the soil via self-purification, but rather accumulate in reservoirs and enter the food chain (Loska and Wiechula 2003). From the environmental point of view, heavy metals are largely immobile in the soil system, explaining why they tend to accumulate and persist in agricultural soils for a long time. The concentration and distribution of heavy metals in the soil often differ from metal to metal, probably as a result of their differential accumulation rates. For instance, a study conducted on Nigerian agricultural soils revealed that heavy metal contamination by lead, zinc and cadmium was minimal whereas copper contamination was very high (Aikpokpodion, Lajide, and Aiyesanmi 2010). According to Toselli et al. (2009), accumulation of copper in Italian soils is due to repeated application of fungicides to control fungal diseases of pear and grapes. A study conducted by Savithri, Joseph, and Poongothai (2003) in India also revealed that high levels of Bordeaux mixture application has resulted in a significant accumulation of copper in surface and subsurface soils. These further show that horticultural operations with a long history of copper fungicide application often have a significant accumulation of copper in surface horizons (Merry, Tiller, and Alston 1986; Alva, Huang, and Paramasivam 2000). There is increased awareness that heavy metals present in soil may have negative consequences on human health and on the environment (Abrahams 2002; Selinus et al. 2005). This situation is gradually taking cocoa soils in the studied area to a condition of deterioration because of the contamination load and the adverse effect on soil biodiversity (Aikpokpodion et al. 2010).

11.3.4 Heavy Metals (Cadmium, Lead, Copper and Arsenic) in Cocoa and Cocoa Products

The levels of cadmium, lead, copper and arsenic in foodstuff are of interest as these metals are generally considered as toxic to human beings. Information on the analysis of these metals in raw cocoa and finished chocolate products is, however, rather scarce. Knezevic (1979, 1980, 1982) investigated the contents of cadmium, copper, lead and arsenic in cocoa beans from various countries as well as in some chocolate products. He reported cadmium content of 0.48–1.83 mg/kg and arsenic content of 0.77 mg/kg in Malaysian cocoa beans. These values are higher than those found in beans from South America, Africa and West Indies. The levels of lead (0.21 to 0.42 mg/kg) and copper (21.5–32.8 mg/kg) in Malaysian beans were comparable to those from other countries. The metal contents of various chocolate products decreased with the fraction of cocoa mass in the products.

Musche and Lucas (1973) studied lead content in some cocoa beans and chocolate products and concluded that there was no lead contamination in the manufacturing process of chocolate products. There is no documented study of cadmium, lead, copper and arsenic levels in locally manufactured chocolate products. The present study reports the determination of the levels of these metals in raw cocoa and intermediate chocolate products in a manufacturing process as well as in some locally manufactured chocolate bars. The suitability of the method of dry ashing followed by analysis by inductively coupled plasma emission spectrometry of these metals in the cocoa samples is also evaluated.

The possibility of the presence of lead (Pb) and cadmium (Cd) in chocolate is a matter of health consideration because the chemical composition of cocoa allows strong binding of Pb and Cd, as discussed by Valiente et al. (1996). The presence of Pb and Cd in chocolate products could be natural or due to processing (Dahiya et al. 2005; Rankin et al. 2005). These can be absorbed directly by the theobroma (cocoa tree), can be introduced during the preparation process or result from contamination via utensils, environmental pollution or transportation and storage, but, regardless, they may be found in the final product (Mesallam 1987). The American Environmental Safety Institute (AESI) claimed that the chocolate manufacturers neither take appropriate measures to remove potentially dangerous levels of lead and cadmium from their chocolate products nor notify consumers of their health risks. On the contrary, world chocolate manufacturers have dismissed claims that their products pose any health hazard due to Pb and Cd.

Chocolates are also considered an important source of nickel as discussed by Smart and Sherlock (1987), as this element is used as a catalyst in the hydrogenation process. Moreover, cocoa butter that is used in chocolates may also contain nickel (Dahiya et al. 2005). Ni contamination in chocolates could be due to the type of containers used for storage and transportation (Mesallam 1987). Analysis of copper content in chocolate is also important because copper compounds are widely used as fungicides in the farming of cocoa (Dos Santos et al. 2005). Copper in the human body plays a role in the mobilisation of tissue iron and the formation of mitochondrial heme (Lucia, Neuza, and Fernando 2005).

12 Cocoa Processing Technology

12.1 COCOA BEAN QUALITY

Quality may be considered as a specification or set of specifications which are to be met within given tolerances or limits. In the context of cocoa quality, it is used to include not just the important aspects of flavour and purity, but also physical characteristics that have direct bearing on manufacturing performance, especially yield of cocoa nib (BCCCA 1996). The different aspects or specifications of quality in cocoa therefore include: flavour, purity or wholesomeness, consistency, yield of edible material and cocoa butter characteristics.

The quality of cocoa beans is an important trade parameter because the quality of chocolate depends to a large extent upon the quality of the cocoa beans used to make the chocolate. After cocoa is harvested, the beans have to be fermented and dried, a process which enables them to develop the characteristic cocoa flavour after they have been roasted. Nearly all exported cocoa is sold on the international markets in London and New York. Because chocolate is sold in a very competitive market, manufacturing companies would like to buy the best quality cocoa. Fine and flavour cocoas have distinctive aroma and flavour characteristics and are therefore sought after by chocolate manufacturers but they represent only 5% of global cocoa production.

12.2 BEAN SELECTION AND QUALITY CRITERIA

The quality of dried cocoa beans that reaches the processor is crucial to the final product quality. Chocolate manufacturers thus follow a strict set of guidelines and quality criteria if they are to produce products that maintain the consumers' loyalty to their products. Before processing, the quality of beans is evaluated using two different methods. With the first technique, the beans are assessed for the following indicators:

1. Degree of fermentation
2. Moisture content (max. 7.5%)
3. Number of defects

4. Number of broken beans
5. Bean count (number per 100 g)
6. Degree of mouldiness
7. Flavour profile
8. Colour
9. Fat content (min 52%)
10. Fat quality relating to % of free fatty acids (as oleic acid – max. 1.75%)
11. Shell content (10–12%)
12. Uniformity of bean size
13. Insect and rodent infestation

Generally, to make good quality chocolate, cocoa beans must have cocoa flavour potential, be free from off-flavours such as smoky and mouldy flavours, should not be excessively acidic, bitter or astringent, the beans should have uniform sizes and on the average weigh 1 gram, should be well fermented, thoroughly dry with a moisture content of between 6 and 8%, have a free fatty acid content of less than 1%, cocoa butter content of 50 to 58%, shell content of less than 11 to 12% and be free from live insects, foreign objects, harmful bacteria, and pesticide residue (Amoa-Awua et al. 2006). The International Cocoa Standards require cocoa of merchantable quality to be fermented, dry, free from smoky beans, free from abnormal or foreign odours and free from any evidence of adulteration. It must be reasonably free from living insects, broken beans, fragments and pieces of shell and foreign matter and be reasonably uniform in sizes. Throughout the world the standards against which all cocoa is measured are those of Ghana cocoa. Cocoa is graded on the basis of the count of defective beans in the cut test (Awua 2002; Amoa-Awua et al. 2006). Figure 12.1 shows typical fermented and dried cocoa beans.

12.2.1 Free Fatty Acid

Generally to make good quality chocolate products the free fatty acid of the cocoa beans should be less than 1%. If the free fatty acid in a sample of beans exceeds 1%, it is likely that the FFA in the cocoa butter derived from them will exceed 1.75%, a limit which applies in EEC countries and any others adopting the recommendations of the Codex Alimentarius Commission. Higher levels of free fatty acids occur in cocoa that is either improperly stored for a long time (i.e., too hot and high moisture) or the activity of bacteria or mould in the cocoa. In particular the action of enzyme lipase (not only introduced by microbiological activity but also present in the natural raw cocoa) acts to break down the triglycerides into

FIGURE 12.1 Fermented and dried cocoa beans.

its separate groups of the fatty acids. In significant amounts their presence gives a rancid flavour to the products.

12.2.2 Bean Count Test

The second technique is evaluated based on the size of the beans using either bean count (number of beans per 100 g) or the weight in grams of 100 beans. On the international cocoa market, different bean sizes attract different prices. Beans with smaller sizes usually contain a proportionately lower amount of nibs, higher shell content, lower fat content and attract lesser prices. Typically, beans from Asian origin have higher shell content than West Africa beans.

12.2.3 Cut Test

The cut test used in the grading of cocoa reveals the presence of certain defects that may cause off-flavours and indicates the degree of fermentation of the beans which has a bearing on the flavour and quality of the beans. It involves randomly selecting a certain number of beans from the sample and carefully cutting them lengthwise to examine the cross-sections physically (Figure 12.2). The number of beans with the following defects is counted separately: mouldy beans, slaty beans, insect-damaged beans, germinated beans, flat beans and purple beans in some cases. According to the International Organisation for Standards (ISO), a minimum of 300 beans must be checked for every tonne of cocoa, and a minimum of 30% of the sacks or bagged cocoa checked to determine the grade.

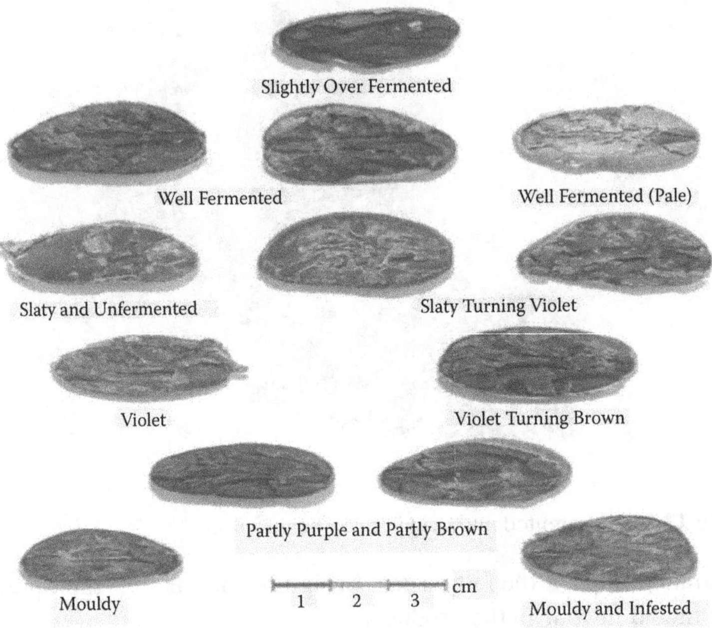

FIGURE 12.2 Visual examination of bean quality from the bean cut tests.

The bean cut test is used to assess defects and the degree of fermentation. In this process, a sample of 300 beans is randomly selected and split open longitudinally. The cut surfaces are then examined and assessed based on the following criteria:

1. Flat and shrunken beans
2. Mouldy beans
3. Slaty beans
4. Germinated beans
5. Degree of insect and rodent infestation

All these factors affect the flavour and taste of the finished products from the beans which would be used. Good cocoa beans should be well fermented, dry and free from insect and rodent infestation, abnormal odours and contaminations/adulterations. Figure 12.2 shows the visual examination of beans quality from the bean cut tests.

12.2.4 Flavour Quality

Flavour quality is another key criterion used to assess the quality of fermented and dried cocoa beans. Here, consideration is usually given to the desired quality of the finished chocolate or products upon which or in which the chocolate would be used. For instance, harsh cocoa and bitter notes are required to contrast a very sweet or heavily flavoured center, using delicately flavoured beans such as Java beans. It is important to note that just because a bean comes from a flavour grade stock does not mean it will automatically improve a product's profile. The overall impact on a particular stock of its inclusion upon the blend has to be carefully assessed. Also noteworthy is the fact that although beans are characteristically typed, flavour quality may vary from year to year, crop to crop, and so on, and therefore require a continuous assessment of availability of the beans before using them in recipe formulations.

12.3 COCOA QUALITY, GRADING AND STORAGE

Cocoa quality is determined by a combination of factors that determine the acceptability of the cocoa to a buyer. These factors include proper fermentation, dried to the proper moisture level, free from abnormal odours and free from mould contamination. The standards against which all cocoa is measured are those of Ghanaian cocoa. Cocoa is graded on the basis of the count of defective beans in the 'cut test'. The cut test reveals the presence of certain defects which may cause off-flavours and indicates the degree of fermentation of the beans which has a bearing on the flavour and quality of the beans. The International Standards Organisation cut test procedure states that for a complete determination of bean quality, cut lengthwise through the middle, so as to expose the maximum cut surface of cotyledons. Both halves of each bean are visually examined. Each defective type of bean is counted separately, and the result for each kind of defect is expressed as a percentage of the beans examined. Defective beans include slaty, insect damaged, flat beans, over-fermented and mouldy beans. According to international standards mouldy beans should not exceed 3%; slaty beans should not exceed 3% and insect damaged, germinated or flat beans 3% by count. Table 12.1 shows the different types of defect which can be seen using the cut test and their specifications that are used to establish the standard or grade of cocoa.

According to the Ghana Cocoa Board, the required quality specifications of dry cocoa beans as stated in the sales contract include the following, among others:

1. Superior Quality/Good Fermented: Cocoa should contain no more than 5% slaty beans and not more than 5% of all other defects.

TABLE 12.1
Cocoa Quality Parameters and Percentages of Defective Beans Used to Determine Grade

Grade–Bean Count/100 g	Mouldy	Slaty	Defects[a]
1A (<+100)	3	3	2.5
1B (101–120)	3	3	2.5
2A (<+100)	4	8	5.0
2B (101–120)	4	8	5.0
Sub-standard (>120)	>4	>8	>5.0

[a] Including insect damage, infested beans and germinated beans.

2. Average Quality/Fair Fermented: Cocoa should contain no more than 10% of 'all other defects'. Mouldy, germinated, flat, insect-attacked beans and the like are considered as 'all other defects'. In all cases, bean count (i.e., the number of beans per 100 gm wt.) should either be
   a. Main Crop beans (i.e., up to 100 beans/100 g)
   b. Light Crop beans (i.e., 101 – 120 beans/100 g)

Other quality parameters covered in the sale contract documents are

1. Uniformly fermented cocoa
2. Dry beans, moisture content of 7.5%
3. Uniform in size
4. Homogeneous in all other respects and the parcel shall be
   a. Fit for the production of a foodstuff
   b. Virtually free from foreign matter and adulteration, contamination, live insects (including mites), rodents or other types of infestation

Great care to achieve optimum quality from harvest to drying must continue during transport and storage. There must be proper humidity control to avoid re-humidification of the beans, which would lead to mould growth. Also, storing on gratings or decking should allow at least 7-cm air space above the floor to avoid rodent and insect pests. Storage must also not be in close proximity to any strong odour. Cocoa should never be stored for a significant period of time before shipment, thus reducing the chance of picking up strong off-odour, increased moisture levels and mould growth. Forced air ventilation, fumigation and good phytosanitary practices all contribute towards optimal storage conditions.

12.4 STEPS IN COCOA PROCESSING

12.4.1 Cleaning, Breaking and Winnowing

Before processing, cocoa beans are passed through the processes of cleaning, breaking and winnowing to obtain nibs of consistent quality. These processes also ensure that the nibs are cleaned (free from dirt and infestation), well broken and properly deshelled. The kernels (nibs) obtained after the process must be of uniform size to achieve constant quality. The process involves, first, sieving the beans and removing all extraneous materials such as stones, strings, coins, wood pieces, soil particles, nails and so on. The cleaned beans are then broken to loosen the shells from the nibs using multiple steps to avoid an excess of fine particles. The products obtained are then sieved into a smaller number of fractions to obtain optimal separation during subsequent winnowing. The fractions are then transported to the winnowing cabinet where the lighter broken shells are removed by a stream of air. The breaking and winnowing steps are vital in separating the essential components of the bean, the nibs from the shells; the shells are then discarded and sold for use as agricultural mulch or as fertilisers. Strong magnets are then used to remove magnetic foreign materials from the nibs, which are then stored, awaiting further processing. Figure 12.3 shows the flow diagram for the general production of cocoa butter, cocoa cake and cocoa powder. The details outlined for the production of cocoa liquor, cocoa butter, cocoa cake and cocoa powder are as shown in Figure 12.4.

12.4.2 Sterilisation

Sterilisation is the technique of exposing the cocoa beans or nibs to sufficiently higher temperatures for sufficiently long times to destroy all micro-organisms in the beans. Depending on the factory and equipment used, this process can either be done before or after the roasting process. The treatment can be done in a batch or continuous process by wetting or heating with steam, eliminating the micro-organisms that might have contaminated the nibs during the post-harvest processes of fermentation, drying, bagging and transportation. The process ensures that the total plate count (TPC) is reduced to less than 500 per gram, and all pathogenic bacteria are destroyed. After sterilisation, the nibs can then be roasted directly (natural process) or can be alkalised first by the Dutch process before roasting. In situations where sterilisation is done after roasting, the heat treatment is used to ensure total destruction of heat-resistant bacteria and spores that might have survived the high temperatures of the roasting process. The procedure is to inject, over a period of

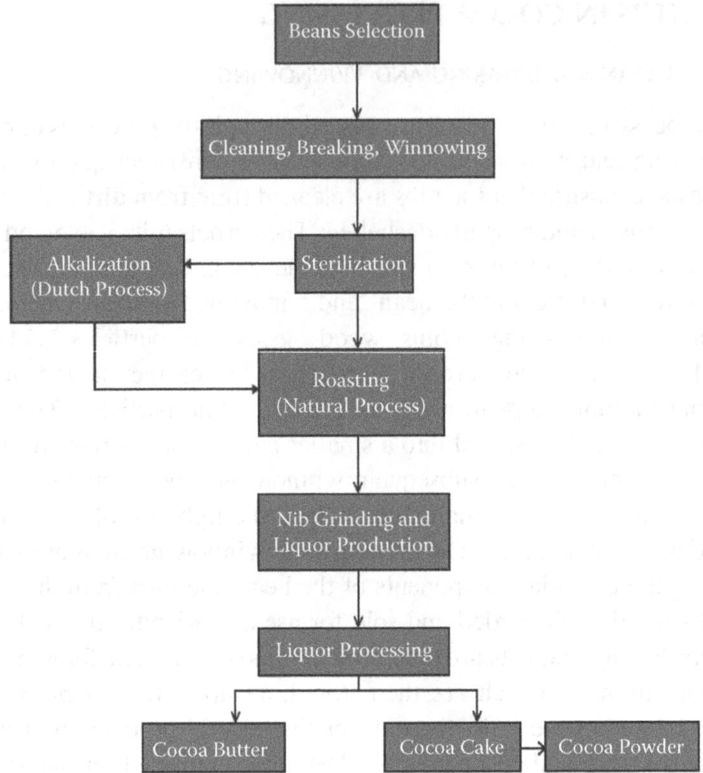

FIGURE 12.3 Flow diagram for the production of cocoa butter, cocoa cake and cocoa powder.

about 20 seconds, a fine water spray of steam into the roasting drum at the end of the roasting period (Awua, 2002). This guarantees a considerable reduction in microbial count in the roasted nibs.

12.4.3 Alkalisation

The technique of alkalisation was first introduced by a Dutchman known as van Houten in 1928 and therefore named as the Dutch process. All cocoa, beans, nibs or liquor that is so treated is described as 'alkalised' or 'Dutched'. This consists of treating the cocoa nibs with an alkali solution such as potassium or sodium carbonate. The alkali is used to raise the pH of the beans or nibs from 5.2–5.6 to near neutrality at 6.8–7.5, depending on the alkali used, and the purposes are primarily to modify the colour and flavour of cocoa powder or cocoa liquor, and also improve dispersibility or suspension of the cocoa solids in water. During the process, the alkali solution is sprayed into the drum after it has been charged with the nibs, which are then slowly dried at a temperature below 100°C (212°F). The

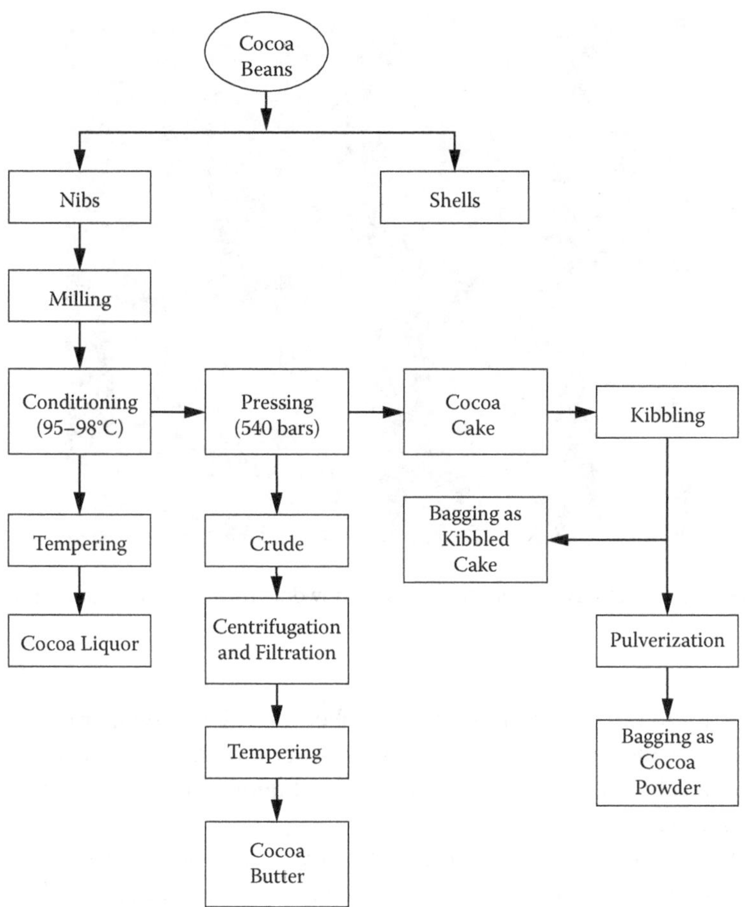

FIGURE 12.4 Flow diagram for the production of cocoa liquor, butter and powder.

alkalisation process can also be used to produce many different colours of cocoa powder (Figure 12.5).

12.4.4 Roasting

Cocoa beans are roasted to develop further the original cocoa flavour that exists in the form of precursors generated during the processes of fermentation and drying of the beans. During roasting of the dried fermented beans, several physical and chemical changes take place, which include the following:

1. Loosening of the shells.
2. Moisture loss from the beans to about 2% final content.
3. The nibs (cotyledons) become more friable and generally darken in colour.

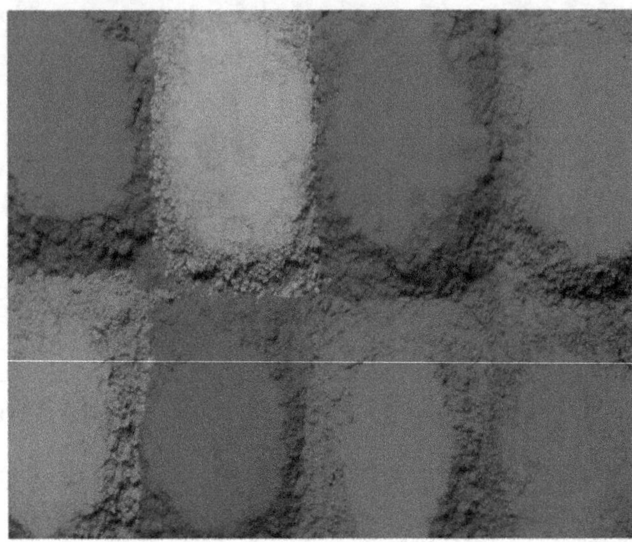

FIGURE 12.5 Different colours of cocoa powder from the alkalization process.

4. Additional reduction in the number of micro-organisms present in the beans. This helps attain food-grade products such as cocoa butter, cocoa powder and cocoa liquor, which have stringent microbiological specifications.
5. Degradation of amino acids takes place and proteins are partly denatured; the natural reducing sugars are almost destroyed during degradation of amino acids.
6. Losses of volatile acids and other substances that contribute to acidity and bitterness. A large number of compounds have been detected in the volatile compounds including aldehydes, ketones, pyrazines, alcohols and esters. The substances that undergo only minimal changes are the fats, polyphenols and alkaloids.

However, the degree of change is related to the time and temperature of roasting and the rate of moisture loss during the process. The roasting temperature varies between 90°C and 170°C depending upon the type of roasting adopted being dry or moist roasting.

Three main methods of roasting are employed within the cocoa processing industry and these include the following:

1 Whole bean roasting
2. Nib roasting
3. Liquor roasting

Whole bean roasting is usually the traditional way of producing cocoa liquor. By this process, the beans are roasted first before winnowing to facilitate removal of the shells which are broken by high-speed impact against metal plates. During the process, the heat causes some of the fat to migrate into the shells, thus resulting in a loss of some cocoa butter. This is particularly important in the case of broken or crushed beans. Nib roasting is done by first removing the shells before roasting and by this many of the limitations of whole bean roasting are overcome. This also makes it possible to treat the nibs with alkaline or sugar solution during roasting to help improve flavour development in certain types of cocoa. In liquor roasting, thermal pre-treatment is often used before winnowing for liquor roasting. The nib is then ground to liquor before roasting. The major disadvantage of both nib and liquor roasting is that the shell must be removed before it has been loosened from the nib by heating and this may result in poor separation, especially with some types of cocoa. As a result, a variety of machines has been developed to pre-treat the beans thermally. These develop a high surface temperature and evaporate the internal moisture, which in turn builds up a pressure within the bean causing the shell to come away from the nib.

12.4.5 Nib Grinding and Liquor Treatment

Nib grinding involves milling of cocoa nibs to form cocoa liquor. The purpose is to produce as low a viscosity as possible to obtain smooth cocoa powder and chocolate taste during subsequent use of the liquor. The nib has a cellular structure containing about 55% cocoa butter in solid form locked within the cells. Grinding of nib cells releases the cocoa butter into liquor with particle size up to 30 μm and for production of cocoa powder, fine grinding is particularly important. The viscosity of the liquor is related to the degree of roasting preceding the grinding and to moisture content of the nib.

Many machines are used for reducing the nibs into liquor and these include stone mills, disc mills, pin or hammer mills and bead or ball mills. The grinding is done in a multi-stage process, and the heat treatment generated during the grinding process causes the cocoa butter in the nib to melt, forming the cocoa liquor. The refined cocoa liquor is heated in storage tanks at a temperature of about 90–100°C for aging and microbial destruction, after which the liquor is packaged for sale (Awua 2002). Typically, approximately 78–90% cocoa butter is collected by pressing; residual lipids may be removed by supercritical fluid extraction (Beckett 2000).

12.4.6 Liquor Pressing

Cocoa butter constitutes about half of the weight of the cocoa nib. This fat is partially removed from the cocoa liquor by means of hydraulic presses applying pressures as high as 520 kg/cm^2 and the larger presses take a charge of up to 113.4 kg per one pressing cycle. Depending upon the pressing time and the settings of the press, the resulting cake may have a fat content of between 10 to 24%. Two kinds of cocoa cake can be obtained by the process, including

1 High fat cake containing between 22–24% residual fat in the pressed cake
2. Low fat cake containing between 10–12% residual fat in the pressed cake

The cocoa butter extracted is discharged into receptacles from which it is pumped into an intermediate tank for further processing.

12.4.7 Cake Grinding (Kibbling)

After pressing, the cakes released are quite big to handle and are therefore passed through kibbling machines to be broken down into smaller pieces, known as kibbled cake. The kibbled cake obtained is stored by fat content and degree of alkalisation, and may be blended before pulverisation to obtain the desired type of cocoa powder.

12.5 COCOA POWDER PRODUCTION

The powder grinding lines usually comprise hammer-and-disc or pin mills which pulverise cocoa cake particles into the defined level of fineness of the cocoa powder. The powder is then cooled after pulverisation so that the fat of the cocoa powder crystallises into its stable form. This prevents any discolouration (fat bloom) and the formation of lumps in the bags after packing, a phenomenon that is caused by insufficient crystallisation of the fat at the moment of filling. The free-flowing powder is then passed through sieves and over magnets prior to packing in bulk containers or four-ply multi-wall paper bags lined with polyethylene.

12.5.1 Cocoa Butter Quality

Cocoa butter is by far the most abundant, about 53–58% (Biehl and Ziegleder 2003; Borges, Tomaś-Barberán, and Crozier 2006; Afoakwa 2010; Afoakwa et al. 2013a,b) of the cotyledon on a dry weight basis. It is the most valuable chemical component of the cocoa beans. The fat contains about 95% triacylglycerols, 2% diacylglycerols, less than 1% monoacylglycerol, 1% polar lipids, and 1% free fatty acids (Biehl and Ziegleder 2003). The quantity of fat and its properties such as melting point and hardness depend on the variety of cocoa and the environmental conditions (ICCO 2012b). The physical and chemical characteristics of the fat provide specific functional properties of great demand in the food, pharmaceutical and cosmetic industries (Howell et al. 2005; Sukha 2003; Beckett 2009), hence the fat sells at a price approximately twice higher than that of cocoa powder (Venter et al. 2007).

The availability of cocoa butter can be unstable due to few countries cultivating cocoa, hence other fat replacers are being developed to care for this situation when they occur. Triglycerides consist of about 37% oleic (O), 32% stearic (S), 27% palmitic (P) and 2–5% linoleic (L) acids (Biehl and Ziegleder 2003; Saldaña, Mohamed, and Mazzafera 2002; Kaphueakngam, Flood, and Sonwai 2009; Quast, Luccas, and Kieckbusch 2011). Trace amounts of lauric acid (C12) and myristic acid (C14) also exist (Kaphueakngam et al. 2009). The oleic fatty acid is always sterified in the central position of the glycerol molecule whereas the palmitic and stearic fatty acids are positioned on carbon one and three (Quast et al. 2011), respectively. Three main symmetric triglycerides are represented by the cocoa butter. These are POP (palmitic–oleic–palmitic), POSt (palmitic–oleic–stearic) and StOSt (stearic–oleic–stearic; Saldaña et al. 2002; Rodrıguez et al. 2009; Quast et al. 2011). Other smaller amounts of palmitic-linoleic-palmitic (PLP), palmitic-oleic-oleic (POO), palmitic-linoleic-stearic (PLS) and stearic-oleic-oleic (SOO) have also been reported (Rodrıguez, Péreza, and Guzmán 2009). The triacylglycerols in the fat contain more than 75% of 1,3-dipalmitoyl-2-oleoylglycerol, 1-palmitoyl-2-oleoyl-3-stearoylglycerol and 1,3-distearoyl-2-oleoylglycerol (Minifie 1989; Quast et al. 2011).

The different favourable characteristics of chocolate such as hardness at room temperature, brightness and fast and complete melting when placed in the mouth (Saldaña et al. 2002) are due to the cocoa butter. Cocoa butter is light-yellow coloured, mainly solid at temperatures below 26.7°C but melting completely at body temperature or at 35°C (Minifie 1989; Kaphueakngam et al. 2009). Biehl and Ziegleder (2003) indicated that cocoa butter melts at a temperature of 31–34°C. This melting behaviour of cocoa butter is due to the high levels of these three main symmetric triglycerides: 1, 3-disaturated and 2-monounsaturated (Rodrıguez et al. 2009). The hardness and crystallisation behaviors depend on the origin, climatic and processing conditions (Saldaña et al. 2002; Chaiseri and Dimick 1989). The climate, mainly the ambient temperature and stress due to heat or drought, affect the biosynthesis and the final composition of the triacylglycerols, the melting and crystallisation characteristics of the cocoa butter (Biehl and Ziegleder 2003; Chaiseri and Dimick 1989). Cocoa butter obtained from fruits grown at low temperature is soft and contains high diunsaturated triacylglycerols (Lehrian and Keeney 1980) and high unsaturated fatty acids, that is, oleic and linoleic acid (Berbert 1976; Chaiseri and Dimick 1989). For example, rainfall patterns have an effect on the composition of the triacylglycerol. Enough rainfall during the fruit-bearing age causes high concentrations of stearic acid and oleic acid, in the triacylglycerol as well as free fatty acids (Fowler 1999).

Sunlight increases the palmitic acid content (Fowler 1999) and the iodine value of cocoa butter (Berbert 1976). Chaiseri and Dimick (1989) and other researchers in the 1980s indicated that cocoa butter exhibits six polymorphs (Table 12.2) in which polymorph I has the least stability and the lowest melting point and polymorph VI the highest stability and melting point.

Evidence indicates that only four cocoa butter polymorphs actually occur inasmuch as polymorph III is a mixture of polymorphs II and IV (Schlichter, Sarig, and Garti 1988; Riiner 1970) and polymorph VI is only a phase differing in composition (Chaiseri and Dimick 1989; Schlichter et al. 1988).

Triacylglycerols in cocoa butter appear in various crystal lattices such as α (hexagonal sub-cell), β′ (orthorhombic sub-cell) and β (triclinic sub-cell) depending on the cooling rate and agitation level (Quast et al. 2011) during industrial processing. The three polymorphs are based on sub-cell structures that define cross-sectional packing modes of the zig-zag aliphatic chain. In addition, the fat content and physical and chemical characteristics of the fat within a cocoa bean change according to the area of

TABLE 12.2
Melting Point and Chain Packing of Cocoa Butter Polymorphic Forms

Polymorphic Forms of Cocoa Butter		Melting Point (°C)	Chain Packing
Form I	β'_2	16–18	Double
Form II	α	21–22	Double
Form III	Mixed	25.5	Double
Form IV	β_1	27–29	Double
Form V	β_2	34–35	Triple
Form VI	β'_1	36	Triple

Source: G. Talbot, *Industrial Chocolate Manufacture and Use*, 3rd edition, Oxford: Blackwell Science, pp. 218–230, 1999. With permision.

cultivation. In general, the nearer the area where the cocoa tree is grown to the equator, the harder is the fat in terms of melting characteristics. This means that Malaysian cocoa butter is relatively hard, whereas most Brazilian cocoa butter is much softer (Beckett, 2008). Ghanaian cocoa beans or cocoa beans from West Africa have cocoa butter hardness in between that of Brazil and Malaysia.

13 Chocolate Manufacturing and Processing Technology

13.1 INTRODUCTION TO CHOCOLATE MANUFACTURE

Chocolate manufacturing involves several physical and chemical processes requiring numerous technological operations and addition of different ingredients to attain the desirable quality and sensory characteristics. The product so formed is a complex emulsion, which could be transformed to a solid product to form several ranges of different branded products. Its worldwide popularity stems from the fact that it is considered a luxury food that during consumption evokes a range of stimuli that activate pleasure centres of the human brain, providing a general sensation of satisfaction, happiness, joy and in some cases sexual stimulation. It is a semi-solid suspension of fine solid particles from sugar, cocoa and milk powder (depending on type), making about 70% total, in a continuous fat phase, mostly of cocoa butter (Figure 13.1).

Central to chocolate quality is an appropriate melting behaviour so that products are solid at ambient temperature and on ingestion melt to undergo dissolution in oral saliva, with a final assessment of texture after phase inversion. Particle size distribution and ingredient composition play important roles in shaping its rheological behaviour and sensory perception (Afoakwa, Paterson, and Fowler 2007). With opportunities for improvements in quality possible through improved and more transparent supply chain management, cocoa varietal breeding strategies, novel product development technologies and development of niche premium quality products, greater understanding of process and product quality variables will be highly beneficial.

13.2 CHOCOLATE MANUFACTURING PROCESSES

Chocolate manufacturing processes (Figure 13.2) generally share common features such as

1. Mixing
2. Refining

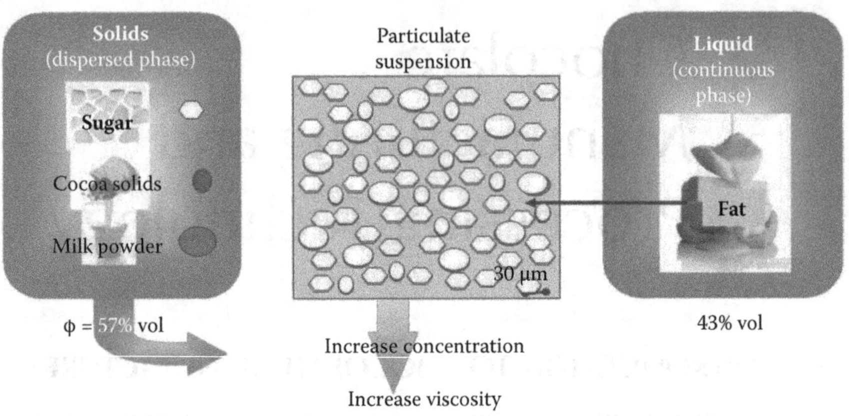

FIGURE 13.1 The chocolate model.

3. Conching of chocolate paste
4. Tempering and depositing
5. Casting and moulding
6. Cooling and demoulding
7. Wrapping and packaging

The outcome sought is smooth texture of product considered desirable in modern confectionery and elimination of oral perceptions of grittiness. Primary chocolate categories are dark, milk and white that differ in content of cocoa solid, milk fat and cocoa butter. The outcome is varying proportions of carbohydrate, fat and protein.

13.2.1 Mixing

Mixing of ingredients during chocolate manufacture is a fundamental operation employed using time–temperature combinations in a continuous or batch mixer to obtain constant formulation consistency. In batch mixing, chocolate containing cocoa liquor, sugar, cocoa butter, milk fat and milk powder (depending on product category) is thoroughly mixed normally for 12 to 15 minutes at 40–50°C. Figure 13.3 shows chocolate manufacturing processes from cocoa to chocolate. Continuous mixing is usually used by large chocolate manufacturers such as Nestlé and Cadbury using well-known automated kneaders, producing a somewhat tough texture and plastic consistency. Figure 13.4 shows mixing of raw materials during chocolate manufacturing.

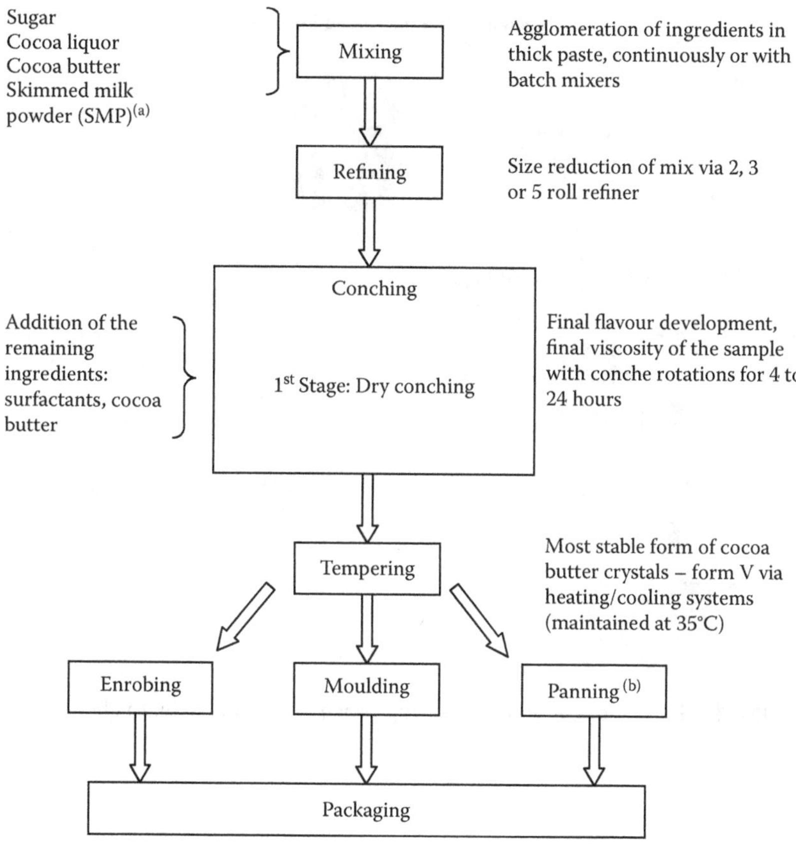

Note: (a) Skimmed milk powder is only used in milk chocolate manufacture;
(b) Panning means that the chocolate is used as coating for hard centres such as nuts.

FIGURE 13.2 Processing steps for chocolate manufacture. (Adapted from E.O. Afoakwa, A. Paterson, and M. Fowler, *Trends in Food Science & Technology*, 18:290–298, 2007. With permission.)

13.2.2 Refining

Refining of chocolate is important to the production of the smooth texture that is desirable in modern chocolate confectionery. Mixtures of sugar, cocoa liquor (and milk solids depending on the type of chocolate) at an overall fat content of 8–24% is refined to particle size <30 μm normally using a combination of two- and five-roll refiners. Final particle size critically influences the rheological and sensory properties. A five-roll refiner (Figure 13.5) consists of a vertical array of four hollow cylinders, temperature controlled by internal water flow, held together by hydraulic pressure. A thin film of chocolate is attracted to increasingly faster rollers, travelling up the refiner until removed by a knife blade. Roller shearing fragments

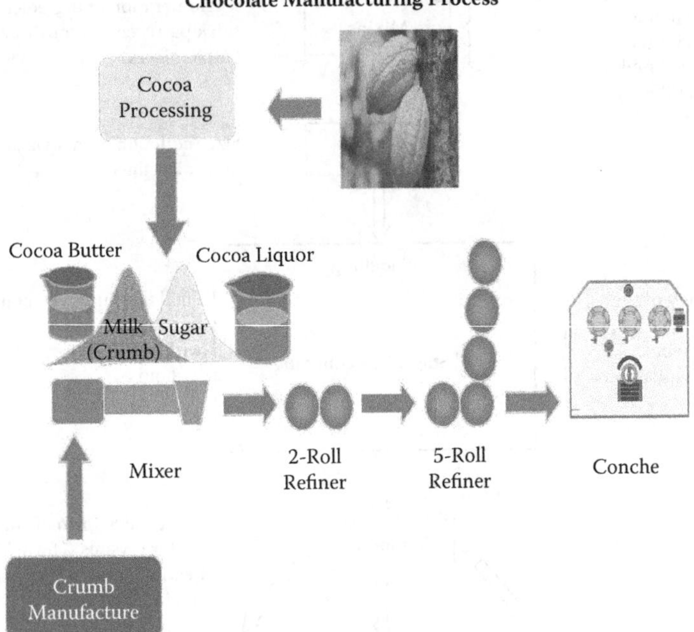

FIGURE 13.3 Chocolate manufacturing processes from cocoa to chocolate.

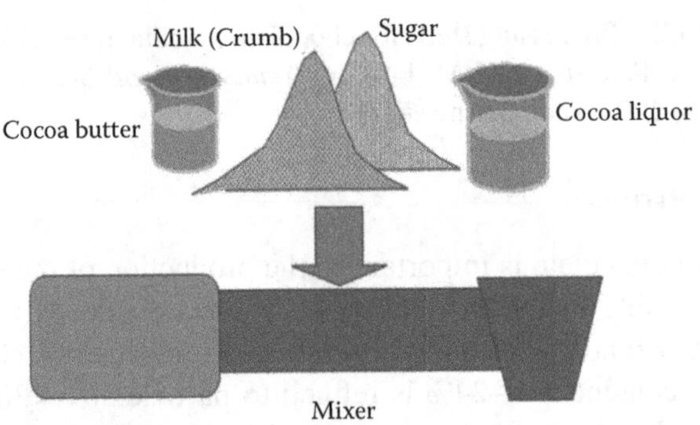

Combines the main chocolate ingredients together:
White Chocolate: Sugar, milk powders, cocoa butter, milk fat
Milk Chocolate: Sugar, milk powders, cocoa liquor, cocoa butter, milk fat
Dark Chocolate: Cocoa liquor, sugar, cocoa butter

FIGURE 13.4 Mixing of raw materials during chocolate manufacture.

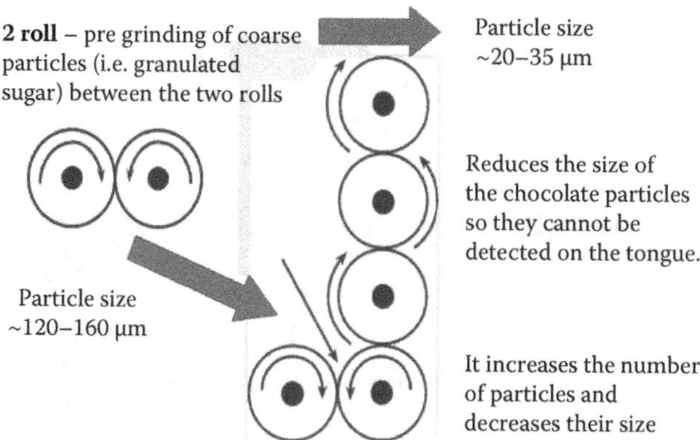

FIGURE 13.5 Two- and five-roll refining processes.

solid particles, coating new surfaces with lipid so that these become active, absorbing volatile flavour compounds from cocoa components.

Texture in milk chocolate appears improved by a bimodal distribution of particles with a small proportion having sizes up to 65 μm. Optimum particle size for dark chocolate is lower at <35 μm although values are influenced by product and composition. Refiners, in summary, not only effect particle size reduction and agglomerate breakdown but distribute particles through the continuous phase, coating each with lipid.

13.2.3 Conching

Conching is regarded as the endpoint or final operation in the manufacture of bulk chocolate, whether milk or dark. It is an essential process that contributes to development of viscosity, final texture and flavour. Conching is normally carried out by agitating chocolate at >50°C for some hours. In the early stages moisture is reduced with removal of certain undesirable flavour-active volatiles such as acetic acid and subsequently interactions between disperse and continuous phase are promoted. In addition to moisture and volatile acid removal, the conching process promotes flavour development due to the prolonged mixing at elevated temperatures, giving a partly caramelised flavour in non-milk crumb chocolate. The process also aids reduction in viscosity of refiner pastes throughout the process, and reduction in particle size and removal of particle edges.

The name of the equipment, the *conche*, is derived from the Latin word for 'shell', as the traditional conche used in chocolate manufacture resembled the shape of a shell. Figure 13.6 is an illustration of the internal mechanics of the Frisse conche. The Frisse conche is a typical example of an overhead conche used in the modern chocolate industry. It consists of a large tank with three powerful inter-meshing mixer blades,

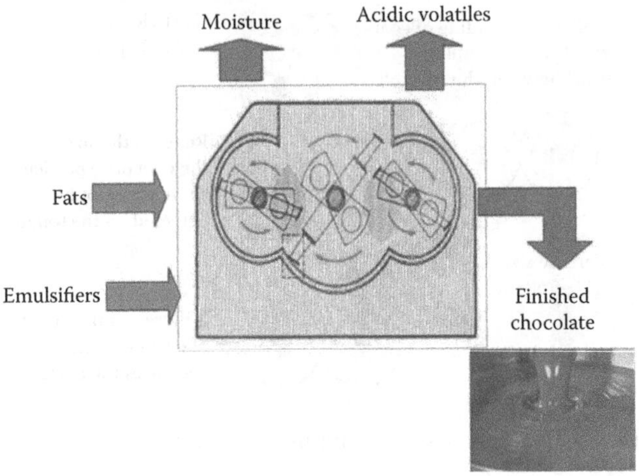

FIGURE 13.6 Internal mechanics of the Frisse conche.

providing shearing and mixing action. The conching process goes through three different phases as shown in Figure 13.7. These include the dry paste phase, the plastic phase and the liquid phase.

Conching times and temperatures vary typically: for milk crumb, 10 to 16 h at 49–52°C; with milk powder products, 16 to 24 h at up to 60°C; and with dark chocolates at 70°C and continuing up to 82°C. Replacing full-fat milk powder with skimmed milk powder and butter fat, temperatures up to 70°C may be used (Awua 2002). To give chocolate a suitable viscosity,

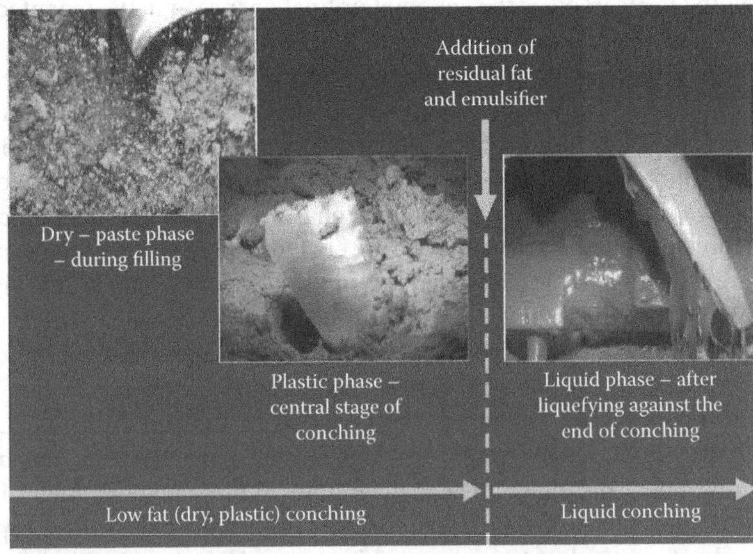

FIGURE 13.7 The three different phases of the conching process.

additional cocoa butter and lecithin can be added towards the end of conching to thin or liquefy the chocolate prior to tempering (Beckett 2000; Whitefield 2005).

13.2.4 Tempering and Lipid Crystallisation

Cocoa butter can crystallise in a number of polymorphic forms as a function of triglyceride composition, with fatty acid composition influencing how liquid fat solidifies (Awua 2002). Cocoa butter has six polymorphic forms (I–VI), the principals being α, β and β' (Figure 13.8). Form V, a β polymorph, is the most desirable form (in general) in well-tempered chocolate, giving a glossy appearance, good snap, contraction and resistance to bloom (Beckett 2000).

If chocolate is poorly tempered, the outcome is the β Form IV which rapidly transforms into Form V. This influences colour as reflected light is disoriented by unstable, disorganised crystal growth (Hartel 2001). Untempered chocolate is soft and not effectively demoulded. In cocoa butter Forms V and VI are the most stable forms. Form VI is difficult to generate although formed on lengthy storage of tempered chocolate accompanied by fat bloom. In addition Form VI has a high melting temperature (36°C), and crystals that are large and gritty on the tongue. The

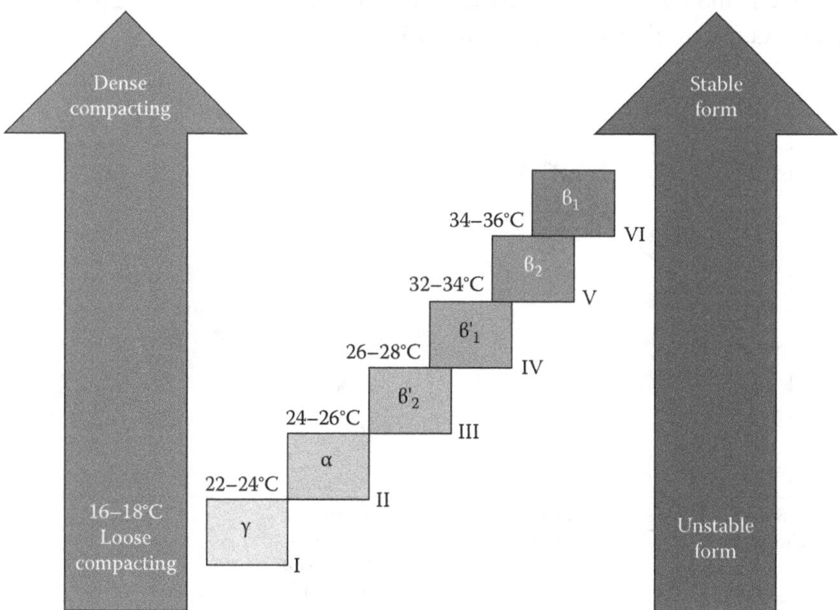

FIGURE 13.8 Polymorphic arrangements of crystalline fat. (Adapted from S.T. Beckett, *The Science of Chocolate*, 2nd edition. London: Royal Society of Chemistry, 2008. With permission.)

unstable Form I has a melting point of 17°C and is rapidly converted into Form II that transforms more slowly into III and IV. Polymorphic triglyceride forms differ in distance between fatty acid chains, angle of tilt relative to plane of chain end methyl group and manner in which triglycerides pack in crystallisation.

Polymorphic form is determined by processing conditions. Fatty acids crystallise in a double- or triple-chain form depending on triglyceride composition and positional distribution. Form IV crystallises in a double-chain form, with Form V in a triple-chain system that enables closer packing and greater thermodynamic stability. Unstable lower polymorphic forms (II and III) transform into higher melting, more stable forms, with closer packing and lower volume. These changes can be observed in terms of overall contraction of the chocolate, appearance of undesirable fat bloom formation at rates dependent on relative stabilities of the polymorphic forms and temperature (Talbot 1999). For chocolate to be in an appropriate polymorphic form, tempering is crucial, influencing final quality characteristics such as colour, hardness, handling, finish and shelf-life characteristics.

Tempering involves pre-crystallisation of a small proportion of triglycerides, with crystals forming nuclei (1–3% total) for the remaining lipid to set in the correct form. Tempering has four key steps: melting to completion (at 50°C), cooling to point of crystallisation (at 32°C), crystallisation (at 27°C), and conversion of any unstable crystals (at 29–31°C). These different stages are illustrated in Figure 13.9.

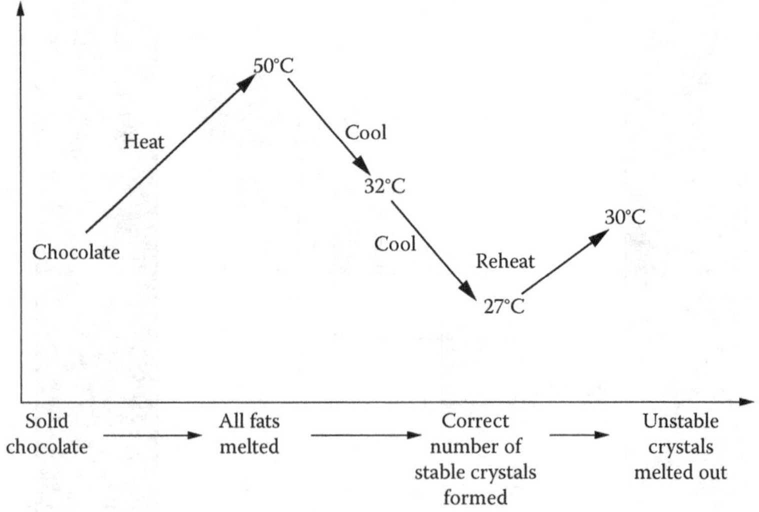

FIGURE 13.9 Tempering process of molten chocolate. (Adapted from S.T. Beckett, *The Science of Chocolate*, 2nd edition. London: Royal Society of Chemistry, 2008. With permission.)

The tempering sequence is a function of recipe, equipment and the final purpose. Before the use of tempering machines, chocolate used to be hand-tempered, and this method is still occasionally used by chocolatiers who produce relatively small quantities of hand-made confections. Current tempering machines consist of multi-stage heat exchangers through which chocolate passes at widely differing rates making it difficult to identify optimum conditions.

Time–temperature combinations are of paramount importance in process design, and in continuous tempering molten chocolate is usually held at 45°C then gently cooled to initiate crystal growth. Working with the Buhler 'Masterseeder', Windhab (ETH Zurich, Switzerland) and Mehrle (Buhler AG, Uzwil, Switzerland) found that high shear seed tempering can be beneficial as the kinetics of fat crystal nucleation and polymorphic transformations ($\alpha \rightarrow \beta_2 \rightarrow \beta'_1$) are strongly accelerated by shear forces acting in high shear flow fields: overall quality of products was better, as fat bloom was reduced. During tempering, the temperatures are precisely controlled and the agitation provided enhances nucleation rates. As the viscosity increases, the chocolate is reheated again in the third stage to prevent runway solidification. In the fourth stage, crystals are matured.

Chocolate can also be tempered by the use of high pressure with molten chocolate compressed to 150 bar. This increases the chocolate melting point and causes it to solidify into solid crystals of all polymorphic forms. When pressure is released, lower polymorphic forms melt, leaving behind tempered chocolate. Subsequent batches can be seeded with stable fat crystals.

A well-tempered chocolate will have the following properties: good shape, colour, gloss, contraction from the mould, better weight control, stable product, harder and more heat resistant (fewer finger marks during packaging) and longer shelf-life. The tempering regime for milk chocolate slightly differs from that for dark due to the influence of milk fat molecules on crystal lattice formation. Milk chocolate contains a proportion of butter fat that causes an eutectic effect, which prevents bloom formation, results in a lower melting point, softening of texture and lowering of temperature to obtain crystal seed for the tempering process (around 29.4°C compared to 34.5°C for plain chocolate). Cocoa butter equivalents (CBEs) and replacers (CBRs) may also find application in the chocolate industry. Although cocoa butter equivalents are compatible with cocoa butter, cocoa butter replacers (CBRs), which do not require tempering, can only be used if almost all the cocoa butter is replaced. These CBRs melt in the same temperature range as cocoa butter, but crystallise only in the β' form. Figure 13.10 shows typical images of (a) white chocolates, (b) milk chocolates and (c) dark chocolates.

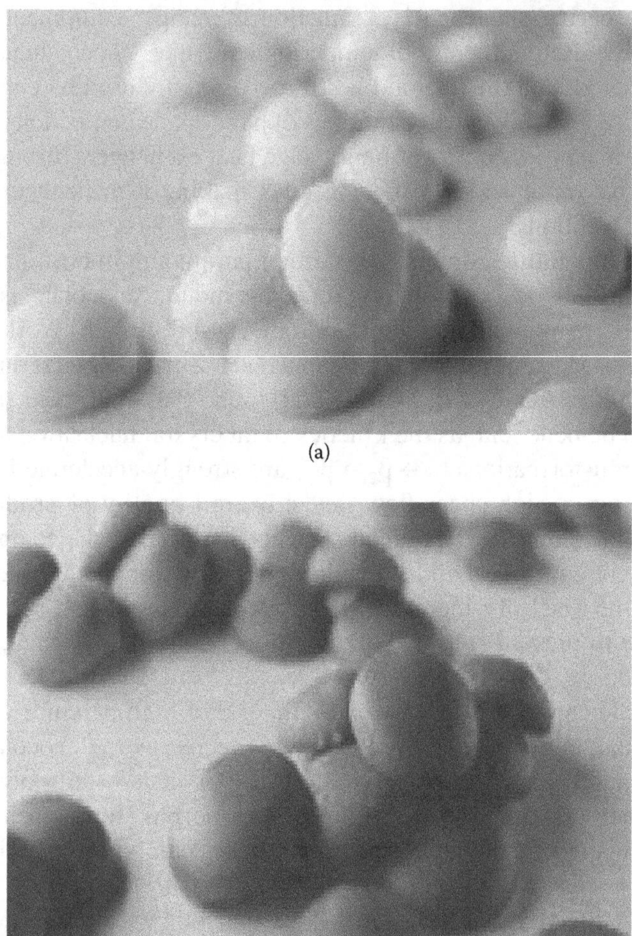

(a)

(b)

FIGURE 13.10 Typical images of (a) white chocolates, (b) milk chocolates. (*Continued*)

13.2.5 Casting and Moulding

The temperer sends the tempered mixture to the Chocomaster, an automated system for moulding and demoulding of chocolate. The mixture first enters the hopper of the depositor which deposits chocolate into 20-g, 50-g and 100-g moulds depending on the selected mould type. The depositor automatically fills molten chocolate into moulds which are of the same temperature as the chocolate (Figure 13.11). If the moulds are too hot detempering occurs, resulting in product sticking in the impressions of the mould, poor gloss and bloom. If they are too cold, poor gloss and sticking in the mould can result with an increase in the number of air bubbles and

(c)

FIGURE 13.10 (CONTINUED) Typical image of (c) dark chocolates.

FIGURE 13.11 Deposition of molten chocolate into a mould.

markings on the finished product. A vibrator shakes each mould to level the liquid chocolate in the moulds.

13.2.6 Cooling

The moulds after vibrations pass through the freezing section which is a multi-tier cooler. This is a mechanised process that passes the moulds gradually layer by layer through the freezer reducing the temperature of the chocolate to about 12–15°C solidifying the chocolate into bars. This type of cooling is usually found in the well-established processing industries. During batch processing by small-scale chocolate manufacturers, cooling is mostly effected using refrigerators. In this situation, the moulds containing the chocolate may be cooled in the refrigerator set at 7–10°C for between 20–30 min.

13.2.7 Demoulding

With optimised tempering and cooling, the demoulding becomes a minor part of the process resulting in good quality product cleaning, leaving moulds that are returned to the start of the process. A small amount of force is needed to part the product from the mould and this is supplied by a hammer, aided by a mechanism that twists the moulds. Product is demoulded onto a belt which conveys chocolate onto plastic trays. These trays are collected onto trolleys and wheeled to nearby wrapping plants. The period between the deposition on trays and wrapping is known as the drying stage where excess moisture on the surface of the products is lost.

13.2.8 Wrapping/Packaging

At the wrapping plants trays of chocolate are emptied onto a conveyor belt which transports chocolate into the machine. The wrapping machine, depending on the grams/size of the chocolate, cuts aluminium foil and picks up a paper wrapper. The aluminium foil initially covers the chocolate before the paper wrapper. The foil provides the best barrier to water vapour and gas transmission, maintaining aroma and cool temperature of the chocolate. The paper material is also chosen because it is strong, easily printed and relatively inexpensive. The machine then labels the chocolate with the batch number as well as the production and expiry date. Figure 13.12 shows some different wrappers and packaging materials used by different chocolate manufacturing industries. Wrapped chocolates without any defects are manually picked and boxed before sending to the warehouse for storage at temperatures between 18–20°C.

FIGURE 13.12 Different wrappers and packaging materials used by different chocolate manufacturing industries.

14 Chocolate Quality and Defects

14.1 CHOCOLATE QUALITY

The International Organisation of Standardization (ISO) defines quality as 'the totality of features and characteristics of a product that bear on its ability to satisfy stated or implied needs' (ISO 1998). Quality may be judged as good or bad depending on the level of adherence to specifications or standards for the products—with regard to raw material input and finished products—and how well it matches consumer preferences. The control of chocolate quality is often determined by rheological measurements of the viscous liquid formed during production, and sensory evaluation of the final solid product.

14.1.1 Rheological Measurements of Chocolate Quality

Chocolate behaves as a non-Newtonian liquid, exhibiting non-ideal plastic behaviour, where shear-thinning occurs once a yield value has been overcome. This is caused by the three-dimensional structure of the material collapsing, and asymmetric particles which align as the shear rate increases, causing a decrease in viscosity until it becomes independent of shear rate at high shear rates. This behaviour is exhibited when solid chocolate is melted into the liquid state at temperatures above 50°C. The rheological properties of the molten chocolate (Figure 14.1) are determined to assess the effectiveness of conching. They are also used to evaluate the flow properties of the molten chocolate during processing (Beckett 2009; Afoakwa 2010). Several rheological models have been used to estimate the yield stress and plastic viscosity of chocolates (ICA 2000; Beckett 2009; Afoakwa et al. 2009). However, the Herschel–Buckley models, as well as the Casson model, are both used as the most popular models to fit non-ideal plastic behaviour. The Casson model was adopted in 1973 as the standard rheological equation for chocolate by the International Office of Cocoa, Chocolate and Sugar Confectionery.

In 2000, another study by the International Confectionery Association (ICA 2000), showed that the mathematical models used to express the whole flow curve by a single equation using only a small set of parameters

FIGURE 14.1 Molten chocolate.

are limited in accuracy, as chocolate flow properties do not exactly fit the Casson equation. It was, therefore, suggested that yield values be measured at low shear rates and viscosities be measured at high shear rates if data are to be compared between different laboratories. As a result, the ICA (2000) now recommends the measurement of stress and viscosity at shear rates between 2 s^{-1} and 50 s^{-1} in 7 minutes, using both up and down curves in shear rate, this being preceded by a pre-shear at 5 s^{-1} lasting for 5 minutes. Unfortunately, most factory-grade viscometers are, however, not accurate at a shear rate of 2 s^{-1}, hence the yield stress at a shear rate of 5 s^{-1} is taken to relate to the yield value of chocolate. As the stress at 5 s^{-1} is, however, a completely different order of magnitude to the shear stress calculated by the Casson model, a relationship can be established by dividing the shear stress at 5 s^{-1} by 10. Furthermore, shearing at 50 s^{-1} does not present results that are representative of the actual chocolate structure and is not always achievable when testing very viscous chocolate using factory-grade viscometers; hence a speed of 40 s^{-1} is chosen. Servais, Ranc, and Roberts (2004) showed that the viscosity at 40 s^{-1} can be considered to be an accurate reference value for the plastic viscosity of chocolate (Afoakwa et al. 2008).

To relate this plastic viscosity to the Casson plastic viscosity, it should be multiplied by a factor of 0.74. This calculation does not mean that one could obtain the same values as using the Casson model, but that one could keep the same order of magnitude as before. Furthermore, chocolate exhibits time-dependent behaviour; in other words, a change of shear stress and viscosity at a given shear rate occurs with time, which can be related to the change in the structure of the material. This decrease of viscosity with time of shearing, followed by recovery of the structure when the stress is removed, is called thixotropy. A well-conched chocolate should, however, not be thixotropic (Servais et al. 2004; Afoakwa et al. 2008).

14.1.2 Sensory Characteristics of Chocolate and Their Measurement

Chocolate has a distinct set of sensory characteristics (appearance, texture and flavour) that dictate its choice and acceptability by consumers. These characteristics originate from precursors present in cocoa beans, from those that are generated during fermentation and post-harvest treatments, and those formed or transformed during chocolate manufacture. In addition to these inherent factors, others include the ingredients and processing techniques used in chocolate manufacture. However, flavour is the most important sensory attribute of chocolates, as it is influenced by aroma, taste and texture during consumption. Nowadays, chocolate is not a rare or privileged product. Recognition of its values also involves influences of previous experiences by the consumer and the expectations created by marketing and package design. However, what makes it so desirable is the perception of its sensory quality. Figure 14.2 shows some moulded chocolate balls.

Chocolate quality may be evaluated in terms of its appearance, taste, mouthfeel, flavour and aftertaste. These characteristics can be determined either subjectively or objectively. Subjective opinions are based on the relative levels of likes and dislikes by consumers. Objective measures use scoring systems, which are independent of likes or dislikes, and need to be determined by a trained panel of sensory assessors. Objective measures

FIGURE 14.2 Moulded chocolate.

can also be obtained by instrumental analysis of properties such as shear and rheological measurements, texture or some key chemical compounds. In all cases, the challenge is to relate the data obtained by instruments (e.g., rheological data) to the sensory quality experienced by trained assessors or consumers, so that evaluations are meaningful (Afoakwa 2010).

Sensory analysis can be of two kinds, analytical and affective. The analytical approach involves the evaluation of products for differences or similarities of prescribed criteria or attributes. It is based on an analytical tool and is usually carried out by a trained panel of 10–20 assessors. The panel is asked to provide objective evaluation, and should not be used to evaluate preference. Affective analysis applies preference or acceptance evaluation or getting opinions to a product. It uses a large number of panellists, which should be representative of the target population. To study the global perception of a food product, descriptive analytical methods in association with scaling may be used.

In descriptive analysis, various techniques can be used to describe the perceived sensory characteristics of chocolate, such as Flavour Profile®, Quantitative Descriptive Analysis® (QDA), Texture Profile Analysis® (TPA) and Sensory Spectrum® (Lawless and Heymann 1998). Descriptive analysis provides a complete description of chocolate's sensory characteristics in the form of words (descriptors). The application of these tests to multiple samples is more informative than evaluating single samples, first by providing a much more complete picture of how products differ one from the other, and second by providing information on more than a single product. Frequently many different basic procedures, such as the duo–trio test, are used to compare products and determine if one is different from another. The qualitative aspects of a chocolate product include appearance, flavour, texture and taste, which distinguish the products. Sensory judges then quantify these product aspects in order to facilitate description of the perceived product attributes.

A major strength of descriptive analysis is its ability to allow relationships between descriptive sensory and instrumental or consumer preference measurements to be determined. Knowledge of 'desired composition' allows for product optimisation and validated models between descriptive sensory and the relevant instrumental or preference measures are highly desirable and increasingly are being utilised within the confectionery industry.

Descriptive sensory analyses are also used for quality control, for the comparison of product prototypes to understand consumer responses in relation to products' sensory attributes and for sensory mapping and product matching. They may also be used to track product changes over time with respect to understanding shelf-life and packaging effects, to investigate the effects of ingredients or processing variables on the final sensory quality of a product and to investigate consumer perceptions of products (e.g., Free Choice Profiling, FCP).

14.2 SENSORY ASSESSMENT OF CHOCOLATES

The sensory qualities of chocolates are mostly evaluated using descriptive sensory tests. In sensory evaluation, these tests are amongst the most sophisticated tools in the arsenal of the sensory scientist (Lawless and Heymann 1998) and involve the detection (discrimination) and description of both the qualitative and quantitative sensory components of a consumer product by trained panels of judges. Recent information by Thamke et al. (2009) suggests that the use and application of descriptive sensory testing has increased rapidly, and will continue to do so in the confectionery industry for many years.

There are several different methods of descriptive analysis that could be used to evaluate various sensory qualities of chocolate, including the Flavour Profile Method, Texture Profile Method, Descriptive Analysis, the Spectrum method, Quantitative Flavour Profiling, Free Choice Profiling and Generic Descriptive Analysis. The specific methods reflect various sensory philosophies and approaches (Lawless and Heymann 1998). However, generic descriptive analysis, which can combine different approaches from all these methods, is frequently employed during practical applications in order to meet specific project objectives.

14.3 MEASUREMENT OF CHEMOSENSORY PROPERTIES OF CHOCOLATES USING ELECTRONIC NOSES AND TONGUES

Electronic noses and tongues are relatively new and modern technologies that could be employed to assess the flavour/odour quality of chocolates. The processes use chemical array sensor systems for flavour, odour and taste classifications. Both technologies can be used to identify flavour varieties and geographical origin, composition, aroma intensity and degree of freshness but, of course, their meaningful application requires appropriate comparison and calibration with sensory panel classifications.

The e-nose uses ultra-high-speed gas chromatography and a solid-state chemical sensor to analyse quickly the chemical components of flavours, aromas, odours or vapours with parts per trillion sensitivity. It mimics the way the system of interconnected receptors and neurons in the human nose interact and responds to volatile molecules based on analysis of the cross-reactivity of an array of semi-selective sensors. The signals are processed via a pattern recognition program. During its operation, an array of sensors, composed of polymers, for example, expands like a sponge when they come in contact with volatile compounds in the headspace of a sample, increasing the resistance of the composite. The normalised change in resistance is then transmitted to a processor to identify the types, quantity and quality of the odours based on the pattern change in the sensor array (Leake 2009). In a review conducted on the use of e-noses in dairy products, Ampuero and Bosset (2003) noted that although e-noses (detection ppb, response time of seconds) are normally less sensitive and less powerful than human noses (detection ppt, response time milliseconds), they offer some significant advantages, for example, in instrumental classifications based on hedonic or sensory analyses and automated on-line monitoring of food volatiles. E-noses could be used for quality control applications in the confectionery industry to detect conformity control of raw materials, processed or end-product quality, batch-to-batch consistency, certification of origin and clean-in-place process monitoring.

Electronic tongues (e-tongues) measure dissolved compounds responsible for taste in liquids. They can be used to detect major tastes such as sweet, sour, bitter, salty and umami, which resembles effects of sub-taste attributes such as spicy and metallic. They could have diverse applications in the confectionery industry, such as to do a complete bitterness/sweetness assessment of new products. They could also be used to measure the taste-masking abilities of new or different ingredients in some confectionery. As mentioned for e-noses, there has as yet been no application of e-tongues to chocolate production

14.4 CHOCOLATE DEFECTS

When a product has defect(s) in quality, it may either be rendered unwholesome due to food-safety concerns or unacceptable in sensory character. In the case of the latter it may be subjected to re-work to meet expected or aspired sensory perceptions. Typically, two main types of defects occur in chocolates during post-processing handling, storage, warehousing and distribution. These include fat and sugar blooms.

14.4.1 Fat Bloom

Fat bloom occurs when fat crystals protrude on the chocolate or chocolate-flavoured coating surface, disturb the reflection of light and appear visible as a whitish film of fat, usually covering the entire surface, making the products unacceptable for marketing and consumption. Although fat-bloomed chocolate does not pose any public health or safety hazards to consumers, the process renders the product unappealing (Figure 14.3), and therefore renders it inedible. Fat bloom can be caused by

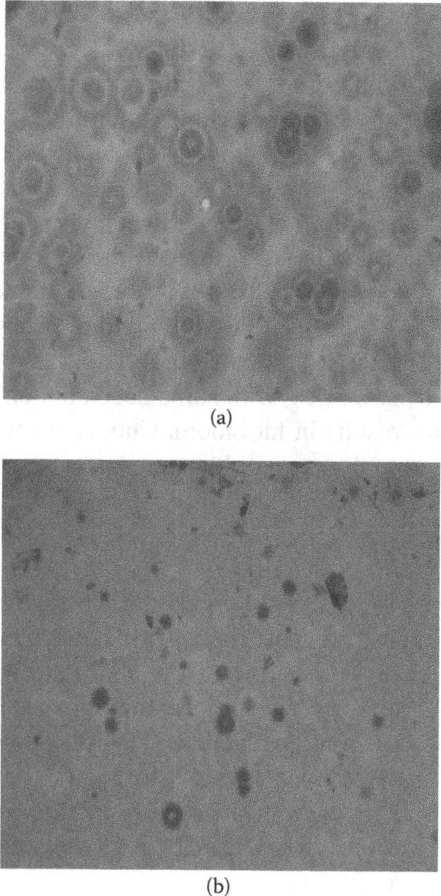

FIGURE 14.3 Fat bloom of (a) milk chocolate and (b) dark chocolates.

1. Insufficient crystallisation during tempering
2. Re-crystallisation without appropriate tempering
3. Inhomogeneity of the chocolate or chocolate flavoured coatings
4. Differences in temperature between the chocolate and the centre
5. Incorrect cooling conditions
6. Fat migration
7. Touch, also known as touch bloom
8. Inappropriate storage conditions, that is, humidity and temperature

When chocolate is poorly tempered there is formation of the soft Form IV that transforms over a period to the denser and stable Form V influenced by temperature. During this transformation some cocoa butter remains in a liquid state as the stable Form (V) solidifies and contracts. This coupled with the release of thermal energy as a more stable Form (V) forms, the liquid fat forces between solid particles and onto the surface where large crystals impart a white appearance to the surface which is recognised as fat bloom (Beckett 2008). Naturally, Form V transforms to the more stable Form VI slowly over an extended period, again influenced by temperature. This process also results in formation of fat bloom (Afoakwa et al. 2009). When optimally tempered products are stored under high temperatures such as exposure to sunlight, chocolate melts, and during re-crystallisation, in the absence of seeding to ensure the direct formation of the stable Form (V), a gradual transition from unstable to stable forms results in fat bloom. A fourth mechanism of fat blooming occurs with chocolates that have centres. Usually liquid fat from the centres migrates and consequently reaches surfaces along with some cocoa butter. Re-crystallisation of this cocoa butter results in fat bloom. Chocolates with nut centres are mostly pre-disposed to this type of bloom.

14.4.2 Sugar Bloom

Sugar bloom occurs through either poor storage conditions (high humidity) or rapid transition of products from an area of low to high temperature. Both conditions result in sweating of the chocolate which consequently dissolves sugar. As the surface water evaporates, sugar crystals remain on surfaces, producing a white appearance. This phenomenon is often confused with fat bloom but is completely different. The difference can be established microscopically or, whichever is simpler, by heating the chocolate to 38°C. Fat bloom disappears at this temperature, whereas sugar bloom remains visible.

15 Effects of Fermentation and Extended Pod Storage on Cocoa Bean Quality

15.1 RESEARCH SUMMARY AND RELEVANCE

Cocoa beans are the principal raw material for chocolate manufacture. Fermentation and drying are critical processes during which the beans develop the flavour precursors that generate into distinctive chocolate flavour notes during subsequent manufacturing processes such as roasting and conching. The technique of pod storage as a method of pulp pre-conditioning has been found to cause reduction in nib acidification and increase in flavour notes in Malaysian cocoa beans. However, the extent to which this technique would influence the quality of Ghanaian beans still remains unknown. This study investigated the effect of pod storage (pulp pre-conditioning) and fermentation on the chemical, physicochemical, biochemical and fermentative quality of Ghanaian cocoa beans (mixed hybrids of *Forastero*). Chemical analysis on the samples revealed that fermentation and increasing pod storage result in decreases in ash, protein and fat content of Ghanaian cocoa beans. Carbohydrate content, however, increased with similar treatments. Amongst the minerals studied, potassium was the most abundant mineral followed by magnesium, phosphorus and calcium in both the fermented and unfermented cocoa beans. Fermentation and increasing pulp pre-conditioning of the beans increased the copper content whereas iron and magnesium levels decreased progressively.

The effects of pod storage and fermentation time on the physicochemical properties, biochemical constituents and polyphenolic compounds were studied, with pod storage times (0, 7, 14, 21 days) and fermentation time (0, 2, 4, 6 days) as variables. Pod storage caused significant increases in the pH of the beans at all times of fermentation. Changes in titratable acidity (TA) during fermentation of the cocoa beans were contrary to observations made with pH for all pod storage treatments. The total sugar and non-reducing sugars decreased consistently with fermentation time whereas reducing sugars increased at all pod storage treatments. The reductions in total and non-reducing sugars with consequential increases in reducing sugars were due to the action of native cocoa seed invertase, which hydrolyses sucrose into glucose and fructose during fermentation. The rate of reducing sugars generation in the cocoa beans was largely affected by fermentation rather than by pod storage.

Polyphenol components were extracted from the cocoa beans from the different pod storage treatments. The total polyphenol contents, *o*-diphenols and anthocyanin content of the control treatment at 7 days pod storage were 148.56 mg/g, 21.17 mg/g and 15.68 mg/kg, respectively. During fermentation, all stored pod samples showed decreased levels of total polyphenols, *o*-diphenols and anthocyanin but the rates of decrease were found to be different and dependent on the rate of polyphenol oxidation and condensation as well as diffusion in the beans. Pod storage significantly reduced the polyphenolic content of the beans. Further to these, changes in the fermentative quality of the beans were studied with varying pod storage and fermentation times. Pod storage was found to be significant in affecting the changes in the degree of fermentation of the beans. Significant increases in the fermentation index (FI) were observed as pod storage days were increased by the fourth day of fermentation. Beans became darker, yellower and less red with increasing pod storage days. A cut test revealed that post-harvest pod storage of 7 and 14 days increased the percentage of brown beans of the control sample by 15 and 38%, respectively, by the sixth day of fermentation. This study revealed that cocoa pods should not be stored beyond 7 days for optimum generation of acids and flavour precursors (reducing sugars and proteins), degradation of polyphenols and improvement in the degree of fermentation.

15.2 INTRODUCTION

In Ghana, cocoa has been labelled 'the golden pod' owing to the pivotal role it plays in the health of the nation's economy. It is cultivated on about 1.5 million hectares of land by some 800,000 families in six out of the ten regions. It is cultivated almost exclusively by smallholder farmers with average farm sizes of about 4.0 hectares and mean production yields of 246.4 kg/hectares. Only the *Forastero* variety of cocoa is planted in Ghana. The main cultivars of this variety cultivated are *Amazon* (34.4%), the *Hybrid* (32.7%), *Amelonado* (13.3%) and a mixture of the two (19.6%; Amoa-Awua et al. 2006).

Cocoa beans are normally processed into chocolate and cocoa products, but other intermediate products are made as well. These are cocoa liquor, cocoa butter, cocoa cake and raw cocoa powder. Cocoa powder is essentially used in flavouring biscuits, ice cream and other dairy products, drinks and cakes and in the manufacture of coatings for confectioners or frozen desserts. It is also used in the beverage industry, for example in the preparation of chocolate milk. Cocoa butter is used in the manufacture of chocolate, confectionery, tobacco, soap and cosmetics (Amoa-Awua et al. 2006). Other by-products include cocoa pulp juice, which is also fermented to produce industrial alcohol and alcoholic beverages such as brandy and wine. The pod husk and shell are used for the preparation of animal feed and fertiliser in Ghana. The unique culture of raw cocoa processing in Ghana ingrained in the peasant farming practices of the growers coupled with rigorous research and quality control programmes embarked upon consistently by successive governments, to date has guaranteed Ghanaian cocoa its premium status on the international market.

Processing of cocoa beans into chocolate bars, milk chocolate and cocoa butter starts with an on-farm fermentation of the beans followed by drying and roasting. These post-harvest processes are very important because they initiate the formation of the precursors of chocolate flavour (Schwan, Rose, and Board 1995). Fermentation is a very important aspect of cocoa processing because it helps to break down the mucilaginous pulp surrounding the beans and causes cotyledon death (Sanchez et al. 1985; Gotsch 1997; Afoakwa et al. 2008). It also helps to trigger biochemical changes inside the beans that contribute to reducing bitterness and astringency, and to the development of flavour precursors (Thompson et al. 2001). Thus, fermentation is necessary to induce the biochemical transformations within the beans that lead to the formation of flavour precursors such as free amino acids, peptides and sugars. Cocoa fermentation is influenced by many factors such as type of cocoa, disease, climatic and seasonal differences (Rohan 1963), turning, batch size (Lehrian and Patterson 1983; Said and Samarakhody 1984), quantity of beans (Said and Samarakhody 1984; Wood and Lass 1985) and also pulp pre-conditioning (Meyer et al. 1989).

Pulp pre-conditioning entails changing the properties of the pulp prior to the development of micro-organisms. The pulp is the substrate metabolised during fermentation by a sequence of bacteria and fungi (Ostovar and Keeney 1973), and because the properties of the substrate determine microbial development and metabolism, changes in the pulp may affect the production of acids by lactic acid bacteria, yeasts and acetic acid bacteria. Three basic processes of pulp pre-conditioning have been evaluated for the treatment of fresh cocoa beans prior to fermentation: pod storage, mechanical or enzymatic depulping and bean spreading (Rohan 1963; Wood and Lass 1985; Biehl et al. 1989; Schwan and Wheals 2004). In West Africa, pods are usually gathered for breaking by a team of workers a few days after harvesting.

The quality and flavour of a batch of cocoa beans depends on complex chemical and biochemical changes, which occur in the cocoa beans during fermentation and drying. Micro-organisms grow in the pulp and produce a diversity of metabolites, along with substantial heat. The metabolites and heat diffuse into the cocoa seeds, killing them and disrupting their cellular integrity. Cellular disruption and seed death initiate various enzymatic and non-enzymatic reactions between seed components. These reactions develop a range of flavour precursors (peptides, amino acids, reducing sugars and polyphenols), and also affect the colour of the beans. During roasting, the precursors undergo further transformations to form the final chocolate flavour compounds (Afoakwa et al. 2008).

15.2.1 Bean Death and Cellular Disruption

Bean death is a critical event during cocoa fermentation which allows the biochemical reactions responsible for flavour development to occur within the cocoa bean. The diffusion of microbial metabolites from the pulp, combined with elevated temperatures, kills the seeds and disrupts their cellular integrity (Lopez and Dimick 1995; Afoakwa et al. 2008; Fowler 2009). Although some reviews confuse it, this process actually comprises two distinct phases.

The production of ethanol during the anaerobic yeast growth phase as well as rising temperatures and increasing acetic acid concentrations during fermentation have been implicated in causing the seed death resulting in loss of germinative power. This prevents certain flavour and quality defects by preventing radical protrusion and the metabolism of lipid and protein stores, hence producing a more stable product (Thompson et al. 2001).

Roughly parallel to the death of the bean of the embryo, the cellular integrity of the cocoa seed disintegrates. Cellular breakdown is indicated by the presence of a fluid within the testa, swelling the cocoa bean, and by the seed becoming softer and spongier in texture (Camu et al. 2008). The breakdown of cellular integrity permits enzymes and substrates to react, initiating various reactions, both enzymic and non-enzymic. The activities of the enzymes in the cotyledon result in significant increases in free amino acid and reducing sugar contents (glucose and fructose). While the temperature of the beans increases, the concentration of organic acids also increases, causing a decrease in pH. All these factors influence the biochemical activities within the bean and have an impact on flavour and quality (Thompson et al. 2001; Afoakwa et al. 2008; Afoakwa and Paterson 2010)

Traditionally, Ghanaian farmers have unknowingly adopted this technique of pod storage by their practice of using family labour to collect the harvested pods into piles 3–5 days before organising friends and neighbours to help break open the pods prior to fermentation (Duncan 1984). It appears that this method of pod storage has a highly beneficial effect on the subsequent development of chocolate flavour, though the precise biochemical pathways, conditions and processes are still poorly understood.

15.2.2 Rationale

The technique of pod storage as a method of pulp pre-conditioning has been found to cause a reduction in nib acidification and increase in flavour notes in Malaysian cocoa beans (Meyer et al. 1989). However, the extent to which this technique would influence the quality of Ghanaian beans still remains unknown. With increasing specialty niche products in chocolate confectionery, greater understanding of factors contributing to variations in the biochemical constituents (flavour precursors) and flavour character would have significant commercial implications.

15.2.3 Main Objective

The main objective of this work was to investigate effects of pulp pre-conditioning (pod storage) on the biochemical constituents and flavour precursor formation and development during fermentation of Ghanaian cocoa beans.

15.2.4 Specific Objective

The specific objectives of this study included the following;

1. To establish the chemical and physical composition of pulp pre-conditioned fermented and unfermented cocoa beans
2. To evaluate the effect of pulp pre-conditioning on biochemical constituents during fermentation of cocoa beans
3. To assess the effect of pulp pre-conditioning on polyphenolic compound concentrations during fermentation of cocoa beans
4. To determine the effect of pulp pre-conditioning on the degree of fermentation

15.3 MATERIALS AND METHODS

15.3.1 Materials

15.3.1.1 Sample Preparation

Ripe cocoa pods from mixed hybrids were harvested from the experimental plots of the Cocoa Research Institute of Ghana (CRIG), Tafo in the

FIGURE 15.1 Pod breaking of cocoa beans.

eastern region of Ghana. The cocoa pods were selected according to their ripeness and maturity level. The beans were pulp pre-conditioned by storing the harvested pods for a period of time before splitting. About 1,200 pods were stored on a concrete floor in the fermentary at ambient temperature (25–28°C) for 0, 7, 14 and 21 days. The pods were then split at these pre-determined times and fermented. Figure 15.1 shows pod breaking of cocoa beans by farmers to start the bean fermentation process.

The heap fermentation method was used (Figure 15.2). About 50 kg of extracted cocoa beans were heaped on and covered with banana leaves and fermented for six days. During the six-day period of fermentation, the fermenting cocoa beans were mixed and sampled at 48 hr intervals by aseptically scooping about 1.5 kg of the beans into a sterile polyethylene bag. The samples were immediately transported to the laboratory for drying. The fermented beans were dried (to moisture content below 8%) by spreading the fermented cocoa beans approximately 5 cm deep on metal trays (40 cm × 60 cm), and placed in a temperature-controlled, forced-air oven for about 48 h at a temperature of 45–50°C. The dried beans were bagged in airtight black plastic bags and stored at ambient temperature (25–28°C) in a dark room free from strong odours.

Samples (fermented and unfermented cocoa beans) for chemical, biochemical and physicochemical analysis were milled using a hammer mill (Model 2A, Christy and Norris Ltd., Chelmsford, England) and the resulting meal packed in black polyethylene bags and used.

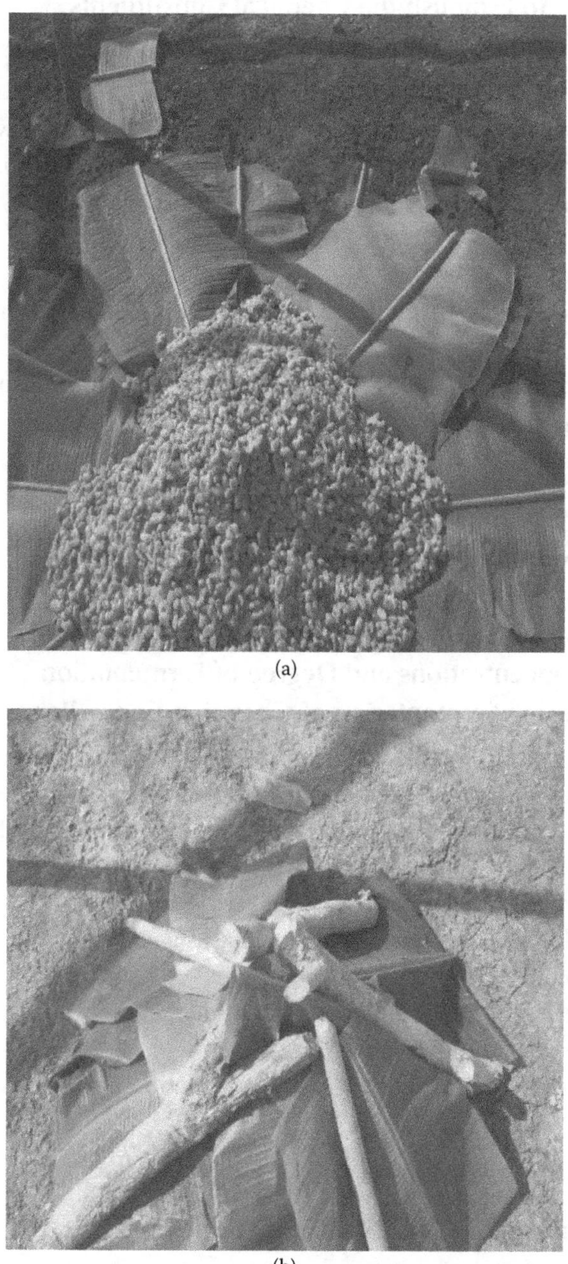

(a)

(b)

FIGURE 15.2 Uncovered (a) and covered (b) cocoa beans to undergo heap fermentation.

15.3.1.2 Experimental Design for Specific Objective 1: To Establish the Chemical Constituents of Pulp Pre-Conditioned Fermented and Unfermented Ghanaian Cocoa Beans

The study was conducted using a 4 × 2 full factorial design with experimental factors as with pod storage (0, 7, 14, 21 days) and cocoa treatment (fermented and unfermented). The samples were analysed for the following indices;

1. Proximate analysis: moisture, fat, protein, ash, carbohydrate content (AOAC 2005).
2. Mineral analysis was determined using an atomic absorption spectrophotometer.
3. Physical analysis: proportion of cocoa nibs, shells and germ.

15.3.1.3 Experimental Design for Specific Objectives 2–4: To Determine the Effect of Pod Storage on the Biochemical Constituents, Polyphenol Concentrations and Degree of Fermentation during Fermentation of Ghanaian Cocoa Beans

These studies were conducted using a 4 × 4 full factorial design with experimental factors as pod storage: (0, 7, 14, 21 days) and fermentation time (0, 2, 4, 6 days). The following indices were analysed on the samples:

1. Biochemical constituents and physicochemical properties: pH, titratable acidity, total sugars, reducing sugars, non-reducing sugars, proteins, free fatty acids
2. Polyphenolic compound concentrations: anthocyanins content, *o*-diphenols, total polyphenols
3. Degree of fermentation: cut test, fermentation index and colour

15.3.2 Analytical Methods

15.3.2.1 Percentage Nib, Shell and Germ

The percentages of nib, shell and germ were determined according to the method described by Wood and Lass (1985). One hundred (100) grams of cocoa bean sample were weighed and the nibs carefully separated from the shells using a sharp knife and weighed separately. The germ was then carefully separated from the nibs and weighed. The percentages of nib, shell and germ were calculated. The analyses were carried out eight times and the mean values reported.

15.3.2.2 Colour Evaluation

The colour of each sample was measured with a Minolta CR-310 Tristimulus Colorimeter (Minolta Camera Co. Ltd, Tokyo, Japan). Measurements were based on the L*a*b* colour system. The colour space parameters L*, a* and b* of the samples were calculated as: L* (L_s–L_0) for lightness (100 = perfect/brightness to 0 = darkness/blackness); a* (a_s–a_o) for the extent of green colour (in the range from negative = green to positive = redness); b* (b_s–b_o) qualifies blue in the range of negative = blue to positive = yellow. The reference white porcelain tile had the following readings: Lo = 97.51, a0 = 0.29 and b0 = 1.88).

15.3.2.3 Proximate Analysis

The moisture, fat, protein, ash and carbohydrate content were determined in accordance with AOAC (2005) methods 931.04, 963.15, 970.22 and 972.15, respectively. Analyses were performed in triplicate and the mean values reported.

15.3.2.4 Mineral Analyses: Wet Digestion

Mineral analyses were determined using AOAC (1990) methods with slight modifications. About 0.5 g of the sample were weighed into a 250-ml beaker. Twenty-five ml (25 ml) of concentrated nitric acid were added and the beaker covered with a watch glass. The sample was digested with great care on a hot plate in a fume chamber until the solution was pale yellow. The solution was cooled and 1 ml perchloric acid (70% $HCLO_4$) added. The digestion was continued until the solution was colourless or nearly so (the evaluation of dense white fumes was regarded to be indicative of the removal of nitric acid). When the digestion was completed, the solution was cooled slightly and 30 ml of distilled water added. The mixture was brought to a boil for about 10 minutes and filtered hot into a 100-ml volumetric flask using a Whatman No. 4® filter paper. The solution was then made to the mark with distilled water.

15.3.2.5 Determination of Ca, Mg, Zn, Fe, Cu, Na and K

The concentrations of Ca, Mg, Zn, Fe, Na and K were determined using a Spectra AA 220FS Spectrophotometer (Varian Co., Mulgrave, Australia) with an acetylene flame. One (1) ml aliquots of the digest was used to determine the Ca, Mg, Zn, Fe, Cu, Na and K content of the samples.

15.3.2.6 Phosphorus Determination

A two (2) ml aliquot of the digest was reacted with 5.0 ml molybdic acid. The molybdic acid was prepared by dissolving 25 ml of ammonium molybdate in 300 ml distilled water; with 75 ml of concentrated sulphuric acid in 125 ml of water to get 0.5 L of molybdic acid. One (1) ml each of 1% hydroquinone and 20% sodium sulphite was added in that sequence, and the solution was made up to 100 ml and allowed to stand for 30 min in order to develop colour, after which the absorption was measured at 680 nm. A standard curve of colorimetric readings versus concentration of phosphorus using portions of standard phosphorus solutions (1 ml, 2 ml and 3 ml) subjected to reactions with molybdic acid, hydroquinone and sodium sulphate solutions were drawn. All readings were corrected by the reading of a blank to eliminate the effect of any colour produced by the reagents.

15.3.2.7 Titratable Acidity (TA) and pH

TA of the nibs was determined according to AOAC method 970.21 (AOAC 2005). Titratable acidity was expressed as a percentage of acetic acid by titrating juice with 0.1N NaOH. Five-gram samples of nibs were homogenised for 30 s in 100 ml of hot distilled water and vacuum filtered through Whatman filter paper No. 4. A 25-ml aliquot was pipetted into a beaker and the pH measured using a pH metre (model MP230 Mettler Toledo MP 230, Geneva, Switzerland). A further 25-ml aliquot was titrated to an endpoint pH of 8.1 with 0.01N NaOH and the values reported as moles of sodium hydroxide per 100 g dry nibs. The analysis was conducted in triplicate and the mean values reported.

15.3.2.8 Free Fatty Acid

Fat from the samples was extracted with petroleum ether (40–60°C) in a Soxhlet apparatus (AOAC 2005 method 963.15). Free fatty acid of the extracted oils was determined using IOCCC method 42-1993 (IOCCC 1996). Five (5) grams of the oil were weighed into a dry 250-ml stoppered conical flask and 25 ml of 95% ethanol/ether (1:1) and phenolphthalein indicator. The solution was titrated with 0.1N NaOH, shaking constantly until pink colour persisted for 30 seconds and the percentage of free fatty acid determined. The analysis was conducted in triplicate and the mean values reported.

15.3.2.9 Total Sugars

Total sugars were determined using the phenol sulphuric acid method (Dubois et al. 1979). Dry weights of the samples were determined by drying the fresh samples overnight in an oven at 105°C.

15.3.2.9.1 Extraction of Sugars

The tissue was boiled in 80% ethanol under reflux for 30–40 min and the supernatant removed. The ethanol was evaporated under reduced pressure in a rotary evaporator. Some water was added to make sure the sugar present was in solution. After removal of the alcohol, alcohol-soluble, water-insoluble substances were precipitated. The precipitate was filtered into a clean flask through Whatman No. 41 filter paper followed by washing of the paper with distilled water.

The extract was clarified by adding 5 ml of the $ZnSO_4$ solution to the extract followed by 4.94 ml of the $Ba(OH)_2$ solution. The solution was allowed to stand for about 5 min and filtered through Whatman No. 41; the filter paper was washed with distilled water to remove any sugar still trapped in the precipitate. A mixture of Zeokarb 225 (H^+), a cation exchange resin, and Deacidite FF (OH) was added to the filtrate, shaken and then filtered. To 1 ml of the extract, 1 ml phenol reagent and 5 ml of H_2SO_4 were added and allowed to stand for an hour to develop colour. The absorbance of the sample was read at 480 nm and the results calculated. The analysis was conducted in triplicate and the mean values reported.

15.3.2.10 Reducing Sugars

The reducing and non-reducing sugars were determined using the Luff–Schoorlf method (Egan, Kirk, and Sawyer 1990). Three grams of fine powdered sample were weighed into a wide-necked 250-ml volumetric flask. Extraction of water-soluble matter was done by shaking the sample with 150 ml of hot distilled water. The solution was clarified by the addition of 5 ml Carrez 1 solution followed by 5 ml Carrez II solution. Content was made to the 200-ml mark with distilled water and filtered.

Twenty-five (25) ml of copper reagent was measured into each of two 250-ml Quickfit Erlenmeyer flasks. To one (the blank), 25 ml of distilled water were added to make up to 25 ml. A few anti-bumping granules were added, transferred to the hot plate and the condenser lowered. This was boiled gently for 10 min and then transferred with tongs to a cooling bath for 5 min. About 3.0 g of potassium iodide were added to the cooled flask, swirled to dissolve and, with constant swirling, 20 ml of 6N HCl were added from a measuring cylinder. The content of the flask was titrated with 0.1N thiosulphate until the iodine colour nearly disappeared. To this, 1.0 ml starch indicator was added and titration continued until the blue colour changed to give a white precipitate of cuprous iodide with no trace of blue. A sample titration was subtracted from the blank to obtain the quantity of thiosulphate used and the results calculated. The analysis was conducted in triplicate and the mean values reported.

15.3.2.11 Non-Reducing Sugar

Non-reducing sugars were completely hydrolysed to reducing sugars by refluxing with 1N HCl. Six (6) ml of clarified sample were measured out into a 250-ml quickfit Erlenmeyer flask and 2 ml of 1N HCl was added and made up to 20 ml using distilled water. Refluxing was done for 10 min on a hot plate. The flask and its contents were allowed to cool and 2 ml of NaOH and 3 ml of water added. The copper reagent was added and preceded as with reducing sugars. The analysis was conducted in triplicate and the mean values reported.

15.3.2.12 Total Polyphenols

Cocoa nibs were defatted by extracting the fat with petroleum ether (40–60°C) in a Soxhlet apparatus. The total contents of polyphenolic compounds in the cocoa bean extracts were determined according to the Folin–Ciocalteau procedure (EEC, 1990). About 0.2 g of dry defatted cocoa nibs was extracted with methanolic HCl (80% MeOH containing 1% HCl) with shaking for 2 h at room temperature. This was centrifuged at 1,000 rpm for 15 min, about 1.0 ml of supernatant was taken to develop colour reaction with 5.0 ml folin–ciocalteau reagent. To the supernatant 4.0 ml of Na_2CO_3 were added and allowed to stand for 1 h at 30°C, then 1 h at 0°C. The absorbance was read at 760 nm. The analysis was conducted in triplicate and the mean values reported. A working standard catechin solution of blank (0), 0.2, 0.4, 0.6, 0.8 and 1 mL was prepared and made up to 1.0-mL volume with distilled water. The colour was developed, the absorbances read at 760 nm and a standard curve drawn. From the standard graph, the amount of polyphenol present in the sample preparation was calculated.

15.3.2.13 *o*-Diphenols

o-Diphenol content was determined with Arnow's reagent (10 g $NaNO_2$, 10 g Na_2MoO_4 in 100 ml H_2O). To 1 ml of extract (ethanol or methanol extract), 1 ml 0.5N HCl, 1 ml Arnow's reagent, 10 ml H_2O and 2 ml 1N NaOH were added. The solution was mixed and the absorbance was read at 515 nm (520) after 30 s. The analysis was conducted in triplicate and the mean values reported. A working standard cathechol solution of blank (0), 0.2, 0.4, 0.6, 0.8 and 1 mL was prepared and made up to 1.0 mL volume with distilled water. The colour was developed, the absorbances read at 520 nm and a standard curve drawn. From the standard graph, the amount of *o*-diphenol present in the sample preparation was calculated.

15.3.2.14 Anthocyanins

Anthocyanin content was determined using a method described by Misnawi et al. (2002). The extract obtained for total polyphenol analysis was filtered using Whatman No. 4 filter paper, and the supernatants were read spectrophotometrically for total absorbance (TOD) at 535 nm. The content of total anthocyanins was calculated as follows: total anthocyanins (mg/kg) = TOD * 1,000/$(AvE_{530})_{1cm}$/10, where TOD is the total optical density (absorbance) and $(AvE_{530})_{1cm}$ is the average extinction coefficient for total anthocyanin when a 1-cm cuvette and 1% (10 mg/ml) standards are used; the value is 982. The analysis was conducted in triplicate and the mean values reported.

15.3.2.15 Cut Test

The cut test (index of fermentation), which relies on changes in the colour of cocoa beans, is the standard test used to assess the suitability of cocoa beans for making chocolate. The cut test was performed according to the international method described by Guehi et al. (2008). A total of 300 beans was cut lengthwise through the middle in order to expose the maximum cut surface of the cotyledons. Both halves were examined in full daylight and placed in one of the following categories: fully brown (fermented), partly brown, partly purple, purple, slaty, insect damaged, mouldy, and germinated.

15.3.2.16 Fermentation Index (FI)

The FI was determined according to the method described by Gourieva and Tserrevitinov (1979). In this procedure, 0.5 g ground cocoa nibs was extracted with 50 ml of 97:3 mixture of methanol:HCl. The homogenate was allowed to stand in a refrigerator (8°C) for 16–19 h and then vacuum filtered. The filtrate was made up to volume in a 50-ml volumetric flask. The fermentation index of the sample was obtained by calculating the ratio of absorbance at 460 nm to the absorbance at 530 nm (LKB Biochrom Novaspec II UV Spectrometer, Birmingham, UK). Three replicate readings were obtained for each sample and the mean values reported.

15.3.3 Statistical Analyses

The data were analysed using Statsgraphics software version 3.0 (STSC Inc., Rockville, MD) for analysis of variance (ANOVA). Least significant difference (LSD) was used to separate and compare the means and significance was accepted at the 5% level ($p < 0.05$). Further analyses were conducted to evaluate the combined effect of pulp pre-conditioning and fermentation time on the physicochemical, biochemical and polyphenol concentrations using response surface methodology. The data obtained were studied using stepwise multiple regression procedures. Models were developed to relate pulp pre-conditioning and fermentation time on the biochemical constituents, physicochemical properties and polyphenolic concentrations in cocoa beans. Later in the chapter, Tables 15.10, 15.11 and 15.13 show the coefficients of the variables in the models and their contribution to the model's variation. R^2 values were used to judge the adequacy of the models. The R^2 of a model refers to the proportion of variation in the response attributed to the model rather than random error. For a good fit of a model, an R^2 of at least 60% was used.

15.4 RESULTS AND DISCUSSION

15.4.1 Effect of Pod Storage on Chemical Composition and Physical Qualities of the Fermented and Unfermented Dried Cocoa Beans

15.4.1.1 Proximate Composition

Table 15.1 shows the proximate composition of unfermented and fermented cocoa beans under different pod storage conditions. The moisture levels of the cocoa beans were considerably lower (3.89–4.95%) than the acceptable limits (6–7%) for long-term storage of cocoa (Wood and Lass 1985; Dand 1997; Fowler 2009), hence the beans were quite brittle in nature. The relatively lower moisture content attained was to ensure that virtually all enzymatic reactions had ceased. Although the fermentation process reduced the water content of the beans there was still a considerable amount of moisture lost during drying. Fermentation introduced significant variation in the moisture levels (Table 15.2). Moisture levels were significantly lower ($p < 0.05$) in all pulp pre-conditioned fermented cocoa beans than in the unfermented beans (Table 15.1); this may be ascribed to the initial higher moisture levels of unfermented samples.

Crude protein content ranged from about 16 to 22% and this was comparable to literature values of 15.2–19.8% (Aremu, Agiang, and Ayatse 1995; Afoakwa et al. 2008). There were general decreases in crude protein with fermentation for all the cocoa samples. Similarly, apparent decreases were observed as pod storage increased (Table 15.1).

The results indicate that protein content was significantly influenced ($p < 0.05$) by pod storage and fermentation time (Table 15.2). Further analysis using least significance difference (LSD) revealed that the decreases amongst the 7- and 14-day pod storage were not significantly different. The protein content of the fermented cocoa beans reduced from 17.80 to 16.52% by 14 days of pod storage. The protein content was significantly reduced after 6 days of fermentation (21.63 to 17.80%) for the beans that

TABLE 15.1
Effect of Pod Storage on Proximate Profile of the Fermented and Unfermented Dried Cocoa Beans

Pod Storage (Days)	Fermentation Condition	Moisture (%)	Protein (%)[a]	Fat (%)	Ash (%)	Carbohydrate (%)[b]
0	Unfermented	4.21 ± 0.02	21.63 ± 0.83	53.21 ± 0.10	3.48 ± 0.11	17.48 ± 0.63
	Fermented	4.01 ± 0.02	17.80 ± 0.56	50.35 ± 0.63	2.82 ± 0.07	25.03 ± 0.08
7	Unfermented	4.42 ± 0.04	19.79 ± 0.05	51.21 ± 1.5	2.90 ± 0.05	21.68 ± 0.72
	Fermented	4.27 ± 0.09	17.16 ± 0.13	49.86 ± 0.05	2.32 ± 0.04	26.40 ± 0.54
14	Unfermented	4.24 ± 0.02	19.64 ± 0.06	52.49 ± 0.04	3.10 ± 0.01	20.53 ± 0.24
	Fermented	4.45 ± 0.03	16.52 ± 0.60	50.51 ± 0.15	2.73 ± 0.18	25.79 ± 0.31
21	Unfermented	4.95 ± 0.01	20.83 ± 0.48	52.27 ± 0.07	3.26 ± 0.05	18.69 ± 0.09
	Fermented	3.87 ± 0.04	18.88 ± 0.07	50.40 ± 0.05	2.92 ± 0.09	23.94 ± 0.11

[a] Protein (N × 6.25).
[b] Carbohydrate was obtained using by the difference method.
Results presented are mean values of triplicate analysis ± standard deviation.

TABLE 15.2
ANOVA Summary Showing F-Ratios for Variations in Proximate Composition of Fermented and Unfermented Cocoa Samples

Variables	Protein	Fat	Carbohydrate	Ash	Moisture
Pod storage (PS)	15.29[a]	166.54[a]	47.29[a]	28.41[a]	2.40
Fermentation time (FT)	16.13[a]	543.32[a]	397.71[a]	115.87[a]	16.93[a]
Interaction (PS × FT)	3.05	15.82[a]	3.32	3.07	23.82[a]

[a] Significant at $p < 0.05$.

were not stored (0-day pod storage) and likewise in all the beans that were pulp preconditioned (7, 14 and 21 days pod storage) and this is consistent with reported literature by Biehl and Passern (1982), Biehl et al. (1985) and Crouzillat et al. (1999). Contrary to this, Aremu et al. (1995) reported a significant increase in bean protein content by the sixth day of fermentation. The observed decreases in protein content with pod storage and fermentation might be due to protein breakdown during the curing process, occurring partly due to hydrolysis to amino acids and peptides and partly by conversion to insoluble forms by the action of polyphenols as well as losses by diffusion (Afoakwa et al. 2008; Afoakwa and Paterson 2010).

Fat content or yield is an important quality index for cocoa processors during purchasing of fermented cocoa beans. In West African fermented and dried cocoa beans, the fat content is between 56 and 58% and most *Forastero* cocoas fall between 55 and 59% (Rohan 1963; Reineccius et al. 1972; Wood and Lass 1985; Afoakwa et al. 2008). The fat content of the beans was slightly lower (49.86–53.21%) than that reported by Afoakwa et al. (2008) and these variations may be attributed to the relatively lower bean weight of cocoa beans used in this study. Variations in the bean sizes could also account for the observed relatively lower fat content. Wood and Lass (1985) and Dand (1997) reported that smaller bean size results in lower fat yield.

Analysis of variance on the data revealed that the fat content of the samples decreased significantly ($p < 0.05$) with fermentation (Tables 15.1 and 15.2) and this corroborates studies carried out by Aremu et al. (1995) in Nigeria where the lipid content of the cocoa beans decreased from 62.9% to 55.7% by the sixth day of fermentation. This suggests that the reductions in fat content in cocoa beans could be avoided by reducing fermentation time.

Carbohydrate content was significantly ($p < 0.05$) higher in fermented samples than in unfermented samples (Table 15.1), with beans stored for 7 days prior to fermentation having the highest carbohydrate content at the start of the fermentation. Even though pod storage influenced the carbohydrate content significantly ($p < 0.05$), samples stored for 0 and 21 days were not significantly different from each other. An apparent inverse relationship appears to exist between the levels of fat and total carbohydrate in fermenting cocoa. Conversion of lipid to carbohydrate via gluconeogenesis, employing the glyoxylate cycle could not be ruled out. It has been indicated that this pathway normally operates in micro-organisms and germinating oil seeds (White et al. 1978).

The ash content of the cocoa beans decreased significantly ($p < 0.05$) with fermentation and was generally comparable to literature values (Rohan 1963; Reineccius et al. 1972; Aremu et al. 1995). ANOVA indicated that the reductions in the ash content due to fermentation and pulp pre-conditioning were significant ($p < 0.05$); however, pod storage of 7 days was significantly lower than the other pod storage days (Table 15.2).

15.4.1.2 Mineral Content

The effect of pulp pre-conditioning on the mineral composition of fermented and unfermented cocoa samples is shown in Table 15.3. Generally, there were decreases in the micronutrients with fermentation and increasing pod storage. The differences in mineral contents for all the different days of pod storage were significant ($p < 0.05$). Also, differences among the unfermented and their corresponding fermented samples were also significant ($p < 0.05$). Iron generally decreased significantly ($p < 0.05$) as pod storage days increased and with fermentation (Tables 15.3 and 15.4). The iron content of unfermented cocoa samples that were not stored prior to fermentation was 2.73 mg/100 g and this decreased significantly by the end of the fermentation to 2.21 mg/100 g (Table 15.3). Similar trends were observed in the beans stored for the other days of pod storage.

Results presented are mean values of triplicate analysis ± standard deviation. Copper content on the other hand increased as fermentation time and pod storage days increased. By 21 days of pod storage, the copper content of cocoa bean samples had increased from 8.75 to 15.49 mg/100 g which is approximately a 100% increase in copper content. This remarkable trend may be explained by the breakdown of anti-nutritional factors such as polyphenols and tannins during fermentation (Svanberg and Lorri 1997). Fermentation is known to provide optimum pH conditions for the enzymatic degradation of polyphenols, which may be present in the cocoa beans in the form of complexes with polyvalent cations such as copper, zinc and proteins thus rendering them unavailable. Reduction in these anti-nutritional factors therefore might have increased the amount of soluble copper by several fold (Chavan and Kadam 1989; Nout and Motarjemi, 1997).

The magnesium content of the cocoa samples was significantly higher ($p < 0.05$) in unfermented samples than in the fermented beans (Tables 15.3 and 15.4). Pod storage, however, had only marginal influence

TABLE 15.3
Effect of Pod Storage on Mineral Content of Fermented and Unfermented Cocoa Beans

Pod Storage (days)	Fermentation Condition	Mineral Content (mg/100 g) Fe	Cu	Mg	Zn	Na	Ca	P	K
0	Unfermented	2.21 ± 0.04	11.06 ± 0.03	286.77 ± 3.19	9.71 ± 0.06	3.35 ± 0.01	140.23 ± 0.60	236.619 ± 23.08	2313.12 ± 6.04
	Fermented	2.73 ± 0.02	8.75 ± 0.01	364.16 ± 1.82	10.58 ± 0.07	2.48 ± 0.16	170.79 ± 0.74	195.824 ± 0.02	2557.92 ± 11.01
7	Unfermented	2.45 ± 0.02	11.53 ± 0.13	318.95 ± 7.27	9.26 ± 0.06	2.46 ± 0.04	141.10 ± 0.60	264.360 ± 184.62	2325.42 ± 12.3
	Fermented	1.80 ± 0.01	13.17 ± 0.05	262.71 ± 3.68	8.21 ± 0.01	2.98 ± 0.01	143.53 ± 0.08	210.509 ± 23.08	2164.22 ± 10.26
14	Unfermented	2.19 ± 0.02	13.71 ± 0.02	331.51 ± 6.89	9.27 ± 0.05	3.29 ± 0.08	158.2 ± 0.38	292.102 ± 23.08	2433.65 ± 16.23
	Fermented	1.52 ± 0.03	15.49 ± 0.06	271.25 ± 1.16	7.48 ± 0.02	2.58 ± 0.06	150.27 ± 0.68	203.982 ± 23.08	2095.55 ± 6.98
21	Unfermented	1.38 ± 0.01	15.31 ± 0.12	349.23 ± 2.98	9.41 ± 0.25	2.69 ± 0.04	142.82 ± 0.07	381.854 ± 46.16	2318.66 ± 3.62
	Fermented	1.23 ± 0.02	17.28 ± 0.07	322.26 ± 5.59	15.64 ± 0.52	2.04 ± 0.06	148.49 ± 0.41	355.744 ± 00	2070.74 ± 5.71

TABLE 15.4
ANOVA Summary Showing F-Ratios for Variation in Mineral Content of Fermented and Unfermented Cocoa Beans

Variables	Ca	Cu	Na	Mg	Fe	Zn	P	K
Pod storage (PS)	29.34[a]	4945.20[a]	51.06[a]	82.02[a]	1341.94[a]	322.82[a]	394.27[a]	1053.69[a]
Fermentation condition (FC)	34.80[a]	2387.31[a]	0.06	577.25[a]	1365.55[a]	36.12[a]	84.51*	6180.26[a]
Interaction (PS × FC)	54.61[a]	13.83[a]	117.99[a]	20.78[a]	80.84[a]	322.77[a]	62.32[a]	118.47[a]

[a] Significant at $p < 0.05$.

on cocoa beans with no precise trends in their observation. Cocoa beans had low sodium content (2.04 to 3.35 mg/100 g) and were not significantly ($p > 0.05$) influenced by fermentation although there were apparent differences observed amongst the samples. On the contrary, increasing pod storage caused general decreases in the sodium content of the samples. ANOVA on the data showed that pod storage had a significant ($p < 0.05$) influence on the sodium content of the cocoa beans (Table 15.4). A multiple comparison test showed that the beans stored for 0 and 14 days prior to fermentation were not significantly different from each other but the observed significant reductions were due to the differences in values from the 0, 7 and 21 days.

Generally, phosphorus content decreased with fermentation at all levels of pod storage (Table 15.3). Contrary to these, increasing pod storage (pulp pre-conditioning) caused consistent increases in the phosphorus content (Table 15.3). ANOVA on the data showed that both pod storage and fermentation had a significant ($p < 0.05$) influence on the phosphorus content (Table 15.4). A multiple comparison test (LSD) suggested that the phosphorus contents at 0 and 21 days of pulp pre-conditioning were significantly different from each other and those from 7 and 14 days.

Cocoa beans had a very high potassium content with values of 2,557.92 and 2,313.12 mg/100 g, respectively, for the fermented and unfermented samples from the unstored pods (Table 15.3). Fermentation of the beans caused a slight reduction in the samples to 2,070.74 mg/100 g after 21 days of pod storage whereas the unfermented samples showed only marginal increases in K content with increasing pod storage. Analysis of variance on the data showed that the K content was significantly higher ($p < 0.05$) by both fermentation and increasing pulp pre-conditioning (Table 15.4).

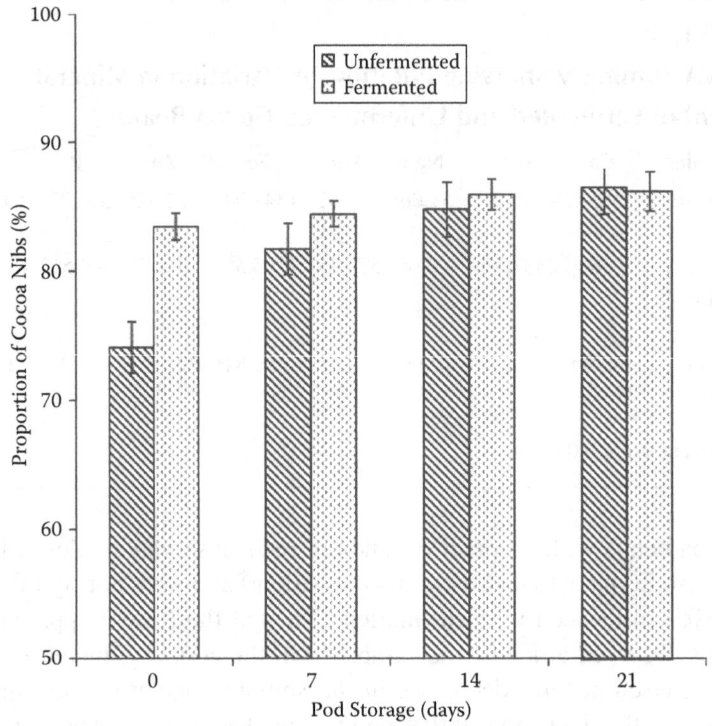

FIGURE 15.3 Changes in proportion of cocoa nibs in fermented and unfermented cocoa beans as influenced by pod storage.

These high values suggest that potassium is the most abundant mineral in Ghanaian cocoa beans and these might have originated from the soil in which the cocoa was planted.

15.4.1.3 Proportion of Cocoa Nibs

The proportion of nibs ranged from 74.1 to 83.5% in the unfermented and fermented cocoa beans that were not stored prior to fermentation and these values were slightly lower than those (86–90%) reported by Rohan (1963), Reineccius et al. (1972) and Afoakwa et al. (2008). These differences in nib content might have resulted from the harvesting season (whether major or minor) as these have been reported to affect the size of the beans (Rohan 1963) and the important determining factors are suspected to be the amount and distribution of rainfall and temperature during the development of the pod (Rohan 1963; Wood and Lass 1985 and Dand 1997). High temperatures and lower rainfall might have accounted for the smaller nib proportion of the cocoa beans used in the study. Generally the proportion of nib was slightly higher in the fermented cocoa samples than in the unfermented samples (Figure 15.3).

TABLE 15.5
ANOVA Summary Showing F-Ratios for Variation in Physical Constituents of Fermented and Unfermented Cocoa Samples

Variables	Cotyledon[a]	Shell[a]	Germ
Pod storage	295.66	281.53	0.73
Fermentation time	275.74	263.66	0.04
Interaction	118.62	113.47	0.61

[a] Significant at $p < 0.05$.

Pulp pre-conditioning and fermentation caused significant increases in the weight of the cocoa nibs (Table 15.5). As illustrated in Figure 15.3, the nib recovery increased with increasing pod storage days as well as fermentation. The amount of pulp on the bean during shell separation would have much influence on the proportions. The amount of nib contained in the bean is of major concern to the cocoa processor because higher nib content results in higher nib recovery and fat yield. The apparent increase in the weight of the nib reflects a decrease in shell content.

15.4.1.4 Proportion of Shells

The shell provides adequate protection of the nib from mould and insects, but the shell percentage should be as low as possible (10–14%) inasmuch as it is removed during processing of the cocoa beans and is of very little value (Rohan 1963; Reineccius et al. 1972; Wood and Lass, 1985; Dand, 1997 and Afoakwa et al., 2008). In the unfermented cocoa beans, shell content for all the pod storage ranged between 25.13 to 12.80%. By the end of the fermentation, the shell content had decreased significantly ($p < 0.05$) for the different pod storage days. Figure 15.4 shows a sharp decrease from 25.13% to 15.77% shell content for 0 days pod storage. The considerably high shell content of all the unfermented samples could be ascribed to the adhering thick mucilaginous pulp immediately surrounding the testa prior to fermentation. Subsequent degradation of pectin by microbial pectinases during fermentation causes the liquefaction and drainage of about 10–50% of the pulp (Outtara et al. 2008) and this might have accounted for the relatively lower shell content in the fermented cocoa beans.

Figure 15.4 also depicts a decrease in shell content as pod storage increased with beans stored for 21 days prior to fermentation having the lowest shell content (12.80%). Reduction in pulp volume by water evaporation occurs during pod storage (Biehl et al. 1989). This phenomenon could

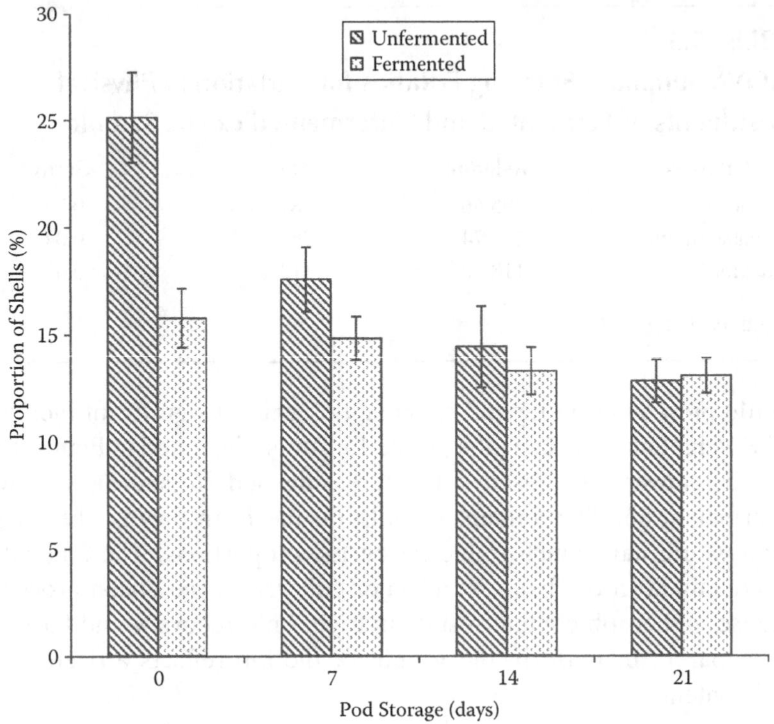

FIGURE 15.4 Changes in proportion of shell in fermented and unfermented beans as influenced by pod storage days.

explain the decrease in shell content with increasing pod storage. The multiple range test (LSD) showed that the beans stored for the different pod storage days (0, 7, 14 and 21 days) were significantly different from each other. The cocoa bean shells make up waste material, thus the lower they are, the more desirable they are to the cocoa processor.

15.4.1.5 Proportion of Germ

The proportion of germ for the cocoa beans ranged from 0.73 to 0.75% for all the cocoa beans (Figure 15.5) and this is similar to results reported by Reineccius et al. (1972; 0.77%). Pulp pre-conditioning and fermentation time did not have any significant effect ($p > 0.05$) on the proportion of germ of the cotyledons (Table 15.5). Pods stored for 21 days had their beans germinating, which accounted for the slightly higher proportion of germs, and this observation may be attributed to pod rotting and the penetration of oxygen during pod storage (Meyer et al., 1989), hence providing favourable conditions for the germination of the beans.

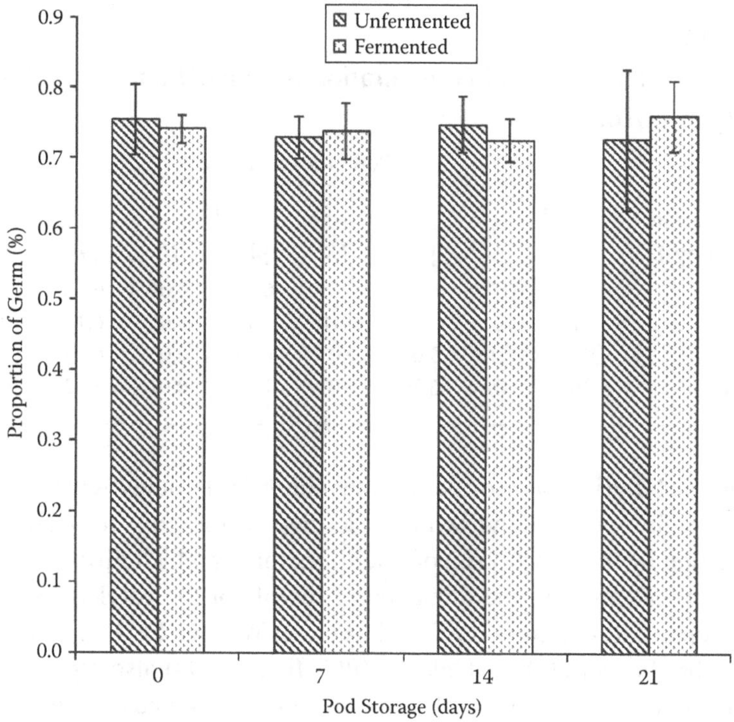

FIGURE 15.5 Changes in percentage of germ as influenced by pod storage and fermentation.

15.4.2 Effect of Pulp Pre-Conditioning on Physicochemical and Biochemical Constituents of Fermented Cocoa Beans

15.4.2.1 pH

The pH of cocoa beans during fermentation is crucial as it dictates the fermentative quality of the beans. Results from this study showed that the pH of the cotyledons in the unstored pods decreased slightly from 6.7 to 4.9 within the first four days of fermentation and then increased slightly again to 5.3 by the sixth day of fermentation (Table 15.5). These observations are in agreement with an earlier report by Biehl (1984), who noted that there was diffusion of acids (predominantly acetic acid) in the fermenting bean leading to decreases in the pH of the cotyledon from 6.5 to 4.6 within the first three days of fermentation and then increasing again to 5.2 by the sixth day of fermentation.

The observed consistent decreases in pH within the first four days of fermentation were primarily due to the diffusion of organic acids produced by lactic acid and acetic acid bacteria into the beans. Several authors have reported that during cocoa fermentation, volatile acids (acetic, propionic,

TABLE 15.6
Effect of Pod Storage and Fermentation on the pH and Titratable Acidity of Cocoa Beans

	Pod Storage (Days)							
Fermentation	0		7		14		21	
Time (Days)	pH	TA	pH	TA	pH	TA	pH	TA
0	6.71	0.02	6.53	0.03	6.43	0.04	6.62	0.03
2	4.91	0.20	5.37	0.10	5.77	0.06	6.11	0.08
4	4.97	0.16	5.64	0.08	6.13	0.05	6.33	0.04
6	5.36	0.09	6.56	0.03	7.33	0.01	7.32	0.01

butyric, isobutyric and isovaleric) and non-volatile acids (citric, lactic, malic, succinic, oxalic and tartaric) develop in the pulp through sugar degradation by the metabolism of micro-organisms which diffuse into the cotyledon and result in the reduction of the pH of the cotyledons (Jinap 1994; Lopez and Dimmick 1995; Schwan and Wheals 2004; Amoa-Awua et al. 2006; Fowler 2009; Afoakwa 2010). Jinap (1994) also reported that the amount of acetic acid absorbed by the nibs corresponds to the concentration found in the pulp after the fourth day of fermentation. As soon as the mass becomes more porous, giving rise to effective air upflow (after about four days of fermentation), acetic acid is strongly reduced in the pulp and in the nibs, accounting for the subsequent increase in pH after the fourth day of fermentation. The gradual pH increase in the pulp during steep acetic acid decay (after about four days) and the steep pH increase during final degradation of acetic acid (six to seven days) are due to the pk_a, (4.7) of this acid (Biehl et al., 1989). Similar findings were also reported by Nazaruddin et al. (2006a) when working on Malaysian cocoa beans.

Pod storage, however, showed different trends in pH at all levels of fermentation. With the exception of the unstored pods which showed a considerable reduction in pH, increasing pod storage caused consistent increases in pH of the cotyledon at all fermentation times (Table 15.6). This explains that increasing pod storage reduces the levels of acidity in the beans at all levels of fermentation. Analysis of variance of the data revealed significant differences ($p < 0.05$) in pH of the fermenting beans with fermentation time and pod storage. In addition, the interaction between fermentation time and pod storage was also significant ($p < 0.05$). A multiple comparison test (LSD) suggested that the pH of the cocoa beans at all levels of pod storage (0, 7, 14 and 21 days) and fermentation (0, 2, 4 and 6 days) was significantly different.

TABLE 15.7
Regression Coefficients and Their Adjusted R^2 Values in the Models for pH and Titratable Acidity of Cocoa Beans

Variables	pH	Titratable Acidity
Constant	6.09228[a]	0.06342[a]
X_1	0.03265[a]	–0.00779[a]
X_2	–0.02956	0.00221
X_1^2	–6.44133	0.00029[a]
X_2^2	0.00018[a]	–1.36990[a]
X_1*X_2	0.00053[a]	–2.36607[a]
R^2	0.818	0.801

[a] Significant at $p < 0.05$; X_1 = Pod storage; X_2 = Fermentation time.

Response surface plots were further developed to study the combined effects of pod storage and fermentation time on the pH of the cotyledons. This was to help broaden our understanding on the extent to which variations in the variables (pod storage and fermentation time) might influence the pH of the beans.

Regression analysis on the data showed significant ($p < 0.05$) influence of the quadratic factor of fermentation time and the linear factor of pod storage on the pH of the cotyledons. There was also significant interaction between pod storage and fermentation time. The model developed could explain over 81% of the variations in pH (Table 15.7). The response surface plot (Figure 15.6) showed a curvilinear relationship between fermentation time and pH whereas pod storage had a linear relationship with pH at all levels of fermentation time. Thus, increasing fermentation time resulted in decreasing pH up to a minimum point within the first 48 h of fermentation, after which consistent pH increases occur at all levels of pod storage (0–21 days) till the end of fermentation. The pH of unfermented cocoa beans showed only slight variations with increasing pod storage. However, as fermentation progressed the pH of the beans increased sharply with increasing pod storage. These trends suggest that increasing pod storage has only a marginal effect on the pH of unfermented beans whereas increasing pod storage causes consistent increases in pH in cocoa beans during fermentation.

Biehl et al. (1985) noted that a pronounced cocoa aroma is obtained after roasting when the pH value does not drop to less than 5.0 during fermentation and an optimum pH of 5.5 is obtained after fermentation. The response curve (Figure 15.6) showed that the unstored beans had their pH dropping slightly below 5.0 within 48 h of fermentation, after which it increased further to the optimum pH ~5.5 after 6 days of fermentation. Further increases

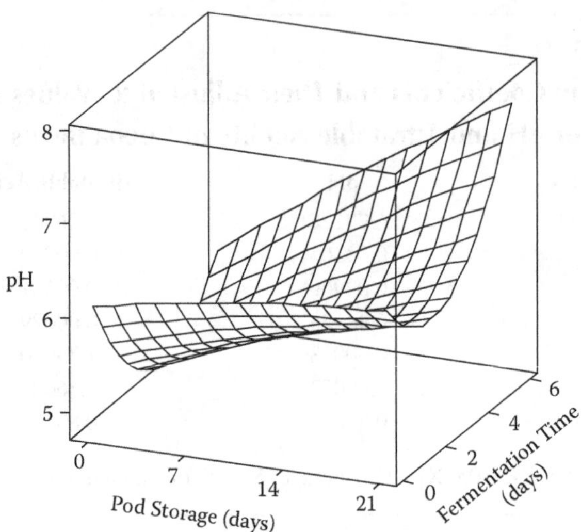

FIGURE 15.6 Response surface plot showing effect of pod storage and fermentation time on the pH of cocoa nibs.

in pod storage beyond 7 days results in consistent increases in pH values above 6.0 at all levels of fermentation. Meyer et al. (1989) reported that increasing pod storage results in low pulp volume and this tends to suppress the anaerobic phase during the initial stages of fermentation and thus account for the reduced acidity in pod-stored beans. For beans stored for 7 days the optimum pH (~5.5) was reached by the fourth day of fermentation.

This suggests that the technique of pod storage (pulp pre-conditioning) could be used to shorten the fermentation time to achieve the required pH and thus the expected fermentative quality of Ghanaian cocoa beans. Beyond seven days of pod storage, however, there is less build-up of acids; hence alkaline cocoa beans are obtained. These observations suggest that the optimum period of pod storage in Ghanaian cocoa beans should not exceed seven days and any further increases might negatively affect the fermentative quality and flavour potential of the beans.

15.4.2.2 Titratable Acidity

Acids are generated in cocoa beans during fermentation as a result of the action of micro-organisms that break down the pulp that surrounds the beans. Their activities result in the conversion of pulp sugars to alcohols and organic acids, predominantly acetic acid, which penetrates into the cotyledon (Lopez and Dimick 1995; Schwan and Wheals 2004; Fowler 2009). Changes in titratable acidity (TA) during fermentation of the cocoa beans were contrary to observations made with pH for all pod storage treatments (Table 15.6). Fermentation caused increases in acidity levels reaching a maximum point within 48 h, after which the titratable acidity decreased considerably till the end of fermentation. This observation was noted with all pod storage treatments. Beans that were not pulp preconditioned had a high TA at the end of the fermentation period (sixth day of fermentation). On the other hand, with the exception of the unstored pods which showed only significant increases in acidity levels, increasing pod storage consistently reduced the levels of acidity in the beans at all fermentation times (Table 15.6). This suggests that pod storage is one of the pulp pre-conditioning techniques that could be effectively employed to reduce the acidity levels of cocoa beans, probably due to reduced pulp and water content in the cocoa beans and thus results in increasing microaeration within the pulp. This increases the sugar respired by yeast rather than fermentation and eventually reduces alcohol fermentation and acetic acid formation in the bean (Said, Meyer, and Biehl 1987).

In an attempt to further understand the combined influences of pod storage and fermentation on titratable acid generation in the beans, response surface methodology was employed. Regression analysis on the data showed a strong and significant ($p < 0.05$) influence of the linear factor of pod storage and quadratic factors of pod storage and fermentation time (Table 15.7). There were also significant ($p < 0.05$) interactions between fermentation time and pod storage days. The model could explain approximately 80.0% of the variations in the titratable acidity. The response surface developed (Figure 15.7) showed that titratable acidity of the cocoa

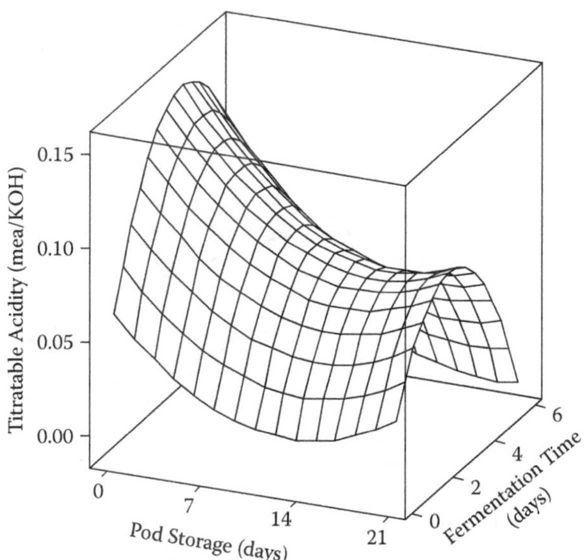

FIGURE 15.7 Response surface plot showing effect of pod storage and fermentation time on the titratable acidity of cocoa nibs.

bean samples at all pod storage days increased to a maximum during the first 48–72 h of fermentation and decreased again till the end of the fermentation. However, the extent of acid generation was more pronounced at lower pod storage periods (0–7 days), and any further increases in pod storage results in drastic reduction in acidity levels at all fermentation times. Work done by Nazaruddin et al. (2006a) also revealed the same trend in titratable acidity levels.

15.4.2.3 Changes in Sugar Concentrations

Changes in sugar concentration during cocoa fermentation at different pod storage times are as shown in Figure 15.8. In unfermented cocoa beans (0 day of fermentation), non-reducing sugars (mainly sucrose) were the main sugars present in significant concentration (18.78 mg/g), which was about 95% of the total sugars. Reducing sugars made up about 2.7%. According to Berbert (1979), the sucrose concentration in the unfermented beans generally comprised about 90% of the total sugars, whereas both fructose and glucose made up about 6%. These relative differences may be attributed to method and time of harvesting, and type and origin of cocoa beans (Reineccius et al. 1972).

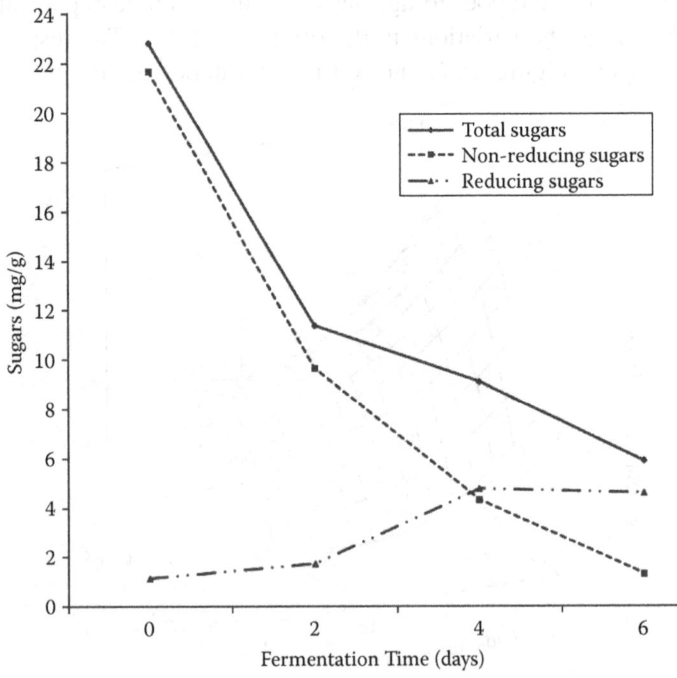

FIGURE 15.8 Effect of fermentation time on total sugars, reducing sugars and non-reducing sugars.

During fermentation, both the non-reducing and total sugars decreased significantly ($p < 0.05$) to 2.03 mg/g and 5.02 mg/g (89% and 75% decrease), respectively. These decreases roughly mirrored the production of reducing sugars (Figure 15.8). However, the rates of decrease in both the total and non-reducing sugars slowed down towards the end of fermentation (day 6). This trend is similar to observations made by previous researchers (Hashim et al. 1998; Ardhana and Fleet 2003). Hansen, del Olmo, and Burri (1998) explained that the reduction in non-reducing sugars is caused by the action of native cocoa seed invertase which hydrolyses sucrose into glucose and fructose during fermentation. In contrast, reducing sugars increased after the death of the beans and reached maximum concentrations after four days of fermentation; thereafter, their concentrations remained steady until the end of the fermentation. Similar trends have been reported by several authors (Rohan and Stewart 1967; Reineccius et al. 1972; Berbert 1979).

15.4.2.4 Changes in Total Sugars

The response surface plot (Figure 15.9) shows the combined effects of pod storage and fermentation time on the total sugar concentrations in cocoa beans. Regression analysis on the data showed that both the linear and quadratic effects of pod storage and fermentation time were significant at $p < 0.05$. However, the interaction effect was not significant ($p > 0.05$; Table 15.8). The R^2 value was 0.946, indicating that the pod storage days and fermentation time could account for 94.6% of the changes in total sugars.

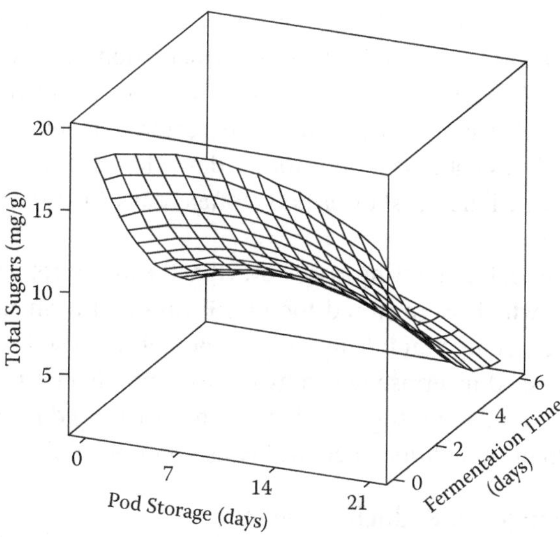

FIGURE 15.9 Response surface plot showing effect of pod storage and fermentation time on the total sugars of cocoa nibs.

TABLE 15.8
Regression Coefficients and Their Adjusted R^2 Values in the Models for Total, Reducing and Non-Reducing Sugars in Cocoa Beans

Variables	Total Sugars	Non-Reducing Sugars	Reducing Sugars
Constant	17.944[a]	17.4849[a]	0.4932[a]
X_1	0.2993[a]	0.2125[a]	0.0912[a]
X_2	−0.1725[a]	−0.2234[a]	0.0498[a]
X_1^2	−0.0205[a]	−0.0168[a]	−0.0038
X_2^2	0.0006[a]	0.0007[a]	−1.6208[a]
$X_1{}^aX_2$	1.0416	0.0007	−7.3645[a]
R^2	0.946	0.973	0.800

[a] Significant at $p < 0.05$; X_1 = Pod storage; X_2 = Fermentation time.

The unfermented beans had the highest levels of total sugars (~18 mg/g) and showed only slight variations with increasing pod storage. Increasing fermentation times caused consistent decreases in total sugars with all pod storage days. However, at all fermentation times, increasing pod storage resulted in only marginal reduction in total sugar concentrations, with the greatest reduction noted between 14 to 21 days during the later days of fermentation (4–6 days; Figure 15.9). These observations explain that the rate of total sugar generation in the cocoa beans is largely affected by fermentation rather than by pod storage.

About 15–18 mg/g of total sugars were present at the initial stages of fermentation (0–2 days). With increasing fermentation time at shorter pod storage days (0–7 days), total sugars slightly fell from about 18 mg/g to 10 mg/g. However, it was reduced significantly after fermentation time was extended up to 6 days at longer pod storage days (14–21 days). Total sugars were found to be higher at shorter pod storage days (0–7 days) at all fermentation times.

The response surface plotting also shows that the lowest total sugars were present in nibs which were stored for 14–21 days and fermented for 4–6 days. These reductions in total sugars were probably caused by the action of native cocoa seed invertase which hydrolyses mainly non-reducing sugars (sucrose) into reducing sugars (glucose and fructose) during fermentation of the pulp pre-conditioned cocoa beans (Hansen et al. 1998).

15.4.2.5 Changes in Reducing Sugars

Figure 15.10 depicts the response surface plot of the effect of pod storage and fermentation time on reducing sugars. Regression analysis on the data showed that both the linear and quadratic effects of pod storage and the

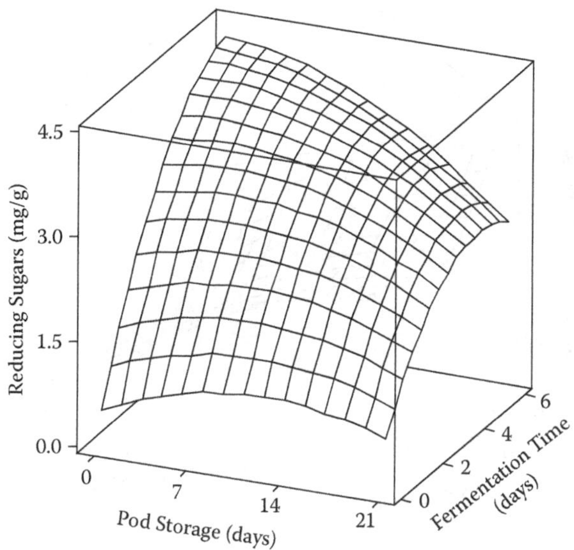

FIGURE 15.10 Response surface plot showing effect of pod storage and fermentation time on the reducing sugars of cocoa nibs.

linear effect of fermentation time were significant at $p < 0.05$. Interaction between pod storage and fermentation time was also significant ($p < 0.05$). The model could explain about 80% of the variations that occurred in the reducing sugars (Table 15.8). Generally, increasing pod storage had no observable influence on the reducing sugars and this observation was noted till the fourth day of fermentation, beyond which increasing pod storage from 0 to 21 caused a slight consistent reduction on reducing sugar concentration and this continued till the end of fermentation. On the other hand, increasing fermentation had a consistent increasing effect on reducing sugars at all pod storage periods (Figure 15.10). The increasing reducing sugar concentration with increasing fermentation might be due to hydrolysis of sucrose to glucose and fructose by the enzyme invertase (Hansen et al., 1998). The highest reducing sugar content was attained in samples that had been pulp pre-conditioned for up to 7 days and fermented for 6 days. Samples that had been pulp pre-conditioned for 14 to 21 days showed a relatively reduced rate of reducing sugar increase which levelled up after 4 days of fermentation till the end of fermentation. This suggests that cocoa beans that have been pulp pre-conditioned for 14 days and beyond impose a decreasing effect on the reducing sugar build-up during the latter stages of the fermentation period and this might negatively influence the flavour precursor formation and development during subsequent drying and roasting.

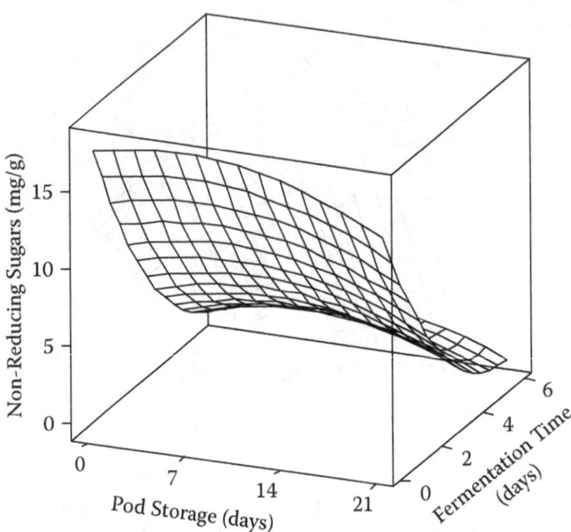

FIGURE 15.11 Response surface plot showing effect of pod storage and fermentation time on the non-reducing sugars of cocoa nibs.

15.4.2.6 Changes in Non-Reducing Sugars

A close examination of the response surface plot (Figure 15.11) shows a similar trend as in the total sugars. Regression analysis on the data showed that both the linear and quadratic effects of pod storage and fermentation time were significant at $p < 0.05$. However, the interaction effect was not significant ($p > 0.05$; Table 15.8). The R^2 value obtained was 0.973, indicating that the pod storage days and fermentation time could account for 97.3% of the changes in non-reducing sugars, suggesting only 2.7% were due to other factors not investigated in this work. The unfermented beans had the highest levels of non-reducing sugars (~16–17 mg/g) and showed only slight variations with increasing pod storage. Increasing fermentation times led to consistent and sharp decreases in non-reducing sugars with all pod storage treatments (Figure 15.11).

However, the rate of decrease was sharp within 4 days of fermentation after which the non-reducing sugar levelled up with no further decreases till the end of fermentation. On the other hand, increasing pod storage resulted in only marginal reduction in non-reducing sugar concentrations at all fermentation times, with the greatest reduction occurring in samples pulp pre-conditioned for 14 to 21 days during the later (4–6) days of fermentation (Figure 15.11). These suggest that the amount of non-reducing sugars generated in cocoa beans is largely influenced by fermentation rather than by pod storage. The sharp decreases in non-reducing sugars with increasing fermentation could be explained by the hydrolyses of

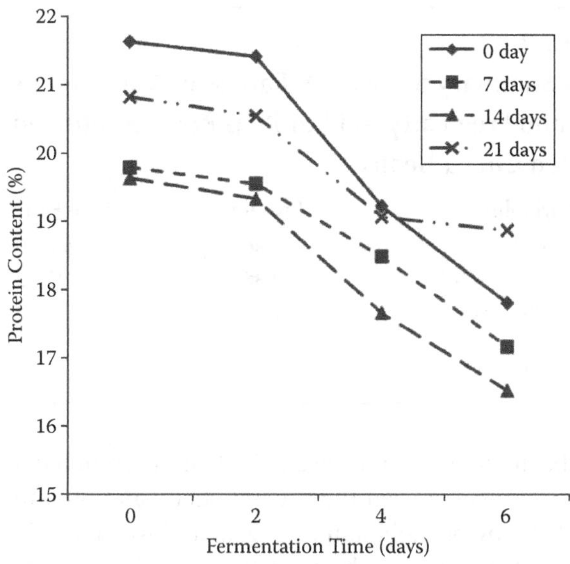

FIGURE 15.12 Effect of pod storage and fermentation time on protein content.

sucrose by the enzyme invertase which is inactivated after 4 days of fermentation (Hansen et al. 1998).

15.4.2.7 Changes in Protein Content

Voigt et al. (1993, 1994a) explained that in addition to reducing sugars, peptides and hydrophobic free amino acids are the cocoa-specific flavour precursors. The sources of these precursors are certainly numerous; however, seed protein degradation and release of peptides and free amino acids are likely to be amongst the most important processes for the formation of flavour precursors.

Figure 15.12 shows the effect of fermentation and pod storage on protein degradation of cocoa beans. Protein content irrespective of all the different pod storage treatments reduced only marginally during the first two days of fermentation, but beyond the second day there were sharp decreases in protein content for all the samples until the end of fermentation. This observation suggests that proteolysis started after the death of the bean, which occurs after 48 hours of fermentation (Voigt et al. 1994a; Afoakwa and Paterson 2010) and is associated with a phase of cellular breakdown, loss of compartmentalisation and water absorption, all of which would be expected to facilitate breakdown of proteins by proteases.

Opinions have differed, however, as to the start of this process (proteolysis), which is suspected to begin before the death of the embryo (Barel 1986), or only afterwards (Biehl et al. 1982a; Crouzillat et al. 1999). The process of decompartmentalization accompanying protein degradation,

TABLE 15.9
ANOVA Summary Showing F-Ratios for Variation in Protein and Free Fatty Acid of Pulp Pre-Conditioned Fermented Cocoa Beans

Variables	Protein	Free Fatty Acid
Pod storage (PS)	16.86[a]	246.50[a]
Fermentation time (FT)	32.36[a]	42.62[a]
Interaction (PS × FT)	1.84	35.37[a]

[a] Significant at $p < 0.05$.

however, has been reported to release polyphenolic compounds, thus making possible reactions between these compounds and proteins, rendering the latter insoluble by phenolic tanning (Zak and Keeney 1976).

Protein content was also reduced with increasing pod storage except for the beans stored for 21 days, which showed relatively higher values (Figure 15.12). The slight reduction in protein content during pod storage (pulp pre-conditioning) is suspected to result from the action of protease enzymes in the pods during storage and thus initiate the process of proteolysis. This observed protein degradation occurring during pulp pre-conditioning of the beans might initiate the release of peptides and free amino acids which could influence the processes for the formation of flavour precursors in the bean during subsequent fermentation and drying. Statistical analysis on the data revealed that both pod storage and fermentation time significantly ($p < 0.05$) affected the protein content of the cocoa beans (Table 15.9).

15.4.2.8 Free Fatty Acid

In the assessment of cocoa quality, the fat yield and physical characteristics of the fat are very important parameters especially for cocoa butter extraction. The physical characteristics of the fat are assessed by measurements of the fat hardness and the free fatty acid content (Knight 2000). The level of free fatty acids (FFA) in the fat of cocoa beans measures the rancidity of the cocoa, and high levels of FFA in cocoa are not acceptable. Fresh cocoa beans should not contain more than 1% and certainly not more than 1.75% of FFA in the dried beans (Dand 1997). This limit has been imposed because higher levels of FFA indicate that hydrolysis of triglycerides occurs which results in softening of the butter.

Generally, it was expected that FFA would increase with increasing fermentation and pod storage but this was not observed. Figure 15.13 showed that samples from the pods stored for 7 and 14 days recorded the lowest amounts (0.43 and 0.44% FFA, respectively). However, the pods stored for

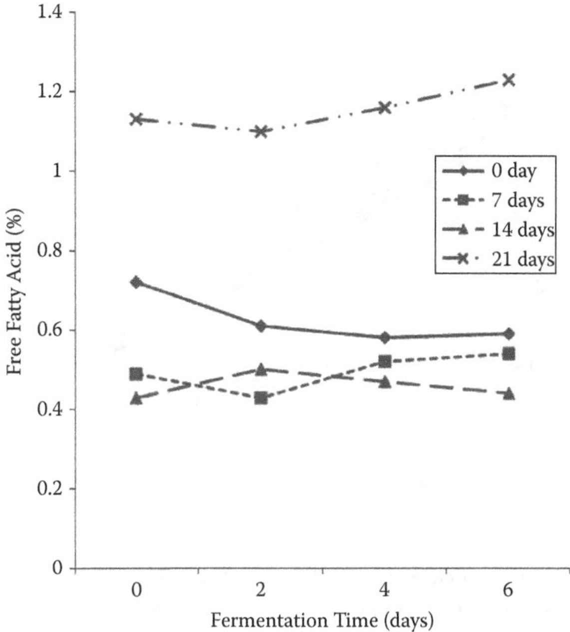

FIGURE 15.13 Changes in free fatty acid as affected by pod storage (pulp pre-conditioning) and fermentation.

21 days recorded the highest FFA value (1.13%) and this high level of FFA could have resulted from prolonged pod storage and mould infestation. The higher FFA values observed in samples pulp pre-conditioned for 21 days might have resulted from the action of lipase enzyme, which is not only introduced into the beans by the microbiological activity but is also present in the natural cocoa and acts to break down the triglycerides into separate groups of the fatty acids and glycerol thereby freeing the fatty acids (Dand 1997), resulting in rancidity of the product.

Fermentation time caused only minimal changes in the free fatty acid levels of the beans with all levels of pulp pre-conditioned treatments (Figure 15.13). Analysis of variance on the data revealed that both pod storage and fermentation time significantly ($p < 0.05$) influenced the free fatty acid levels of the cocoa beans (Table 15.9).

15.4.3 Effect of Pulp Pre-Conditioning and Fermentation on Polyphenolic Compound Concentrations

15.4.3.1 Total Polyphenols

Cocoa bean polyphenols, comprising 12–18% of the whole bean weight, have long been associated with the flavour and colour of chocolate (Kim

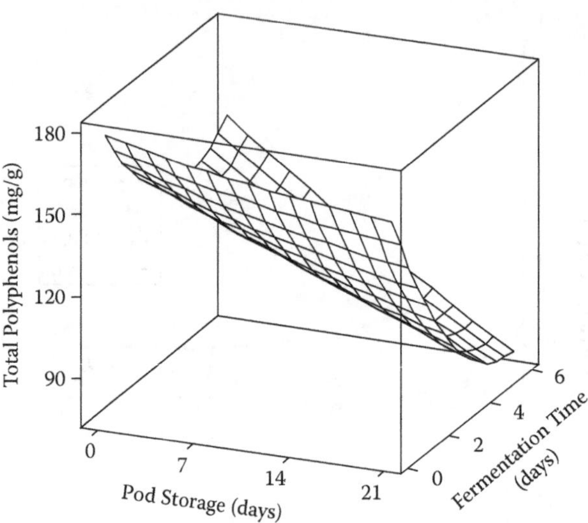

FIGURE 15.14 Response surface plot showing effect of pod storage and fermentation time on the total polyphenols of cocoa nibs.

and Keeney 1984). Polyphenols impart an astringent taste to cocoa beans, which is perceived as a dry feeling in the mouth along with a coarse puckering of the oral tissue. This taste has been associated with the effect of polyphenol–protein interaction in the saliva to form precipitates or aggregates. In this study, the total polyphenolic content of the unstored cocoa samples ranged from 190.87 mg/g to 140.34 mg/g.

A graphical representation of the regression model is as shown in Figure 15.14. Polyphenol content reduced slightly with pod storage at the beginning of fermentation. However, as fermentation time increased, the total polyphenol content decreased sharply with increasing pod storage. Fermentation of cocoa beans is crucial for the development of the precursor for chocolate flavour. Complex interactions among polyphenols to form high molecular weight tannins and their interactions with protein have an impact on the overall quality of fermented cocoa beans for chocolate production. During fermentation of cocoa beans, polyphenols diffuse with cell liquids from their storage cells and are oxidised enzymatically by the polphenol oxidase to condensed high molecular, mostly insoluble tannins. The sharp decreases observed (Figure 15.14) reflect the onset of these phenomena during fermentation. Thus, astringency and bitterness associated with polyphenols may be reduced drastically by the combined effect of pod storage and fermentation prior to chocolate manufacturing inasmuch as roasting and conching, as well as tempering and other processes during chocolate manufacturing, cannot remove the excessive amount of astringency in cocoa beans (Fowler 1995; Afoakwa et al. 2008).

TABLE 15.10
Regression Coefficients and Their Adjusted R^2 Values in the Models for Total Polyphenols, o-Diphenol and Anthocyanins in Cocoa Beans

	Coefficients		
Variables	**Total Polyphenols**	***o*-Diphenols**	**Anthocyanins**
Constant	177.817[a]	24.2500[a]	17.0108[a]
X_1	−1.1937[a]	−0.00958[a]	−0.17877
X_2	−0.8028[a]	−0.09981[a]	−5.14831[a]
X_1^2	0.02143	0.00438	0.00497
X_2^2	0.0042[a]	0.00054[a]	0.536797[a]
X_1*X_2	−0.0182[a]	−0.00380[a]	−0.00964
R^2	0.851	0.855	0.963

[a] Significant at $p < 0.05$; X_1 = Pod storage; X_2 = Fermentation time.

The model developed to predict the polyphenol content had an R^2 of 0.85 implying that it could explain 85% of the variations in polyphenol content. The regression coefficients showed that the linear terms of pod storage and fermentation time as well as the quadratic term of fermentation time had a significant influence on polyphenol content (Table 15.10). There was also a significant ($p < 0.05$) interaction between pod storage and fermentation time (Table 15.10). The implication of this finding is that total polyphenol content at every fermentation time depends on the duration of pod storage.

15.4.3.2 *o*-Diphenols

Polyphenol oxidase is a copper-dependent enzyme that, in the presence of oxygen, catalyses two different reactions: the hydroxylation of monophenols to *o*-diphenols (monophenolase activity) and the oxidation of *o*-diphenols to *o*-quinones (diphenolase activity) (Wollgast and Anklam 2000). In living tissues, the phenolic substrates and the enzymes are separated within the cell, but upon cell damage (in cocoa beans due to acetic acid penetration) the enzyme and substrate may come into contact, leading to rapid oxidation of *o*-diphenols (catechins and epicatechins) and further complexation of *o*-quinones (Camu et al. 2008).

There was a significant ($p < 0.05$) influence of the linear factors of pod storage and fermentation time, but only the quadratic factor of fermentation time was significant ($p < 0.05$) on the *o*-diphenol content of the beans. The model could explain 85% of the variations in *o*-diphenol content meaning only 15% of the variations were due to other factors which were not included in the model (Table 15.10).

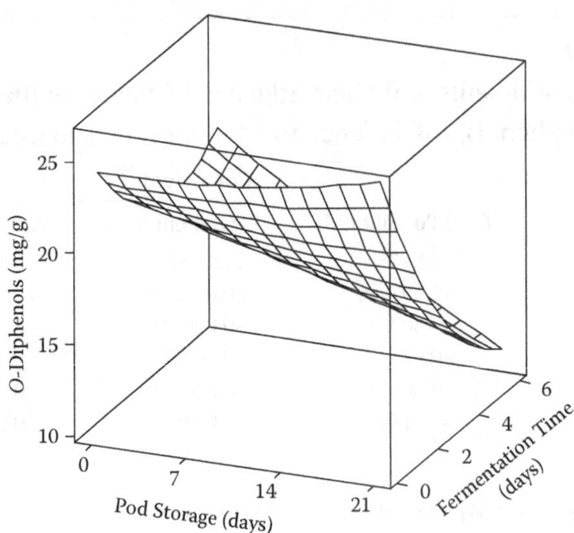

FIGURE 15.15 Response surface plot showing effect of pod storage and fermentation time on *o*-diphenols of cocoa nibs.

The response surface plot generated (Figure 15.15) showed that pod storage and fermentation time all had a significant effect on *o*-diphenol content of the cocoa beans with significant interaction between these factors. It showed a curvilinear plot with fermentation time. This implies that the *o*-diphenol content reduced considerably during the fermentation and increased slightly by the end of the fermentation. Increased pod storage and increased fermentation time caused drastic reductions in the content of *o*-diphenol compounds (catechin and epicatechin) in the beans. Nazaruddin et al. (2006b) reported that the reduction in pulp volume during pod storage may have facilitated the oxidation and polymerisation of *o*-diphenols and its oxidation products as clearly demonstrated by the beans from 14- and 21-day pod storage. The occurrence of condensation reactions is confirmed by the sharp decrease of *o*-diphenol content between the second and sixth day of fermentation for beans stored for 14 and 21 days (Figure 15.15). Epicatechin and catechin content are reduced to approximately 10–70% during fermentation. This is not only due to the oxidation process but is also caused by diffusing of polyphenols into fermentation sweatings (Kim and Keeney 1984).

An earlier study on the effect of pod storage on the content of (−)-epicatechin from different clones also demonstrated almost the same pattern (Clapperton et al. 1992). The reduction in the *o*-diphenols might help to lower the level of astringency in the cocoa beans. Previous work by other

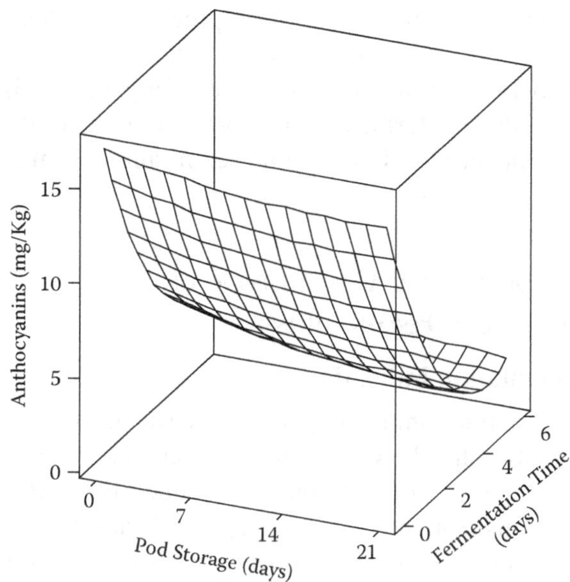

FIGURE 15.16 Response surface plot showing effect of pod storage and fermentation time on anthocyanin content of cocoa nibs.

researchers also shows a similar trend (Meyer et al. 1989; Biehl et al. 1990; Nazaruddin et al. 2006b).

15.4.3.3 Anthocyanins

Among the polyphenols, anthocyanins give the purple colour to unfermented *Forastero* beans, but are absent from *Criollo* beans (Rohan 1963; Wood 1983). The model developed for predicting anthocyanin content could explain 96.3% of the variation observed in this index. The variables which significantly affected the anthocyanin content of the cocoa bean were the linear and quadratic terms of fermentation time (Table 15.10). Significant interaction between pod storage and fermentation time was not found and these findings give an indication that anthocyanin content of pulp pre-conditioned fermented beans is dependent only on fermentation time.

The response surface plots generated (Figure 15.16) showed that the anthocyanin content of the pulp pre-conditioned cocoa beans decreased drastically with increasing fermentation at all pod storage levels. This drastic reduction in anthocyanin content occurred till the fourth day of fermentation, beyond which no considerable changes in anthocyanin content were observed for all the pod storage treatments. Lange and Fincke (1970), Pettipher (1986), Wollgast and Anklam (2000) and Nazaruddin et al. (2006a) have all reported that anthocyanins usually disappear rapidly during the fermentation process (93% loss after four days). The anthocyanins are hydrolysed to anthocyanidins with the latter compounds being polymerised along with simple catechins to form complex tannins during fermentation. Pod storage caused only marginal and insignificant ($p > 0.05$) decreases in anthocyanin content at all periods of fermentation.

15.4.4 Effect of Pulp Pre-Conditioning on the Degree of Fermentation

15.4.4.1 Fermentation Index (FI)

The hydrolysis of anthocyanin is important in fermentation because there is an inverse relationship between flavour development and the degree of purple colour retained after fermentation (Lopez 1986; Misnawi et al. 2003). Anthocyanins usually disappear rapidly during the fermentation process. Thus, anthocyanin content has been considered as a good index for determination of the degree of cocoa bean fermentation (Lange and Fincke 1970; Pettipher 1986; Wollgast and Anklam 2000). The fermentation index can be used to assess the fermentation level of cocoa beans. Being a quantitative measure unlike the cut test, the FI is unbiased by human factors. Table 15.11 shows the mean data of FI from different days of pod storage and fermentation. A cocoa bean is considered under-fermented when it yields an FI of less than 1.0 and any values above this are considered fully fermented (Gourieva and Tserrevitinov, 1979).

Results of spectral measurements (Table 15.11) showed that cocoa beans for all the pod storage treatments prior to fermentation (unfermented beans) were dominated with compounds having maximum absorption at 530 nm attributed to the presence of anthocyanin pigments. As fermentation progressed, more and more condensation products of anthocyanins, such as cyanidin-3-β-D-galactosid and cyanidin-3-α-L-arabinosid were formed (Kim and Keeney 1984), as shown by increasing absorbance at 460 nm but decreasing absorbance at 530 nm (Table 15.11). These oxidation products of the polyphenol oxidase activities contribute the brown pigments in the cocoa beans.

Furthermore, the brown pigments may also be produced from complexation of condensed tannin, a high molecular weight product of flavonoid polymerisation, with protein, via hydrogen bonding (Shamsuddin and Dimmick 1986). The table also shows a rapid change in FI during the first 4 days of fermentation, which subsequently slowed down because the condensation product became less soluble. Analysis of variance showed that pod storage and fermentation time as well as the interaction of these factors had a significant ($p < 0.05$) effect on the fermentation index of the cocoa beans (Table 15.12). The multiple range test revealed that the different pod

TABLE 15.11
Effect of Fermentation Time and Pod Storage on Colour Fractions Absorbance Value and Fermentation Index

Pod Storage (Days)	Fermentation Time (hr)	Colour Fractions Absorbance Value		Fermentation Index (460/530)
		Fraction I (530 nm)	Fraction II (460 nm)	
0	0	1.70	0.63	0.372
	2	0.80	0.56	0.706
	4	0.65	0.50	0.774
	6	0.44	0.46	1.050
7	0	1.54	0.61	0.399
	2	0.57	0.42	0.751
	4	0.52	0.41	0.779
	6	0.39	0.41	1.052
14	0	1.65	0.61	0.368
	2	0.60	0.40	0.661
	4	0.40	0.41	1.025
	6	0.23	0.29	1.278
21	0	1.54	0.59	0.382
	2	0.58	0.36	0.619
	4	0.28	0.29	1.040
	6	0.29	0.28	0.962

storage treatments were significantly ($p < 0.05$) different from each other. Overall, pod storage of more than 7 to 14 days was seen to have a FI value of more than one. Compared to other treatments, this treatment is capable of producing well-fermented beans.

TABLE 15.12
ANOVA Summary Showing F-Ratios for Variation in Fermentation Index and Colour of Pulp Pre-Conditioned Fermented Samples

Variables	Fermentation Index	L[a]	a[a]	b[a]
Pod storage	1766.27[a]	116598.79[a]	48375.59[a]	862404.00[a]
Fermentation time	231111.63[a]	662340.54[a]	335066.84[a]	1237338.00[a]
Interaction	7487.42[a]	42428.22[a]	103.41[a]	155220.22[a]

[a] Significant at $p < 0.05$.

TABLE 15.13
Colour of Cocoa Beans in Relation to Pod Storage Days and Fermentation Time

Pod Storage (Days)	Fermentation Time (Days)	L*	a*	b*
0	0	46.23 ± 0.35	12.10 ± 0.43	6.45 ± 0.52
	2	45.35 ± 0.11	11.95 ± 0.08	9.03 ± 0.03
	4	44.79 ± 0.41	9.23 ± 0.09	8.19 ± 0.14
	6	44.09 ± 0.48	8.69 ± 0.114	9.00 ± 0.11
7	0	46.80 ± 0.54	12.09 ± 0.17	7.90 ± 0.09
	2	44.39 ± 0.14	12.91 ± 0.04	9.11 ± 0.03
	4	44.26 ± 0.04	10.12 ± 0.02	9.82 ± 0.05
	6	41.19 ± 0.23	6.89 ± 0.04	13.26 ± 0.04
14	0	46.53 ± 0.53	11.54 ± 0.03	5.86 ± 0.09
	2	44.60 ± 0.66	9.80 ± 0.05	10.61 ± 0.20
	4	43.60 ± 0.01	8.90 ± 0.02	11.69 ± 0.02
	6	40.73 ± 0.05	5.88 ± 0.02	16.50 ± 0.02
21	0	46.40 ± 0.34	11.07 ± 0.06	8.16 ± 0.18
	2	44.84 ± 0.25	9.98 ± 0.03	13.23 ± 0.38
	4	41.27 ± 0.10	9.39 ± 0.05	17.44 ± 0.05
	6	38.92 ± 0.42	5.80 ± 0.04	16.02 ± 0.06

15.4.4.2 Colour

Colour is one of the most important quality attributes of a food. This is because no matter how nutritious, flavoured or well-textured a food product is, it is unlikely to be accepted unless it has the right colour (Serna-Saldivar, Gomez, and Rooney 1990). Results of the tristimulus colour measurements on cocoa powders obtained from the pulp pre-conditioned fermented cocoa beans (Table 15.13) showed that the lightness of the cocoa samples ranged from 44.09 to 46.23 for the cocoa beans that were not stored prior to fermentation. Generally, the samples became darker as fermentation increased and lightness also decreased progressively with increasing pod storage. This may probably be due to the destruction of the anthocyanins by enzymic hydrolyses, which is accompanied by the bleaching of the beans and a subsequent browning reaction. Multiple range tests showed that the different pod storage treatments (0, 7, 14, 21 days) were statistically different from each other with beans stored for 21 days being the darkest.

The a-value, however, decreased progressively during fermentation, and higher decremental rates were observed during the fourth to sixth day of fermentation for all the pod storage treatments. This decrease could be attributed to the breakdown of anthocyanins which normally impart the purple colour to under-fermented beans. It can also be observed (Tables 15.2 and 15.3) that as pulp pre-conditioning increased, the redness of the samples decreased significantly ($p < 0.05$) as a result of the diffusion of the polyphenols along with cell liquids from their storage cells during pod storage.

The b-value of the cocoa beans increased as fermentation days increased and as pod storage days increased. Increase in b-value during fermentation could be due to the presence of oxidised polyphenols, as a result of enzymatic oxidation by polyphenol oxidase in the beans. For the four different pod storage treatments it can be observed from Table 15.13 that at the end of fermentation, cocoa beans became darker, more yellowish but less reddish. However, these changes became more prominent as the pod storage days increased, thus it can be inferred that pulp pre-conditioning influenced the fermentation of cocoa beans and resulted in the formation of darker beans. This agrees with the spectral measurement (Table 15.11) in which the spectral changes were higher in the pulp pre-conditioned fermented beans than the unfermented beans.

15.4.4.3 Cut Test

The cut test is the standard method of assessing quality as defined in grade standards and can be used to estimate two major off-flavours (mouldy and unfermented beans). It also identifies other defects which can affect the keeping quality. In addition the cut test measures the degree of fermentation (Wood and Lass 1985). The cut test was carried out on cocoa beans that had been fermented for four and six days based on the results obtained for the fermentation index, which illustrated that beans were fermented between the fourth and sixth day.

A number of defects were detected in the beans that were stored for 14 and 21 days and fermented. Germination occurred in beans stored for 14 and 21 days prior to fermentation; it was, however, lesser in the 14-day pod storage. The incidence of germination occurred because of the prolonged storage of the pods which resulted in the rotting of pods and consequently penetration of oxygen into the pods creating optimum conditions for growth of the beans.

By 21 days of pod storage, the proportion of germinated beans had increased considerably and this increased further during fermentation. It was expected that the heat produced during fermentation coupled with the diffusion of some metabolites (ethanol and acetic acid) into the bean could result in the death of the bean, hence arresting germination; however, enough acids were not generated in the beans as shown in the physicochemical analysis of the beans (Figures 15.6 and 15.7).

This suggests that beans stored for 21 days were not adequately fermented. Germinated beans are considered a defect because the hole left by the emerging radicle provides an easy entrance for insects and moulds. They are also considered to lack good chocolate flavour (Wood and Lass 1985). Slaty beans were absent in the fermented beans for all the pod storage days.

Mouldy beans (18%) were also detected in the beans stored for 21 days before fermenting. According to Wood and Lass (1985), internal moulds are the most important causes of off-flavours during manufacture, and samples of beans with as little as 4% of beans with internal moulds can produce off-flavours. Beans stored for 21 days exceeded this limit; this could be ascribed to the invasion of mould species (*phytophthora palminovora, Botryodiplodia theobrommae* etc.) during the prolonged storage, and beans may produce off-flavours when roasted. Moulds inside the beans can also increase the free fatty acid (FFA) content of the cocoa butter (Wood and Lass 1985) and this may have accounted for the high FFA values obtained for the beans stored for 21 days (Figure 15.13).

In order to assess the degree of fermentation, the cut beans were divided into four categories, fully fermented partly brown/partly purple, fully fermented and slaty. The proportion of partly purple and brown beans did not exceed 50% for all the pod storage days and this gives an indication that the beans were adequately fermented and may not give rise to bitter and astringent flavours. Beans described as 'partly brown, partly purple' are not defective and should be present at least to the extent of 20% (Wood and Lass 1985). The 14 and 21 days pod storage had their beans falling within this range whereas the 0 and 7 days pod storage were higher (27–38%); however, this was within the acceptable range (30–40%). For good cocoa flavour development the degree of fermentation (percentage of fully brown beans) should be above 60%. Beans stored for 7 days and beyond fell within this range and this is in agreement with spectral and tristimulus measurements (Table 15.14).

15.5 CONCLUSIONS

Based on the findings of the study, the following conclusions were made:

1. Pod storage and fermentation influenced to varied levels the nutritional and physical composition of cocoa beans. Increasing pod storage and fermentation caused increases in copper and carbohydrate content and reductions in Mg, K and proteins. Amongst the minerals studied, potassium was the most abundant mineral followed by magnesium, phosphorus and calcium

TABLE 15.14
Effect of Fermentation Time and Pod Storage on Dried Cocoa Beans Surface Colour and Cut Test Score

Pod Storage (Days)	Fermentation Time (Days)	Purple (%)	Purple/ Brown (%)	Brown (%)	Slaty (%)	Germinated (%)	Mouldy (%)
0	4	21	38	41	—	—	—
	6	10	32	58	—	—	—
7	4	11	30	59	—	—	—
	6	6	27	67	—	—	—
14	4	1	21	77	—	2	—
	6	—	20	80	—	—	—
21	4	—	1	88	—	11	3
	6	—	1	76	—	18	5

in both the fermented and unfermented cocoa beans. Proportion of cocoa nibs (fermentation yield) also increased with increasing pod storage and fermentation.

2. Fermentation caused a slight reduction in pH of the bean within the first 48 h after which the pH increased consistently until the end of fermentation at all pod storage periods. This was contrary to the trends observed for acidity. Increasing pod storage led to consistent increases in pH with concomitant decreases in acidity at all fermentation times. For optimum acid generation Ghanaian cocoa pods should not be stored beyond 7 days prior to fermentation as this leads to reduced acidity beyond acceptable limits.
3. The biochemical constituents (protein, total sugars and non-reducing sugars) decreased consistently during fermentation, whereas reducing sugars increased. With the exception of protein which reduced, pod storage did not have any effect on the sugars. The rate of total and non-reducing sugars degeneration with concomitant generation of reducing sugars in the cocoa beans is largely affected by fermentation rather than by pod storage.
4. Polyphenolic composition (total polyphenols, *o*-diphenol and anthocyanin) reduced consistently with pod storage and fermentation time, suggesting that the combined effects of pod storage and fermentation could be employed to drastically reduce the astringency and bitterness in cocoa and cocoa products.
5. Fermentation index of ~1.0 (well-fermented level) was attained by all pod stored samples after 6 days of fermentation. Pods stored for 14 and 21 days, however, attained the fermentation index of 1.0

after 4 days. Cocoa beans pulp pre-conditioned for 14 and 21 days were over 60% fully brown by the fourth day of fermentation suggesting that pulp pre-conditioning by pod storage for over 14 days could be used to reduce fermentation time of cocoa beans from 6–7 days to 4 days without any effects on the physical qualities of the dried beans.

16 Effects of Fermentation and Reduced Pod Storage on Cocoa Pulp and Cocoa Bean Quality

16.1 RESEARCH SUMMARY AND RELEVANCE

Cocoa fermentation and drying are influenced by factors such as pod storage and duration of the fermentation and drying processes. Understanding the factors contributing to variations in the qualities of cocoa beans during short periods of pod storage and subsequent fermentation and drying would have significant commercial implications. This study investigated changes in physicochemical constituents, flavour precursors and polyphenolic constituents during fermentation and drying of pulp pre-conditioned Ghanaian cocoa beans. A 4 × 3 full factorial experimental design was used with pod storage (0, 3, 7 and 10 days) and fermentation time (0, 3 and 6 days) being the principal factors investigated. The study also used a 4 × 3 full factorial design with pod storage (0, 3, 7 and 10 days) and drying time (0, 3 and 7 days) being the principal factors investigated. The physicochemical constituents and the mineral composition of the pulp were studied. Again, the physicochemical and polyphenolic constituents of the beans were studied. The fermentation index (FI) and cut test were done to evaluate the fermentative quality of the beans. All analyses were conducted using standard methods.

The pH of the pulp increased significantly ($p < 0.05$) with increasing pod storage and fermentation with a consequential decrease in titratable acidity. Pod storage and fermentation also decreased the reducing sugars and total solids of the pulp significantly ($p < 0.05$). The most abundant mineral in unfermented cocoa pulp was calcium, followed by potassium and sodium with values of 316.92 mg/100 g, 255.12 mg/100 g and 103.26 mg/100 g, respectively. Zinc was the mineral with the least concentration of 1.04 mg/100 g, and iron and magnesium had appreciable values of 4.26 mg/100 g and 32.52 mg/100 g, respectively. Pod storage and fermentation, however, showed variable trends in the studied mineral content.

Increasing pod storage also caused consistent increase in pH of the nibs at the end of fermentation with consequential decrease in titratable acidity. Fermentation significantly ($p < 0.05$) decreased the non-reducing sugars, total sugars and protein content of the beans whereas reducing sugars increased. Pod storage caused marginal reductions in total and non-reducing sugars with a consequential increase in reducing sugars whereas protein content was reduced significantly. Protein, reducing sugars, non-reducing sugars and total sugars decreased significantly ($p < 0.05$) with increasing duration of drying at all pod storage periods.

Pod storage and fermentation significantly ($p < 0.05$) reduced total polyphenols, *o*-diphenols and anthocyanins content of the beans and were subsequently reduced significantly ($p < 0.05$) during drying. Pod storage, fermentation and drying significantly ($p < 0.05$) increased the free fatty acids (FFAs) of the nibs. FFAs of all the nibs were, however, below the acceptable limit of 1.75 oleic acid equivalent. Increasing pod storage and fermentation as well as drying significantly ($p < 0.05$) influenced the fermentative quality (cut test and fermentation index) of the beans. The fermentation index of the beans increased significantly with pod storage and fermentation but decreased slightly during drying. The fermentation index of all the beans was, however, above 1.0 at the end of drying for all pod storage treatments. The cut test revealed that storage of pods for 3, 7 and 10 days increased the percentage of brown beans by 66%, 94% and 72%, respectively, by the sixth day of fermentation. The percentage of brown beans decreased to 61%, 76% and 63%, respectively, for pods stored for 3, 7 and 10 days at the end of drying (7 days). Pod storage between 3–7 days with 6 days of fermentation and 7 days drying could be used to produce cocoa beans with higher concentrations of flavour precursors and also reduced nib acidification.

16.2 INTRODUCTION

Raw cocoa beans have an astringent and unpleasant taste and have to be fermented, dried and roasted to obtain the characteristic cocoa taste and flavour. The quality and flavour of cocoa beans depends on complex chemical and biochemical changes which occur in the beans during fermentation and drying. Fermentation generates flavour precursors, namely free amino acids and peptides from enzymatic degradation of cocoa proteins, and reducing sugars from enzymatic degradation of sucrose in cocoa (Misnawi 2008). In addition to the formation of flavour precursors, phenolic compounds are oxidised and polymerised to insoluble high molecular weight compounds (tannins) leading to a significant reduction of its concentration and thus reducing the bitterness and astringency of the final product to acceptable levels.

Traditionally, cocoa bean fermentation is a spontaneous process initiated by micro-organisms naturally occurring at fermentation sites, including yeasts, lactic and acetic acid bacteria, bacilli, and filamentous fungi (Schwan 1998). Yeasts and lactic acid bacteria consume pulp sugars and organic acids, producing ethanol and lactic acid. Acetic acid bacteria then oxidise the ethanol produced by the yeasts into acetic acid through an exothermal process which gradually increases the temperature of the fermenting seed mass, which can reach values close to 50°C (Schwan and Wheals 2004). Acetic acid diffuses into the seeds and, in combination with the high temperature, causes the death of the seed embryos, disrupting their cellular integrity (Voigt et al. 1994a) and inducing the complex chemical and biochemical changes inside the beans leading to well-fermented cocoa beans. After fermentation, the beans are then dried to reduce the moisture content from about 60% to between 6–8% (Nair 2010) to prevent mould infestation during storage and also allow some of the chemical changes which occurred during fermentation to continue and improve flavour development (Kyi et al. 2005).

Cocoa fermentation is, however, influenced by many factors such as type of cocoa, disease, climatic and seasonal differences (Afoakwa 2010), turning, batch size (Lehrian and Patterson 1983), quantity of beans (Said and Samarakhody 1984; Wood and Lass, 1985) and also pulp pre-conditioning (Meyer et al. 1989). Ostovar and Keeney (1973) reported that the pulp is the substrate metabolised by a sequence of micro-organisms during fermentation, and Afoakwa et al. (2011b) concluded that inasmuch as the properties of the substrate determine microbial development and metabolism, changes in the pulp may affect the production of acids by lactic acid bacteria, yeasts and acetic acid bacteria.

Pulp pre-conditioning therefore involves changing the properties of the pulp prior to the development of micro-organisms in fermentation (Afoakwa et al. 2011b). These changes may be in the form of altering the moisture content of the pulp, sugar content, and volume of pulp per seed as well as pH and acidity of the pulp. Pulp pre-conditioning can be done in three basic ways prior to fermentation and these are pod storage, mechanical or enzymatic depulping and bean spreading (Wood and Lass 1985; Biehl et al. 1989; Schwan and Wheals 2004).

Pod storage is basically storing the harvested cocoa pods for a period of time before opening the pods and fermenting the beans. Harvested cocoa pods are still living tissues undergoing certain physiological processes such as respiration which involves the breakdown of sugars to produce energy. The pods also undergo transpiration leading to the loss of water. Pod storage therefore serves to reduce the pulp volume per seed due to water evaporation and inversion of sucrose; reduces total sugar content and increases micro-aeration within the pulp and eventually reduces alcohol fermentation and acetic acid formation (Biehl et al. 1989; Sanagi, Hung, and Yasir 1997). Pod storage is reported to have a high beneficial effect on the chemical composition of cocoa beans and subsequent development of chocolate flavour (Afoakwa et al. 2011b). Pulp pre-conditioning by post-harvest storage of Malaysian cocoa pods is reported to cause the reduction of nib acidification during subsequent fermentation, reduction of the polyphenol content and an increase in cocoa flavour of the fermented beans (Meyer et al. 1989; Nazaruddin et al. 2006a). Afoakwa et al. (2011a) also reported that increasing pod storage (PS) consistently decreased the non-volatile acidity with a concomitant increase in pH during fermentation of the beans.

16.2.1 Rationale

Cocoa is one of Ghana's major export commodities, contributing 26.3% of export revenue and 26% of agricultural growth in 2006 (Baah and Anchirinah 2011). The Institute of Statistical, Social and Economic Research (ISSER; 2008) reported the subsector also grew quite significantly at 6.5% in 2007 compared to only 2.0% in 2006. Currently, Ghana is ranked as the second highest world producer of cocoa but in terms of quality cocoa beans, Ghana's cocoa beans are rated premium quality and are used as standards against cocoa beans from other producing countries. Cocoa fermentation and drying are crucial for the development of quality beans and these are influenced by factors such as pod storage and duration of the fermentation and drying process. Understanding the factors contributing to variations in the qualities of cocoa beans during pod storage and subsequent fermentation and drying would have significant commercial implications.

Preliminary work has reported on the influence of pod storage and fermentation on the chemical and physical quality characteristics, as well as total polyphenolic content and anthocyanin concentration in Ghanaian cocoa beans (Afoakwa et al. 2011a,b, 2012). However, the extent to which this technique of pulp pre-conditioning would influence the physicochemical constituents of the pulp during fermentation, formation of flavour precursors, changes in polyphenolic constituents and fermentative quality of Ghanaian cocoa beans during fermentation as well as during drying still remains unclear.

The main objective of this study was to investigate changes in physicochemical constituents, flavour precursors and polyphenolic constituents during fermentation and drying of pulp pre-conditioned Ghanaian cocoa beans.

16.2.2 Specific Objectives

The specific objectives of the study were

1. To investigate changes in physicochemical constituents and mineral composition of cocoa pulp during fermentation of pulp preconditioned cocoa beans.
2. To investigate changes in physicochemical constituents and flavour precursors during fermentation of pulp pre-conditioned cocoa beans.
3. To investigate changes in polyphenolic constituents and free fatty acids content during fermentation of pulp pre-conditioned Ghanaian cocoa beans.
4. To investigate the effects of pulp pre-conditioning and fermentation on the fermentative quality of Ghanaian cocoa beans.

16.3 MATERIALS AND METHODS

16.3.1 Material

Ripe cocoa pods (mixed hybrids) were obtained from the Cocoa Research Institute of Ghana (CRIG), Tafo-Akim, Eastern Region.

16.3.1.1 Sample Preparation

Cocoa pods of uniform ripeness were harvested by traditional methods (under ambient temperature during the day; 28–30°C) and transported to the fermentary where they were stored. The pods were divided into four (4) piles each containing about 300 pods and stored for 0, 3, 7 and 10 days under ambient conditions prior to fermentation. After these predetermined storage times, the cocoa pods were then cut open using a cutlass to extract the beans. The beans were withdrawn with the hands wearing gloves.

The basket fermentation method was used. About 30 kg of extracted cocoa beans were placed in woven baskets lined with banana leaves. The surfaces were also covered with banana leaves and fermented for 6 days with consecutive opening and turning every 48 h. Samples were taken at 0, 3, and 6 days into a sterile polyethylene bag and oven-dried for about 48 h at a temperature of 45–50°C until moisture content was between 7–8%. The dried beans were then bagged in airtight black plastic bags and stored at ambient temperature (25–28°C) in a dark room free from strong odours and used for analyses. Random sampling was done at the same time of the day and depth in the mass (40 to 80 cm from upper surface).

16.3.1.2 Preparation of the Pulp Samples

The pulp was manually separated from the beans by rubbing the beans (with adhering pulp) between fingers and squeezing the pulp into a clean sample bag. The pulp was then stored at –20°C prior to analyses.

16.3.1.3 Drying of Fermented Cocoa Beans

The fermented cocoa beans were dried in the open sun on raised platforms using the traditional process (Afoakwa 2010). Drying started at 8 a.m. and ended at 5 p.m. each day for 7 days. The beans were stirred four times each day and were covered with palm mats in the evening till the next morning. Samples were taken at 0 days (undried samples or immediately after fermentation), 3 days and 7 days of drying. The samples were then packaged in airtight plastic bags and taken to the laboratory for analyses. All the treatments were conducted in duplicate.

16.3.1.4 Experimental Design

A 4 × 3 full factorial experimental design was used for specific objectives 1–4. The principal factors investigated were

1. Pod storage time: 0, 3, 7 and 10 days
2. Fermentation time: 0, 3 and 6 days

The analyses conducted were

a. pH and titratable acidity of cocoa pulp
b. Reducing sugars and total solids of cocoa pulp
c. Mineral composition of cocoa pulp
d. pH and titratable acidity of cocoa beans
e. Total, reducing and non-reducing sugars of cocoa beans
f. Protein and free fatty acids content of cocoa beans
g. Total polyphenols, *o*-diphenols and anthocyanins content of cocoa beans
h. Fermentation index and cut test of the beans

16.3.2 Methods

16.3.2.1 pH and Titratable Acidity

The pH and titratable acidity of both the pulp and the beans were done as described by Nazaruddin et al. (2006a) with slight modifications. Ten grams of the pulp (or powdered cocoa beans) were homogenised in 90 ml of hot distilled water, stirred manually for 30 s and filtered using Whatman No. 4® filter paper and cooled to 20–25°C. A 25-ml aliquot of the resulting filtrate was pipetted into a beaker and the pH was measured using a pH metre (model MP230 Mettler Toledo MP 230, Mettler Company Limited, Geneva, Switzerland) calibrated with buffers at pH 4.01, 7.00 and 9.21. A further 10-ml aliquot was used to determine acidity by titration to an endpoint pH of 8.1 with 0.1 N NaOH solution and the values reported as moles of sodium hydroxide per 100 g sample. The analysis was conducted in triplicate and the mean values reported.

16.3.2.2 Determination of Reducing Sugars

Reducing sugars of both the pulp and cocoa beans were determined using the phenol sulphuric acid method as described by Brummer and Cui (2005) with slight modifications. About 0.5 g of defatted cocoa powder (or the pulp) was boiled in 30 ml 80% ethanol under reflux for 30 min. The supernatant was decanted into another round-bottom flask and the process repeated twice. The collected supernatant was concentrated (not to dryness) under reduced pressure using the rotary evaporator. After the removal of ethanol, the extract was then clarified using 7.2 ml of 5% $ZnSO_4$ and 10 ml of 0.3 N barium hydroxide octahydrate [$Ba(OH)_2$, $8H_2O$] to precipitate proteins, colour, and other organic substances out of the solution and allowed to stand for about 5 minutes, then filtered. A mixture of Zeokarb 225 (H^{+}), a cation and anion exchange resin and deactivated $Fe(OH)_2$ was added to the filtrate to rid it of ions, shaken and filtered. One-ml phenol and 5-ml H_2SO_4 reagents were added to 1 ml of the extract and allowed to stand for an hour and absorbance read at 480 nm. A standard glucose solution of 20, 40, 60, 80 and 100 ppm was prepared and the absorbance read at 480 nm and a standard curve drawn. From the standard graph, the amount of reducing sugars present in the samples was calculated and results expressed as mg/g of cocoa beans. The analysis was conducted in triplicate and the mean values reported.

16.3.2.3 Determination of Non-Reducing Sugars

Non-reducing sugars were determined using the phenol sulphuric acid method as described by Brummer and Cui (2005) with slight modifications. To the remaining residues (from the ethanol extraction in Section 16.3.2.2), 20 ml of 1.5 N H_2SO_4 were added and the mixture digested for 1 hr, allowed to cool, filtered and neutralised with barium carbonate. The mixture was then centrifuged at 10,000 rpm for 30 min and supernatant decanted. The supernatant was then clarified using 7.2 ml of 5% $ZnSO_4$ and 10 ml of 0.3N barium hydroxide octahydrate [$Ba(OH)_2$, $8H_2O$] to precipitate proteins, colour, and other organic substances out of the solution and proceeded as described for the reducing sugars. The analysis was conducted in triplicate and the mean values reported.

16.3.2.4 Determination of Total Sugars

Total sugars were determined using the phenol sulphuric acid method (Brummer and Cui 2005) by adding the values of reducing and non-reducing sugars obtained from Sections 16.3.2.2 and 16.3.2.3.

16.3.2.5 Determination of Free Fatty Acids (FFAs)

Fat from the samples was extracted with petroleum ether (40–60°C) using the Soxhlet extraction method (AOAC 2005 method 963.15). FFA of the oils extracted was determined using the IOCCC (1996) method 42-1993. Five grams of the oil were weighed into a dry 250-ml stoppered conical flask and 25 ml of 95% ethanol/ether (1:1) and phenolphthalein indicator were added. The solution was titrated with 0.1N NaOH by shaking constantly until pink colour persisted for 30 s and the percentage of FFA was determined. The analysis was conducted in triplicate and the mean values reported.

16.3.2.6 Determination of Protein Content

Protein content of the defatted cocoa powder was determined by the Kjeldahl method using the AOAC (2005) method 970.22. The percentage of protein was calculated by multiplying the percentage of nitrogen by the conversion factor 6.25. The analysis was conducted in triplicate and the mean values reported.

16.3.2.7 Total Solids of Cocoa Pulp

Total solids of the pulp were determined using the method described by Javaid et al. (2009) with slight modifications. Two (2) grams of pulp sample were weighed into a pre-weighed flat-bottom dish and transferred to a hot-air oven at 101°C for 2 h. Dried samples were transferred to a desiccator having silica gel as desiccant. After 1 h, the dish was weighed and kept in an oven for further drying (30 min). The heating, cooling and weighing processes were repeated until constant weight was achieved. Total solids content was calculated by the following formula:

$$\text{Total solids (\%)} = \text{weight of dried pulp sample/weight of fresh pulp sample} \times 100$$

16.3.2.8 Mineral Analyses: Wet Digestion

Mineral analyses were determined using AOAC (2005) methods with slight modifications. About 0.5 g of the sample was weighed into a 250-ml beaker. Twenty-five ml (25 ml) of concentrated nitric acid was added and the beaker covered with a watch glass. The sample was digested with great care on a hot plate in a fume chamber until the solution was pale yellow.

The solution was cooled and 1 ml perchloric acid (70% $HCLO_4$) added. The digestion was continued until the solution was colourless or nearly so (the evaluation of dense white fumes was regarded to be indicative of the removal of nitric acid). When the digestion was completed, the solution was cooled slightly and 30 ml of distilled water added. The mixture was brought to boil for about 10 min and filtered hot into a 100-ml volumetric flask using a Whatman No. 4 filter paper. The solution was then made to the mark with distilled water.

16.3.2.9 Determination of Ca, Mg, Zn, Fe, Na and K

The concentrations of Ca, Mg, Zn, Fe, Na and K of the pulp were determined using a Spectra AA 220FS Spectrophotometer (Varian Co., Mulgrave, Australia) with an acetylene flame. One-ml aliquots of the digest were used to determine the Ca, Mg, Zn, Fe, Na and K content of the samples.

16.3.2.10 Total Polyphenols: Extraction of Phenolic Compounds

First, cocoa nibs were finely ground using a kitchen blender. Then 10 g of the resultant cocoa powder were weighed into a thimble and the fat fraction removed by Soxhlet extraction (8 h) using petroleum ether (40–60°C) according to the AOAC (2005) method 963.15. The phenolic fraction was then extracted from the defatted cocoa powders. About 0.2 g of the resultant cocoa powder was homogenised in 30 ml of 80% methanol: 1% HCl for 2 h in falcon tubes using an orbital shaker at 420 rpm. The filtrate was decanted into fresh falcon tubes. This extract was used for determination of total polyphenols, anthocyanins and *o*-diphenols concentrations.

Total polyphenol was measured using the Folin–Ciocalteu assay (Othman et al. 2007) with slight modifications. One ml of the filtered sample was diluted with 49.0 ml of 80% methanol and 0.5 ml of this solution further diluted with 0.5 ml 80% methanol into test tubes making 1 ml of solution. Folin–Ciocalteu's phenol reagent was diluted to 10%, and then 5.0 ml of the 10% Folin–Ciocalteu's phenol reagent added to the 1.0-ml solution. This was followed by the addition of 4.0 ml saturated aqueous Na_2CO_3 solution and the mixture incubated at room temperature for 60 min and for another 60 min at –17°C. The samples were taken from the freezer after the 60 min incubation and left to stand to attain a temperature of 30°C. The absorbance at 760 nm was recorded. Results were expressed as catechin equivalents using a standard catechin curve (0–100 µg/ml). The analysis was conducted in triplicate and mean values reported.

16.3.2.11 *o*-Diphenols

o-Diphenol content was determined with Arrow's reagent (10 g $NaNO_2$ and 10 g Na_2MoO_4 in 100 ml distilled water). To 1 ml of the methanol extract, 1 ml of 0.5 N HCl, 1 ml Arrow's reagent, 10 ml distilled water and 2 ml of 1 N NaOH were added. The absorbance of the solution was read at 520 nm after 30 s. The analysis was conducted in triplicate and the mean values reported. A working standard catechol solution of 20, 40, 60, 80 and 100 ppm was prepared and the absorbance read at 520 nm and a standard curve drawn. From the standard graph, the amount of *o*-diphenols present in the samples was calculated.

16.3.2.12 Anthocyanins

Anthocyanin content was determined using the method described by Misnawi et al. (2002). The extract obtained for total polyphenol analysis was filtered using Whatman No. 4 filter paper and the supernatants were read spectrophotometrically for total absorbance (TOD) at 535 nm. The content of total anthocyanins was calculated as

$$\text{Total anthocyanins (mg/kg)} = \text{TOD}/(\text{AvE}_{535})^{1\%}{}_{1\text{cm}}/10 \times 1000/1$$

where TOD is the total optical density (absorbance) and $(\text{AvE}_{535})^{1\%}{}_{1\text{cm}}$ is the average extinction coefficient for total anthocyanins when a 1-cm cuvette and 1% (10 mg/ml) standard are used; the value is 982 (Fuleki and Francis, 1968).

16.3.2.13 Fermentation Index (FI)

The fermentation index was determined using the method described by Gourieva and Tserrevitinov (1979) with slight modifications. About 0.1 g ground cocoa nibs was extracted with 50 ml of 97:3 mixture of methanol: HCl. The homogenate was allowed to stand in a refrigerator (8°C) for 20 h and then vacuum filtered. The filtrate was read in a Spectrophotometer (LKB Biochrom Novaspec II UV Spectrometer, Birmingham, UK) at 460 nm and 530 nm absorbance. The fermentation index of the sample was obtained by calculating the ratio of absorbance at 460 nm to the absorbance at 530 nm. Three replicate readings were obtained for each sample and the mean values reported.

16.3.2.14 Measurements of Cut Test

The cut test was performed using the international method described by Guehi et al. (2007). A total of 300 beans were cut lengthwise through the middle in order to expose the maximum cut surface of the cotyledons. Both halves were examined in full daylight and placed in one of the following categories: purple, pale purple, brown, slaty, germinated and mouldy.

16.3.3 Statistical Analyses

Statgraphics software version 3.0 (STSC, Inc., Rockville, MD) was used to analyse the data for analysis of variance (ANOVA). Least significant difference (LSD) was used to separate and compare the means, and significance was accepted at the 5% level ($p < 0.05$). Again, the combined effects of pulp pre-conditioning and fermentation time and drying time on the studied parameters were studied using the response surface methodology. Models were developed to relate pulp pre-conditioning and fermentation time and also pulp pre-conditioning and drying time on the studied parameters. The coefficients of the variables in the models and their contribution to the model's variation were reported. The R^2 values were used to judge the adequacy of the models. The R^2 of a model refers to the proportion of variation in the response attributed to the model rather than random error. For a good fit of a model, an R^2 of at least 60% was used. All analyses were conducted in triplicate and the mean values reported.

16.4 RESULTS AND DISCUSSION

16.4.1 Changes in Physicochemical Constituents and Mineral Composition of Cocoa Pulp during Fermentation of Pulp Pre-Conditioned Cocoa Beans

16.4.1.1 Changes in pH Profile of Cocoa Pulp

The pulp is the main substrate metabolised by a sequence of microorganisms during the fermentation process. The pH of the pulp during fermentation is thus crucial as it dictates this sequence. The pH of unfermented cocoa pulp has been reported to range between 3.3–4.0, primarily due to a high concentration of citric acid (Thompson, Miller, and Lopez 2001; Ardhana and Fleet 2003; Schwan and Wheals 2004). The pH of the freshly harvested unfermented cocoa pulp was highly acidic (3.88) which increased gradually to 4.02 after 10 days of pod storage (Figure 16.1). The gradual increase in pH of the pulp during pod storage might be due to the breakdown of the pulp sugars, which has been reported to reduce the pulp volume per seed leading to the decrease in citric acid concentrations (Biehl et al. 1989; Sanagi et al. 1997), hence increasing the pH of the pulp.

Fermentation caused consistent increases in the pH of the pulp on all the pod storage days (Figure 16.1). During fermentation, pH increased from 3.88–3.96 for the unstored pods, 3.98–5.04 for pods stored for 3 days, 4.01–5.23 for 7 days pod storage and 4.02–5.24 for pods stored for 10 days at the end of the sixth day of fermentation. These findings were in agreement with reports by previous investigators. Yusep et al. (2002) found the pH of cocoa pulp to increase progressively from 3.80 to 4.80 at 3 days

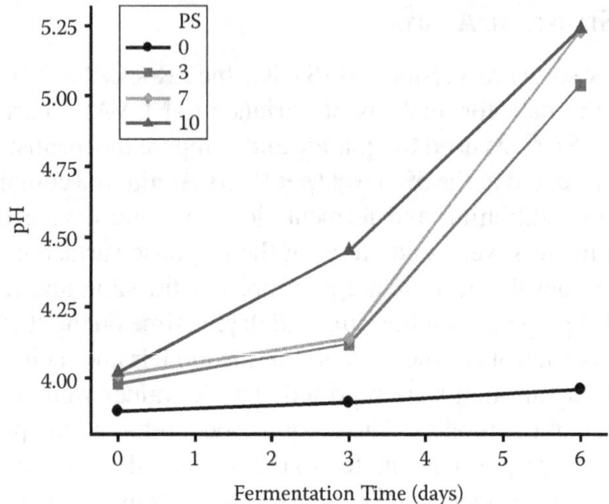

FIGURE 16.1 Changes in pH of cocoa pulp during fermentation and pod storage (PS).

of fermentation. Ardhana and Fleet (2003) found the pH of cocoa pulp to increase from between 3.7–3.9 at the start of fermentation to between 4.8–4.9 by the end of fermentation. Nielsen et al. (2007b) also recorded an increase in pH of the pulp from the starting value of 3.94–4.12 to 4.28–4.69 after 96 hours of fermentation. These gradual increases in pH of the pulp during fermentation are suspected to be due to the reported decline in citric acid concentration (Schwan and Wheals 2004; Jespersen et al. 2005).

During fermentation, yeasts and lactic acid bacteria break down the citric acid in the pulp to metabolise the pulp sugars leading to an increase in the pH from 3.5 to 4.2 (Schwan, Rose, and Board 1995; Schwan 1998; Schwan and Wheals 2004; Jespersen et al. 2005). Nielsen et al. (2007b) reported a decline in the citric acid concentration to low or even non-detectable levels during the first 12 h of fermentation. Statistical analysis on the data showed that both pod storage and fermentation significantly ($p < 0.05$) affected the pH of the pulp (Table 16.1).

16.4.1.2 Changes in Titratable Acidity (TA) of Cocoa Pulp

During fermentation of cocoa beans, micro-organisms break down the sugars in the pulp resulting in the production of alcohols and organic acids, predominantly acetic acid which then diffuses into the beans. Production of acids in the pulp is important in cocoa fermentation as these acids diffuse into the beans and subsequently induce the important biochemical reactions leading to well-fermented cocoa beans. However, high acid production in the pulp is detrimental as it leads to excessive acid diffusing into the beans resulting in the production of acidic beans.

TABLE 16.1
ANOVA Summary Showing F-Ratios of the Physicochemical Properties of Cocoa Pulp during Fermentation and Pod Storage

Variables	pH	Titratable Acidity	Reducing Sugars	Total Solids
Pod storage (PS)	15.10[a]	179.18[a]	807.78[a]	4.58[a]
Fermentation (FT)	54.02[a]	686.74[a]	25545.21[a]	9.71[a]
Interaction (PS X FT)	5.51[a]	18.47[a]	291.80[a]	0.42

[a] Significant at $p < 0.05$.

Changes in titratable acidity of the pulp for all the pod storage treatments and fermentation time are shown in Figure 16.2. Fermentation caused significant ($p < 0.05$) increases in the acidity levels in the pulp reaching a maximum at day 3 of fermentation after which the titratable acidity decreased considerably till the end of fermentation, and this was noted at all pod storage treatments. The acidity level was highest at three days of fermentation as the majority of the pulp sugars were probably degraded into alcohols which were then oxidised to acetic acid by acetic acid bacteria within three days of fermentation.

Ardhana and Fleet (2003) reported a high concentration of 10 mg/g acetic acid in cocoa pulp at 72 h (3 days) of fermentation. Acidity levels decreased after day 3 of fermentation because at that time most of the

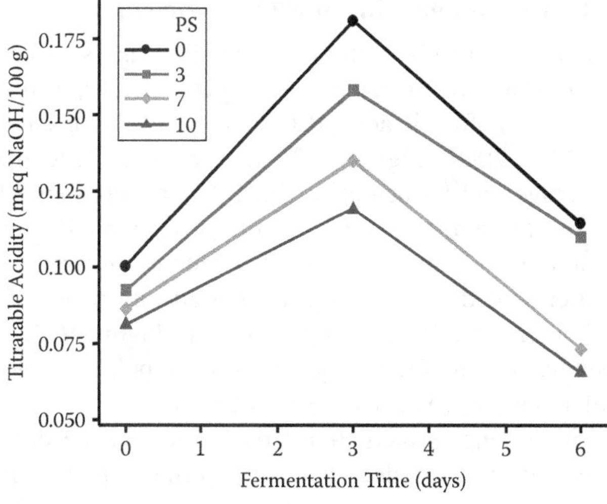

FIGURE 16.2 Changes in titratable acidity of cocoa pulp during fermentation and pod storage (PS).

acid produced diffused into the beans. Again, as the pulp volume reduced, there was improvement in aeration in the fermenting mass leading to the evaporation of volatile acids such as acetic acid. Even though the pH of the pulp increased within 72 h of fermentation for all pod storage periods, TA also increased within 72 h of fermentation for all pod storage periods. This is because the predominant acid in the unfermented pulp is citric acid and as fermentation progresses (after 48 h), the predominant acid in the pulp becomes acetic acid due to the oxidation of alcohol by acetic acid bacteria. Hence the pH of the pulp recorded at 72 h of fermentation is that of acetic acid and that recorded at the onset of fermentation is citric acid and these two acids have different pKa values.

Increasing pod storage significantly ($p < 0.05$) reduced titratable acidity of the pulp (Table 16.1). The interaction between fermentation time and pod storage also had a significant ($p < 0.05$) effect on the acidity levels of the pulp (Table 16.4). Pulp from the unstored pods had the highest titratable acidity at day 3 of fermentation whereas pulp from pods stored for 10 days had the least titratable acidity at day 3 of fermentation. This might be due to the fact that pod storage reduced pulp volume per seed, reduced pulp sugar content and thus increased micro-aeration within the pulp. This decreased the sugar metabolised by yeasts during subsequent fermentation and eventually reduced alcohol fermentation and acetic acid formation in the pulp (Said, Meyer, and Biehl 1987; Biehl et al. 1989; Sanagi et al. 1997). This suggests that pod storage as a means of pulp pre-conditioning could be effectively employed to reduce acidity levels of cocoa beans during fermentation.

16.4.1.3 Changes in Reducing Sugars of Cocoa Pulp

Cocoa pulp is reported to be rich in fermentable sugars, notably glucose and fructose, and has a relatively low initial pH (3.3–4.0), primarily due to a high concentration of citric acid (Pettipher 1986; Thompson et al. 2001; Ardhana and Fleet 2003). Figure 16.3 shows changes in reducing sugars during pod storage and fermentation time. The results showed that both pod storage and fermentation time significantly ($p < 0.05$) affected the reducing sugars of the pulp (Table 16.1). Reducing sugars in the freshly harvested unfermented cocoa pulp decreased drastically from 75.72 mg/g to 48.13 mg/g after 10 days of pod storage (Figure 16.3). This might be due to the breakdown of reducing sugars in the pulp into energy for the physiological and metabolic activities of the beans.

Similarly, increasing fermentation time decreased the concentrations of the fermentable sugars (reducing sugars) in the pulp at all pod storage periods. The concentrations of the reducing sugars decreased from 75.72 mg/g at the onset of fermentation to 7.29 mg/g at the end of the six days of fermentation for the unstored pods (Figure 16.3). Similar decreasing trends

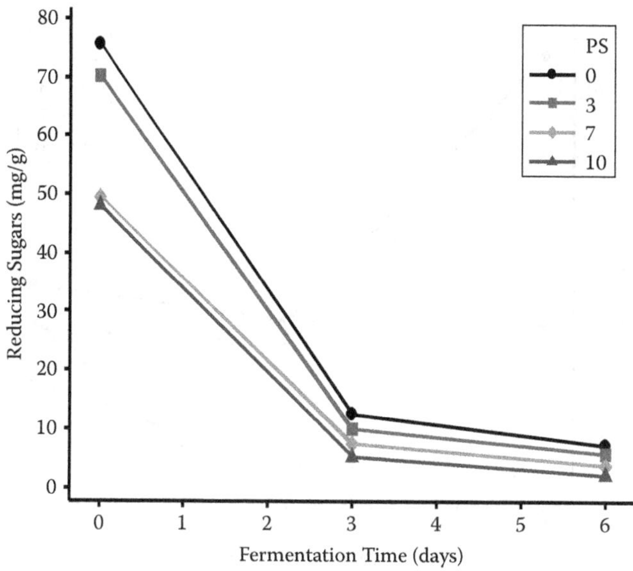

FIGURE 16.3 Changes in reducing sugars of cocoa pulp during fermentation and pod storage (PS).

were observed for all the pods that were pre-conditioned. The decrease in the concentrations of reducing sugars in the pulp during fermentation was probably due to the activities of yeasts and lactic acid bacteria. The high sugar content, the low pH and low oxygen tension in the pulp favour the growth of yeasts at the onset of fermentation. The yeasts metabolise the fermentable sugars in the pulp to ethanol which in turn is oxidised by acetic acid bacteria to acetic acid (Thompson et al., 2001). Ardhana and Fleet (2003) observed a similar trend of decrease in the concentrations of pulp fermentable sugars during fermentation. They reported decreases in the concentrations of fructose and glucose from 62–11 mg/g and 41–7 mg/g, respectively, after 120 h (five days) of fermentation for *Forastero* cocoa and from 42–9 mg/g and 24–5 mg/g, respectively, after 72 h (three days) of fermentation for *Trinitario* cocoa. There was also significant ($p < 0.05$) interaction between pod storage and fermentation on the reducing sugars of the pulp.

16.4.1.4 Changes in Total Solids of Cocoa Pulp

Changes in total solids in the pulp during fermentation at all pod storage periods are shown in Figure 16.4. The results showed that fermentation significantly ($p < 0.05$) influenced the total solids in the cocoa pulp at all pod storage periods. Total solids decreased drastically within the first 3 days of fermentation from 20.5–16.6%, 19.9–16.2%, 17.6–15.5% and 16.6–14.3%, respectively, for pods stored for 0, 3, 7 and 10 days. The decrease

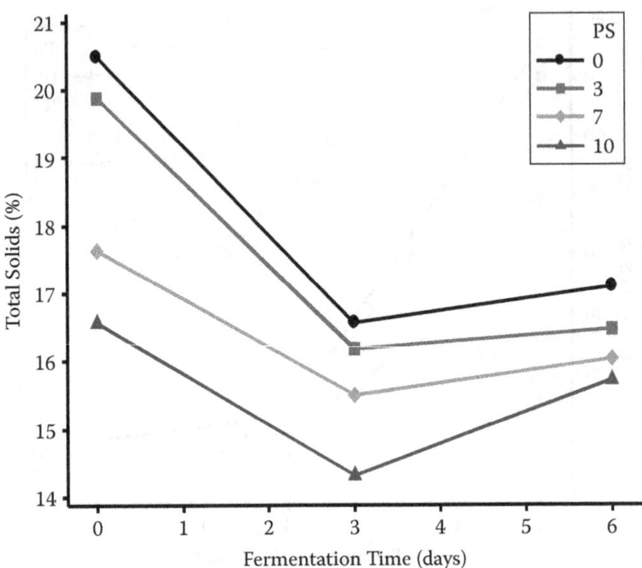

FIGURE 16.4 Changes in total solids of cocoa pulp during fermentation and pod storage (PS).

in total solids during the first 3 days of fermentation might largely be due to the breakdown of sugars in the pulp by yeasts and lactic acid bacteria. Yeasts and lactic acid bacteria have been reported to metabolise the fermentable sugars in cocoa pulp during fermentation to produce ethanol and lactic acid, respectively (Thompson et al. 2001; Ardhana and Fleet 2003). Total solids, however, increased slightly towards the end of fermentation for all pod storage treatments. This slight increase in total solids might be due to the accumulation of some microbial metabolites.

Total solids in the pulp decreased significantly ($p < 0.05$) with increasing pod storage (Figure 16.4). Total solids in the unfermented pulp decreased from 20.5% for the non-pre-conditioned pulp to 16.6% for pulp pre-conditioned for 10 days. A similar trend of decrease was observed for all fermentation times. At the end of fermentation, total solids of the fermented pulp decreased from 16.6% for the non-pre-conditioned pulp to 14.3% for pulp pre-conditioned for 10 days. The observed changes might be due to changes in the fermentable sugars and other constituents of the pulp, thereby reducing the total solid content in the pulp during pod storage. There was, however, no significant ($p > 0.05$) interaction between pod storage and fermentation on the total solids in the pulp (Table 16.1).

16.4.1.5 Changes in Mineral Composition of Cocoa Pulp

Changes in the mineral composition of cocoa pulp during fermentation for all pod storage periods are presented in Table 16.2. The results

TABLE 16.2
Changes in Mineral Composition (mg/100 g) of Cocoa Pulp during Fermentation of Pulp Pre-Conditioned Cocoa Beans

Pod Storage (Days)	Fermentation Time (Days)	Na	Fe	Ca	Mg	Zn	K
0	0	103.26 ± 1.00	4.26 ± 1.00	316.92 ± 0.47	32.52 ± 1.08	1.04 ± 0.04	255.12 ± 1.97
	3	48.12 ± 2.00	3.56 ± 0.58	347.50 ± 0.50	33.98 ± 1.03	2.22 ± 0.89	226.06 ± 2.65
	6	17.48 ± 2.08	2.58 ± 0.43	395.80 ± 1.15	148.32 ± 0.84	4.58 ± 0.47	643.36 ± 0.57
3	0	40.60 ± 1.15	3.42 ± 0.58	248.42 ± 2.08	35.44 ± 0.58	2.44 ± 0.53	155.46 ± 1.10
	3	33.60 ± 0.58	2.90 ± 0.57	275.78 ± 2.08	134.46 ± 0.50	2.68 ± 0.57	408.64 ± 0.45
	6	9.54 ± 2.08	1.42 ± 0.57	374.50 ± 1.92	156.70 ± 2.08	2.78 ± 0.48	569.06 ± 1.00
7	0	20.18 ± 1.00	3.58 ± 0.36	190.12 ± 1.00	29.56 ± 0.48	0.64 ± 0.00	252.52 ±0 .50
	3	12.70 ± 1.53	2.04 ± 1.00	260.00 ± 0.58	97.96 ± 0.91	1.16 ± 0.20	507.16 ± 0.92
	6	9.66 ± 1.53	1.90 ± 0.57	341.14 ± 1.00	144.42 ± 0.54	1.20 ± 0.10	621.10 ± 1.00
10	0	14.15 ± 2.00	5.26 ± 0.87	307.68 ± 1.53	36.10 ± 0.00	0.30 ± 0.01	289.62 ± 1.38
	3	11.72 ± 5.63	2.96 ± 0.46	301.40 ± 1.49	71.02 ± 1.00	0.56 ± 0.02	392.36 ± 1.53
	6	8.68 ± 0.58	2.30 ± 0.00	189.20 ± 1.65	165.66 ± 0.58	0.98 ± 0.02	815.06 ± 2.00

Mean values ± standard deviation.

showed that the most abundant mineral in the freshly harvested and unfermented cocoa pulp is calcium, followed by potassium and sodium with values of 316.92 mg/100 g, 255.12 mg/100 g and 103.26 mg/100 g, respectively (Table 16.2). Zinc is the mineral with the least concentration of 1.04 mg/100 g, whereas iron and magnesium had appreciable values of 4.26 mg/100 g and 32.52 mg/100 g, respectively. Pod storage and fermentation, however, showed variable trends in the studied mineral content (Table 16.2).

Both pod storage and fermentation caused consistent decreases in the composition of sodium and iron. Sodium concentration decreased drastically from 103.26 mg/100 g at the start of fermentation to 17.48 mg/100 g at the end of fermentation (six days) for the unstored pods. Similar decreasing trends in sodium concentration were observed for all pod storage treatments (Table 16.2). Pod storage also led to decreases in iron concentrations at all fermentation times with values decreasing from 4.26 mg/100 g at the start of fermentation to 2.58 mg/100 g at the end of fermentation for the unstored pods (Table 16.2). Similar decreasing trends were observed for all pod storage treatments (Table 16.2). The decrease in the concentration of sodium and iron might be due to the utilisation of these minerals by the different micro-organisms involved in fermentation for their physiological and metabolic activities.

Increasing fermentation time increased the composition of calcium, magnesium, zinc and potassium at all pod storage periods (Table 16.2). Potassium levels in the unfermented pulp increased from 255.12 mg/100 g to 643.36 mg/100 g by the sixth day of fermentation for the unstored pods. Similar increasing trends were observed for the different pod storage treatments. Pod storage also increased potassium concentration significantly at all fermentation times (Table 16.2).

Calcium and magnesium contents in the pulp also increased significantly with increasing fermentation time for all pod storage periods, with values increasing from 316.92 and 32.52, respectively, at the onset of fermentation to 395.80 and 148.32 at the end of the six days of fermentation for the unstored pods. Similar trends were observed at all the pod storage treatments. Increasing pod storage also caused decreases in the concentrations of Ca at all fermentation times and increases in Mg content were noted. Zn composition also increased with fermentation at all pod storage periods (Table 16.2).

The observed increases in Ca, Mg, Zn and K during fermentation might be due to the synthesis of these minerals by the micro-organisms involved in the fermentation of the pulp. Statistical analysis on the data indicated that both pod storage and fermentation time significantly ($p < 0.05$) influenced the mineral composition of cocoa pulp with significant

TABLE 16.3
ANOVA Summary Showing F-Ratios for Variation in Mineral Content of Cocoa Pulp during Fermentation of Pulp Pre-Conditioned Cocoa Beans

Variables	Na	Fe	Ca	Mg	Zn	K
Pod storage (PS)	824.25[a]	6.35[a]	7956.22[a]	2375.77[a]	63.94[a]	17170.88[a]
Fermentation (FT)	701.38[a]	32.67[a]	5344.20[a]	49984.73[a]	31.24[a]	278743.47[a]
Interaction (PS X FT)	228.62[a]	1.64	6371.9[a]	2064.91[a]	11.63[a]	12647.75[a]

[a] Significant at $p < 0.05$.

interaction observed for sodium, calcium, magnesium, zinc and potassium (Table 16.3).

16.4.2 Changes in Physicochemical Constituents and Flavour Precursors during Fermentation of Pulp Pre-Conditioned (Pod Storage) Cocoa Beans

16.4.2.1 Changes in pH Profile of Cocoa Beans

The pH of cocoa beans during fermentation is critical with regard to the biochemical reactions taking place in the beans. During the fermentation process, sugars in the adhering pulp surrounding the beans are metabolised by micro-organisms, leading to the production of organic acids which diffuse into the beans to alter the pH of the beans (Jinap 1994; Lopez and Dimick 1995; Schwan and Wheals 2004; Fowler 2009; Afoakwa 2010). Biehl and Voigt (1999) found the rate of diffusion of organic acids into the cotyledons during fermentation, timing of initial entry, duration of the optimum pH and final pH to be crucial for optimum flavour formation.

With the exception of the unstored pods which showed a continuous reduction in pH during fermentation, the pH of the beans in all the stored pods decreased by the third day of fermentation and then increased by the end of fermentation (Figure 16.5). This was in agreement with Biehl (1984) who found the diffusion of acids (predominantly acetic acid) to decrease the pH of the cotyledon from 6.5 at 0 h to 4.6 at 72 h and an increase to 5.2 at the end of fermentation. The observed consistent decrease in pH within the first three days of fermentation was primarily due to the diffusion of organic acids (predominantly acetic acid) into the beans produced by acetic acid bacteria. The observed continuous reduction in pH for the unstored pods during fermentation might be due to the fact that the unstored pods had a lot of pulp adhering to the beans and thus more

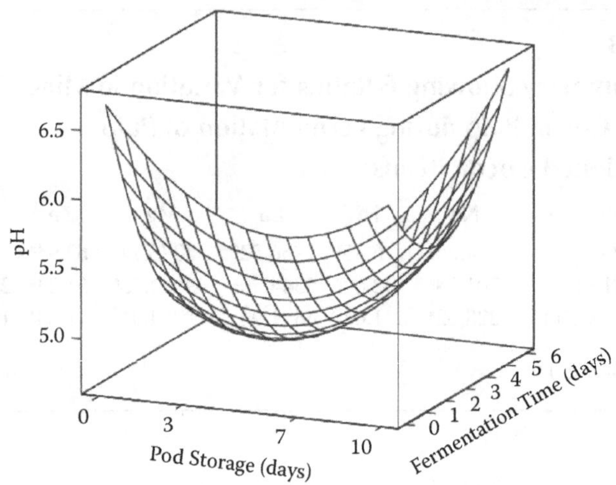

FIGURE 16.5 Response surface plot showing effect of pod storage and fermentation time on the pH of cocoa beans.

pulp sugars were metabolised, producing more organic acids which diffused into the beans even after 72 h of fermentation. Towards the end of fermentation, an increase in pH was observed, possibly due to evaporation of volatile acids such as acetic (Afoakwa et al. 2011b).

Increasing pod storage caused consistent increases in pH of the nibs at the end of fermentation (Figure 16.5). The pH of the nibs increased from 4.80 for the unstored pods to 7.01 for pods stored for 10 days at the end of fermentation. The pH of the beans during fermentation is crucial as it determines the rate of enzyme activity responsible for the production of flavour precursors as well as the development of the characteristic brown colour of cocoa beans (Jinap, Siti, and Norsiati 1994; Hansen, del Olmo, and Burri 1998; Sakharov and Ardila 1999); most of these enzymes are reported to have pH optima of 4.5–5.5 (Biehl et al. 1989). Work done by Biehl et al. (1985) and Biehl and Voigt (1994) reported that fermented cocoa beans with pH between 5.0–5.5 produce higher flavour potentials whereas fermented beans with pH 4.0–4.5 give low flavour potential. The pH at the end of fermentation was 5.10 and 5.36 for pods stored for 3 and 7 days, respectively. The pH was, however, very low for the unstored pods (4.80) and very high (7.01) for pods stored for 10 days at the end of the fermentation. Findings from this study suggest that pod storage between 3 and 7 days and 6 days of fermentation could produce cocoa beans with higher concentrations of flavour precursors compared to unstored pods and pods stored for 10 days.

Regression analysis of the data showed significant ($p < 0.05$) influence of the linear factor of pod storage and fermentation time (FT) and quadratic

TABLE 16.4
Regression Coefficients and Their R^2 Values in the Models for pH, Titratable Acidity, Sugars and Protein of Cocoa Beans

Variables	pH	Titratable Acidity	Reducing Sugars	Non-Reducing Sugars	Total Sugars	Protein
Constant	4.8520[a]	0.23372[a]	10.2891[a]	9.6062[a]	19.8953[a]	25.3221[a]
X_1	0.2957[a]	–0.03744*	1.1356[a]	–1.2547[a]	–0.1191	–1.7408[a]
X_2	–0.3128[a]	0.04937[a]	4.1586[a]	–12.2253[a]	–8.0668[a]	–2.0429[a]
X_1^2	0.5332[a]	–0.06567[a]	–0.3578	0.0071	–0.3507	0.4242
X_2^2	0.7172[a]	–0.08519[a]	–1.8574[a]	7.6370[a]	5.7795[a]	–0.4779
$X_1 . X_2$	0.5495[a]	–0.05461[a]	0.4964[a]	0.1787	0.6751	0.0347
R^2	82.2%	78.4%	94.9%	98.7%	95.0%	87.6%
R^2 (adjusted)	80.1%	75.8%	94.3%	98.5%	94.5%	86.1%

[a] Significant at $p < 0.05$; X_1 = Pod storage; X_2 = Fermentation time.

factor of both PS and FT on the pH of the cotyledons. There was also significant interaction between PS and FT. The model developed could explain about 82% of the variations in the pH of the cotyledons (Table 16.4).

16.4.2.2 Changes in Titratable Acidity of Cocoa Beans

Changes in the titratable acidity (TA) of the cocoa beans during fermentation for all pod storage treatments are shown in Figure 16.6. Titratable acidity for the unstored pods increased continuously from 0.057 meq NaOH/100 g at the start of fermentation to 0.268 meq NaOH/100 g by the end of fermentation. However, with pulp pre-conditioned pods, fermentation caused an increase in acidity levels reaching a maximum within day three after which titratable acidity decreased considerably till the end of fermentation. Similar observations were made by Nazaruddin et al. (2006a) and Afoakwa et al. (2011b). The observed increases in acidity might be due to the fact that volatile acids (acetic, propionic, butyric and isovaleric) and non-volatile acids (citric, lactic, malic, succinic and tartaric) that develop in the pulp through sugar degradation by the metabolism of micro-organisms during the fermentation process diffuse into the cotyledon (Jinap 1994) to cause a gradual increase in acidity of the beans within the first 72 hours of fermentation. Acid production is necessary for killing the seed and to induce the necessary biochemical reactions in the seed, but excessive acid production and the resulting low cotyledon pH will result in the formation of low precursor type peptides and amino acids (Lopez and Dimick 1995).

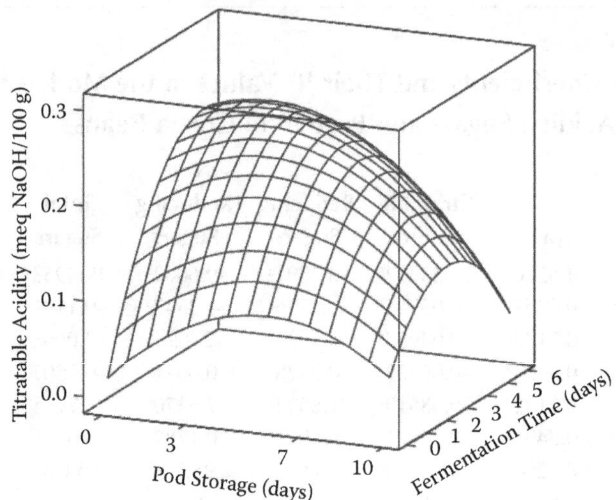

FIGURE 16.6 Response surface plot showing titratable acidity of cocoa beans as affected by pod storage and fermentation time.

Excessive acid in the beans would also result in acidic beans which is deleterious to the flavour quality of the fermented beans.

Again, increasing pod storage caused consistent reduction in nib acidity levels at the end of fermentation (Figure 16.6). Titratable acidity reduced from 0.268 meq NaOH/100 g for the unstored pods to 0.030 meq NaOH/100 g for pods stored for 10 days. This suggests that pod storage could be effectively used to reduce the acidity levels of cocoa beans, probably due to reduced pulp volume per seed and reduced pulp sugar content during the pod storage period which results in increasing micro-aeration within the pulp and reduced the alcohol fermentation and acetic acid production (Afoakwa et al. 2011b).

The model developed to predict the effect of pod storage and fermentation on titratable acidity (TA) of cocoa beans had an R^2 of 78% (Table 16.4). This implies that the model developed could explain about 78% of the variations in the titratable acidity of the nibs and the remaining 22% could be due to other factors not investigated in this work. Regression coefficients showed significant ($p < 0.05$) influence of the linear and quadratic factors of both pod storage and fermentation time on the titratable acidity of the cotyledons. There was also significant ($p < 0.05$) interaction between pod storage and fermentation time.

16.4.2.3 Changes in Reducing Sugars

Reducing sugars are carbonyl aroma precursors in fermented cocoa beans, which are mainly produced through the hydrolysis of sucrose by the action of invertase (Rohan and Stewart 1967). Fructose and glucose are the main

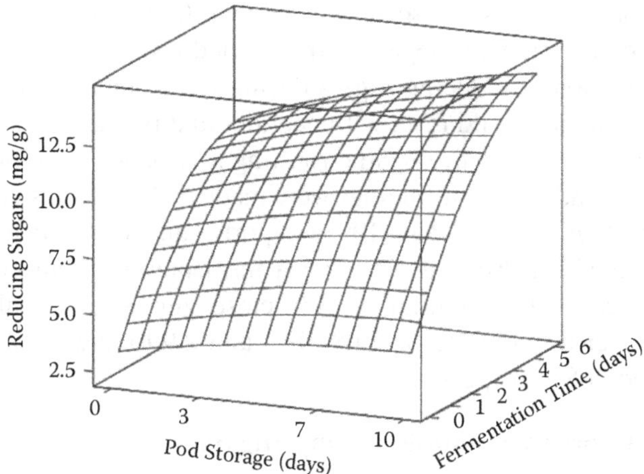

FIGURE 16.7 Response surface plot showing effect of pod storage on reducing sugars of cocoa beans during fermentation.

reducing sugars in cocoa beans; however, reducing sugars can be formed from enzymatic hydrolysis of anthocyanins to yield arabinose and galactose by the action of glycosidase (Said 1989; Hoskin and Dimick 1994).

The reducing sugars of the unfermented cocoa beans increased marginally during the pod storage period (Figure 16.7), from 3.57 mg/g at the start of pod storage to 4.52 mg/g by 10 days of pod storage. This gradual increase in the concentrations of reducing sugars during pod storage might be as a result of the hydrolysis of sucrose in the beans during the storage period.

Fermentation, however, caused significant ($p < 0.05$) increases in reducing sugars at all pod storage periods (Figure 16.7). Reducing sugars increased from 3.57 mg/g at the start of fermentation to 10.69 mg/g at the end of fermentation for the unstored pods, and increased from 3.89–11.94 mg/g, 4.29–13.34 mg/g and 4.52–13.56 mg/g for pods stored for 3, 7 and 10 days, respectively. The increase in the amounts of reducing sugars during fermentation is reported to be the result of enzymatic reactions promoted by invertase, β-galactosidase, α-arbinosidase and α-mannosidase (Hansen et al. 1998). Pod storage also increased the reducing sugars of the fermented beans. At the end of fermentation (6 days), reducing sugars increased by 50%, 60%, 70% and 80%, respectively, for pods stored for 0, 3, 7 and 10 days.

The model developed to predict the effect of pod storage and fermentation on reducing sugars of cocoa beans had an R^2 of 95% (Table 16.4) implying that about 95% of the variations in the reducing sugars of the cotyledons could be explained by the model whereas 5% were due to other factors not investigated in this work. Regression coefficients showed that the linear term of both pod storage and fermentation time and the quadratic term of FT had significant ($p < 0.05$) influence on the reducing sugars of the cotyledons. There was also significant ($p < 0.05$) interaction between PS and FT on the reducing sugars of the cotyledons.

The production of reducing sugars during fermentation is very important as these sugars would react with peptides and free amino acids in the Maillard reaction during drying and roasting to produce the typical cocoa flavour compounds. Similar findings were made by Reineccius et al. (1972) and Berbert (1979). Puziah et al. (1998a) reported an increase in total reducing sugars of about 208% during cocoa fermentation. Reineccius et al. (1972) detected final concentrations of fructose, glucose and total reducing sugars after fermentation to be 1.68, 0.43 and 2.99 mg/g, an increase of 405, 106 and 208%, respectively.

16.4.2.4 Changes in Non-Reducing Sugars

Non-reducing sugars in cocoa beans comprise mainly sucrose (Reineccius et al. 1972; Berbert 1979; Puziah et al. 1998a). Changes in non-reducing sugar concentrations during pod storage and fermentation of cocoa beans are shown in Figure 16.8. Results from this study showed that non-reducing sugars were the major sugar present in significant concentrations in all the unfermented cocoa beans. The concentration of non-reducing sugars in the unfermented beans ranged from 27.95–30.56 mg/g (81–90% of the total sugars). The concentration of non-reducing sugars in the unfermented beans was highest in the unstored pods (accounting for about 90% of the total sugars in the unfermented beans) and decreased marginally with increasing pod storage. The marginal decrease of non-reducing sugars in the unfermented

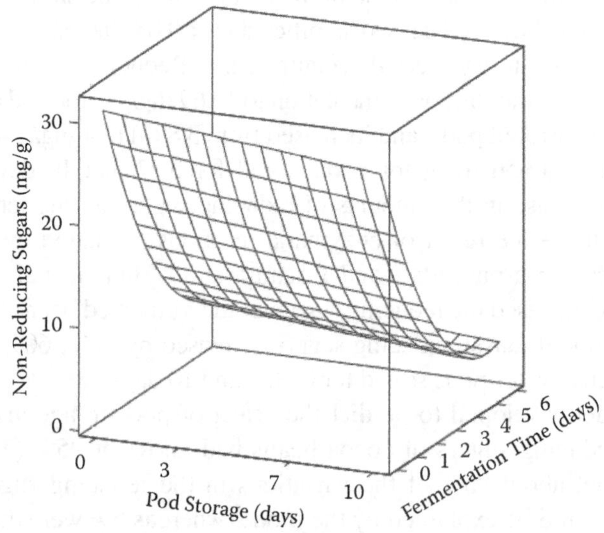

FIGURE 16.8 Response surface plot showing changes in non-reducing sugars of cocoa beans during fermentation of pulp pre-conditioned cocoa beans.

beans with increasing pod storage might be due to some invertase activity in the beans during the storage period. The high concentrations of non-reducing sugars in the unfermented cocoa beans have been reported by several researchers. Puziah et al. (1998a) found that sucrose was the only sugar present in abundant concentration (18.78 mg/g) which was about 95% of total sugars in unfermented cocoa beans. Reineccius et al. (1972) found that the fresh unfermented *Trinidad* cocoa beans contained 15.80 mg/g sucrose and trace amounts of penitol, fructose, sorbose, mannitol and inositol. Berbert (1979) also reported that sucrose concentration of the unfermented beans comprised about 90% of the total sugars (24.80 mg/g), whereas both fructose and glucose made up about 6% (0.90 and 0.70 mg/g, respectively), other sugars (<0.50 mg/g) and total sugars (27.10 mg/g).

During the fermentation process, sucrose was hydrolysed by cotyledon invertase to glucose and fructose (Lopez, Lehrian, and Lehrian 1978). The concentrations of non-reducing sugars decreased significantly ($p < 0.05$) with increasing fermentation time for all pod storage periods (Figure 16.8). It decreased from 30.56 mg/g to 6.11 mg/g (80% decrease) in the unstored pods at the end of the fermentation (6 days). A similar trend of decrease was observed for all pod storage treatments at the end of the fermentation. There were about 83%, 85% and 84% decreases in the concentration of non-reducing sugars at the end of fermentation for pods stored for 3, 7 and 10 days, respectively. Reduction in the non-reducing sugars during fermentation has been reported by several researchers (Reineccius et al. 1972; Berbert 1979). Puziah et al. (1998b) found a significant decrease in the concentration of sucrose during fermentation to 2.0 mg/g which was equal to the 89% decrease.

The rate of decrease was high within the first 3 days of fermentation but slowed down towards the end of the fermentation process. This might be due to high cotyledon invertase activity within the first 3 days of fermentation. Hansen et al. (1998) observed cotyledon invertase activity to decrease within 24 h of heap fermentation. Invertase activity was, however, reported to be insignificant from day 2 of fermentation till the end of fermentation. In the course of the fermentation, enzyme inactivation is caused by generated heat and high concentrations of acetic acid, ethanol and polyphenols (Hansen et al. 1998). This might explain why the reduction in non-reducing sugars slowed down towards the end of the fermentation.

Regression analysis of the data showed significant ($p < 0.05$) influence of the linear factor of both pod storage and fermentation time and the quadratic factor of FT on the non-reducing sugars of the cotyledons. However, interaction between PS and FT was not significant ($p > 0.05$). The model developed could explain about 99% of the variations in the non-reducing sugars of the cotyledons, suggesting only 1% was due to other factors not investigated in this work (Table 16.4).

16.4.2.5 Changes in Total Sugars

Total sugars of cocoa beans comprise non-reducing sugars and total reducing sugars (Puziah et al. 1998a). Changes in the concentrations of total sugars during fermentation for all pod storage treatments are shown in Figure 16.9. Total sugars of the unfermented beans decreased marginally during pod storage. It decreased from 34.13 mg/g in the unstored pods to 32.47 mg/g in the pods stored for 10 days. The marginal decrease in total sugars during pod storage might be due to the marginal breakdown of sucrose in the seeds by invertase (Puziah et al. 1998a).

Total sugars of the beans, however, decreased significantly ($p < 0.05$) during fermentation at all pod storage periods. Total sugars decreased from 34.13 mg/g at the start of fermentation to 16.81 mg/g at the end of fermentation for the unstored pods. It also decreased from 33.72–17.01 mg/g for pods stored for 3 days, 33.84–17.77 mg/g for pods stored for 7 days and 32.47–18.03 mg/g for pods stored for 10 days. This finding was in agreement with Puziah et al. (1998b) who found total sugars to decrease significantly during fermentation to concentrations of 5.0 mg/g representing a 75% decrease. The drastic reduction in the concentrations of total sugars during fermentation was largely due to the breakdown of non-reducing sugars (sucrose) in the cotyledons by invertase. Sucrose is hydrolysed by cotyledon invertase to glucose and fructose during the fermentation (Lopez et al. 1978).

Regression analysis of the data also showed significant ($p < 0.05$) influence of the linear and quadratic factors of fermentation time on the total

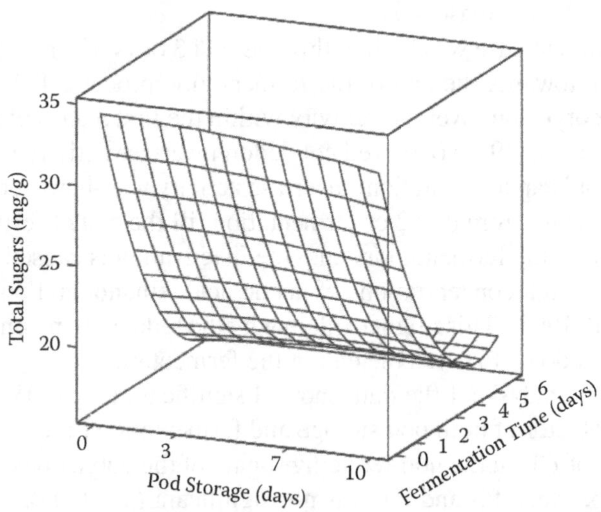

FIGURE 16.9 Response surface plot showing effects of pod storage on total sugars of cocoa beans during fermentation.

sugars of the cotyledons. Both the linear and quadratic factors of pod storage as well as the interaction between pod storage and fermentation time did not significantly ($p > 0.05$) influence the total sugars of the cotyledons. The model developed had an R^2 of 95%, implying that the model could explain about 95% of the variations in the total sugars of the cotyledons, whereas the remaining 5% was due to other factors not investigated in this work (Table 16.4).

16.4.2.6 Changes in Protein during Fermentation

Changes in protein concentration during fermentation for all the pod storage treatments are shown in Figure 16.10. Protein concentration decreased significantly at $p < 0.05$ as fermentation progressed from day 0 to day 6. It decreased from 28.9% at the start of fermentation to 28.1% on day 3 of fermentation for the unstored pods. It further decreased from 28.1% to 24.7% at the end of fermentation (6 days). A similar trend of decrease was observed for all pod storage treatments. This result was in agreement with those previously reported by Dimick and Hoskin (1981), Lopez (1986), Jinap et al. (2008) and Afoakwa et al. (2011b). The decrease in protein content might be caused by the endogenous breakdown of cocoa bean proteins to oligopeptides and free amino acids (Jinap et al. 2008). Findings by Voigt et al. (1994a,b,c) showed that the oligopeptides and free amino acids represent specific cocoa aroma precursors produced during fermentation. The reduction in protein content might also be due to the formation of a complex between polyphenols and the proteins. Polyphenols undergo

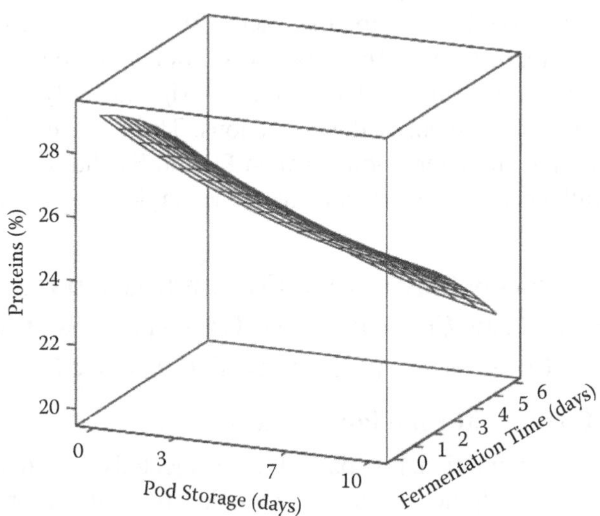

FIGURE 16.10 Response surface plot showing changes in protein content during fermentation of pulp pre-conditioned cocoa beans.

biochemical modification during cocoa fermentation through polymerisation and can complex with protein. This can lead to a reduction in protein content (Bonvehi and Coll 1997; Nazaruddin et al. 2006a).

Cotyledon protein degradation into peptides and free amino acids appears central to flavour formation (Afoakwa 2010). Ziegleder (2009) reported that proteolysis in the seeds mainly takes place within 24 h after destruction of the cells and acidification by acetic acid. Voigt et al. (1994a) and Afoakwa and Paterson (2010) also observed that proteolysis begins after the death of the bean and this occurs after 48 h of fermentation. During fermentation, seed proteins, notably the vicilin class globulins, are degraded by endogenous proteases to peptides and free amino acids.

Again, protein content reduced significantly ($p < 0.05$) with increasing pod storage at all fermentation times (Figure 16.10). The protein content of the unfermented cocoa beans decreased from 28.9% (unstored pods) to 25.8% (10 days pod storage). The reduction in protein content during pod storage (pulp pre-conditioning) is reported to be due to the action of protease enzymes in the pods during storage and thus initiating the process of proteolysis (Afoakwa et al. 2011b). This observation suggests that pod storage might have initiated the release of peptides and free amino acids which could influence the processes for the formation of flavour precursors in the bean during subsequent fermentation and drying.

The model developed to predict the effect of pod storage and fermentation time on the protein content of cocoa beans had an R^2 of 88% (Table 16.4). This implies that the model developed could explain about 88% of the variations in the protein content of the cotyledons whereas 12% of the variations could be due to other factors that were not investigated in this work. The regression coefficients also showed that the linear factor of both fermentation time and pod storage had a significant ($p < 0.05$) influence on the protein content of the cotyledons. The quadratic factor of PS and FT as well as the interaction between PS and FT did not significantly ($p > 0.05$) influence the protein content of the cotyledons (Table 16.4).

16.4.3 Changes in Polyphenolic Constituents and Free Fatty Acids Content during Fermentation of Pulp Pre-Conditioned (Pod Storage) Ghanaian Cocoa Beans

16.4.3.1 Changes in Total Polyphenols

During cocoa fermentation, polyphenols are subjected to biochemical modification through polymerisation and complexation with proteins, hence decreasing concentrations, solubility and astringency (Bonvehi and Coll 1997; Nazaruddin et al. 2006a) and also give rise to the brown colouration of the beans that is typical of well-fermented cocoa beans. Polyphenol oxidase catalyses the hydroxylation of monophenols to diphenols such as hydroquinone, and in a second step, the oxidation of colourless diphenols to highly coloured *o*-quinone, which is an extremely reactive intermediate (Kyi et al. 2005). These *o*-quinones can react in a number of different ways with the vast variety of compounds that are formed during the preceding stages and may also polymerise to form diphenols and diphenol-quinones (Lopez and Dimick 1995).

Total polyphenols decreased with increasing fermentation for all pod storage treatments as shown in the response surface plot (Figure 16.11). It decreased from 171.54 mg/g at the start of fermentation to 153.58 mg/g at the end of the fermentation process for the unstored pods. It also decreased from 169.09–148.77 mg/g for pods stored for 3 days, 149.24–119.43 mg/g for pods stored for 7 days and 123.24–83.48 mg/g for pods stored for 10 days. Results from this study suggested that fermentation of cocoa beans leads to gradual loss of total polyphenols in fermented cocoa beans. Aikpokpodion and Dongo (2010) observed that polyphenol content of cocoa beans decreased from 16.11% (wt/wt; 161.1 mg/g) on day 0 to 6.01% (wt/wt; 60.1 mg/g) after 6 days of fermentation. Nazaruddin et al. (2006a) also reported that epicatechin and catechin content, respectively, were reduced to approximately 10–70% during fermentation. Studies have also shown that loss of polyphenols during fermentation is not only due to the oxidation process but also caused by diffusion of polyphenols into fermentation sweatings (Kim and Keeney 1984; Hansen et al. 1998; Wollgast and Anklam 2000). Forsyth (1952) reported loss of polyphenols caused by

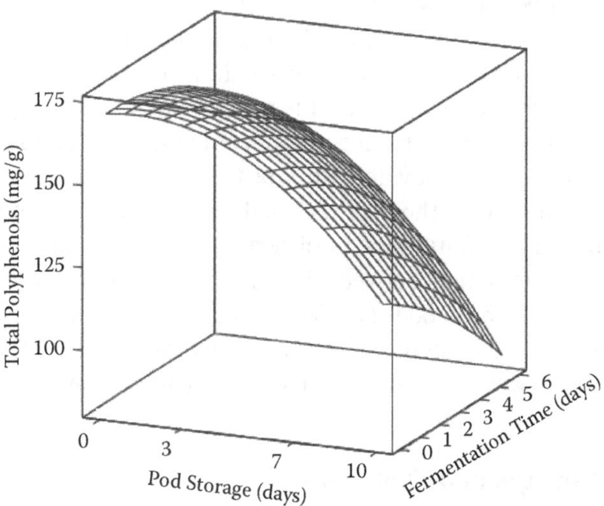

FIGURE 16.11 Response surface plot showing changes in total polyphenols during fermentation of pulp pre-conditioned cocoa beans.

diffusion into the fermentation sweatings which was later confirmed by microscopic studies carried out by de Brito et al. (2000).

Again, results (Figure 16.11) showed that increasing pod storage consistently reduced the levels of total polyphenols in the beans at all fermentation times. Total polyphenols in the unfermented cocoa beans decreased during pod storage. It decreased from 171.54 mg/g at the start of pod storage to 123.24 mg/g by 10 days of pod storage. At the end of fermentation (6 days), total polyphenols decreased by 12%, 14%, 25% and 48% for the pods stored for 0, 3, 7 and 10 days, respectively. Similar observations were made by other researchers (Meyer et al. 1989; Nazaruddin et al. 2006a). Polyphenol oxidase is the major oxidase in cocoa beans responsible for catalysing the oxidation of polyphenols during fermentation. This enzyme is reported to become active during the aerobic phase of the fermentation as a result of oxygen permeating the cotyledon (Thompson et al. 2001). Among the factors that facilitate the activity of this enzyme were reduction in the amount of seed pulp, seed death, subsequent breakdown of subcellular membranes and aeration of the bean by turning of the fermenting bean mass. Pod storage reduced the pulp volume per seed due to water evaporation and inversion of sucrose causing an increase in micro-aeration within the pulp and the fermenting mass (Biehl et al. 1989; Sanagi et al. 1997). This served to enhance the activity of polyphenol oxidase resulting in the oxidation of polyphenols.

Nazaruddin et al. (2006a) also reported that the reduction in pulp volume as a result of pod storage might facilitate the oxidation and polymerisation of (–)-epicatechin and its oxidation products. Said et al. (1988) also observed experimentally that there was significant reduction in the content of polyphenolic compounds, especially (–)-epicatechin in the pod storage period and degradation of the (–)-epicatechin and (+)-catechin was significant after 5 days fermentation. This suggests that pod storage could be effectively employed to reduce the polyphenol content of cocoa beans and hence reduce the astringency and bitter taste of cocoa beans.

Regression analysis of the data revealed significant ($p < 0.05$) influence of the linear and quadratic factors of fermentation time and pod storage as well as the interaction between PS and FT on the total polyphenols of the cotyledons. The model developed could explain about 96% of the variations in the total polyphenols of the cotyledons, suggesting that 4% of the variations were due to other factors not investigated in this work (Table 16.5).

16.4.3.2 Changes in *o*-Diphenols

Polyphenol oxidase, which is a copper-containing enzyme, catalyses the aerobic regioselective oxidation of monophenols to *o*-diphenols followed by dehydrogenation to *o*-quinones (Yağar and Sağiroğlu 2002). During

TABLE 16.5
Regression Coefficients and Their R^2 Values in the Models for Polyphenols, *o*-Diphenols, Anthocyanins, Free Fatty Acids and Fermentation Index of Cocoa Beans

Variables	Total Polyphenols	*o*-Diphenols	Anthocyanins	Free Fatty Acids	Fermentation Index
Constant	152.936[a]	22.252[a]	5.6314[a]	0.47400[a]	1.17296[a]
X_1	–29.450[a]	–2.565[a]	–1.6742[a]	0.05690[a]	0.11583[a]
X_2	–13.480[a]	–2.491[a]	–1.8928[a]	0.09094[a]	0.35506[a]
X_1^2	–15.487[a]	–2.652[a]	0.1686	–0.02629[a]	–0.04400
X_2^2	–4.157[a]	–1.521[a]	0.1136	–0.05656[a]	–0.09588[a]
X_1 x X_2	–5.515[a]	–1.377[a]	0.4271[a]	–0.01724[a]	–0.00884
R^2	96.2%	93.0%	93.4%	87.8%	87.2%
R^2 (adjusted)	95.7%	92.2%	92.7%	86.4%	85.7%

[a] Significant at $p < 0.05$; X_1 = Pod storage; X_2 = Fermentation time.

polyphenol oxidation, monophenols are first oxidised to *o*-diphenols followed by dehydrogenation to *o*-quinones which can react in a number of different ways with the vast variety of compounds that are formed during the preceding stages and may also polymerise to form diphenols and diphenol-quinones (Lopez and Dimick 1995).

The response surface plot (Figure 16.12) showed that *o*-diphenols content in the beans decreased significantly ($p < 0.05$) with increasing fermentation time for all pod storage periods. With the exception of the unstored pods, the *o*-diphenol content of the beans recorded a continuous decrease with increasing fermentation time. It decreased from 23.65 mg/g at the start of fermentation to 18.79 mg/g at the end of fermentation (6 days) for the pods stored for 3 days. It also decreased from 22.05 mg/g to 16.72 mg/g for pods stored for 7 days and from 19.56 mg/g to 11.59 mg/g for pods stored for 10 days at the end of fermentation. The continuous reduction in the levels of *o*-diphenols in the cotyledons with fermentation might be due to the dehydrogenation of *o*-diphenols to *o*-quinones catalysed by polyphenol oxidase (Lopez and Dimick 1995). However, with the unstored pods, the *o*-diphenols content increased from 21.47 mg/g at the start of fermentation to 22.38 mg/g at day 3 and then decreased to 19.70 mg/g at the end of the fermentation. This might be due to high monophenol content in the unstored pods prior to fermentation compared to the unfermented stored pods. Hence the monophenols were first oxidised to *o*-diphenols to increase the content of the *o*-diphenols and then subsequent dehydrogenation to *o*-quinones.

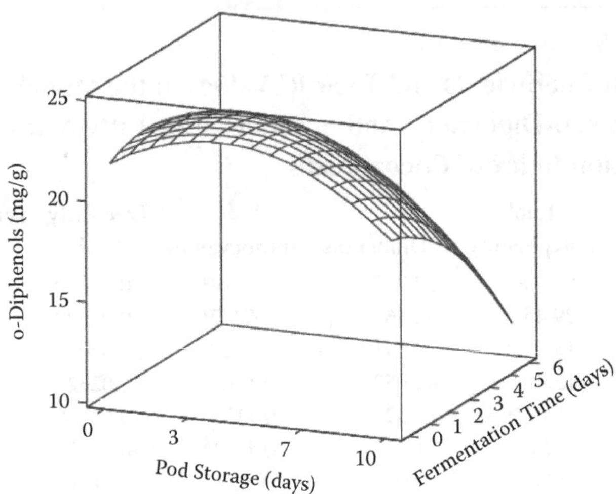

FIGURE 16.12 Response surface plot showing effect of pod storage time and fermentation time on the *o*-diphenols of cocoa beans.

The response surface plot (Figure 16.12) also showed that *o*-diphenol content in the beans decreased significantly ($p < 0.05$) with increasing pod storage for all fermentation times. The *o*-diphenol content of the unfermented cocoa beans decreased from 21.47 mg/g for the unstored pods to 19.56 mg/g after 10 days of pod storage. The observed reduction in *o*-diphenol content suggests the dehydrogenation of *o*-diphenols to *o*-quinones during pod storage. The *o*-diphenol content of the unfermented cocoa beans, however, increased slightly from 21.47 mg/g (unstored pods) to 23.65 mg/g on the third day of pod storage and then decreased to 19.56 mg/g after 10 days of pod storage. This might be due to high monophenol content in the beans of freshly harvested pods prior to storage, hence the monophenols were first oxidised to *o*-diphenols to increase the content of the *o*-diphenols and then subsequent dehydrogenation to *o*-quinones.

Regression analysis of the data also showed significant ($p < 0.05$) influence of the linear factor of fermentation time and pod storage and quadratic factor of PS and FT on the *o*-diphenols of the cotyledons. There was also significant ($p < 0.05$) influence of the interaction between PS and FT on the *o*-diphenols of the cotyledons. The model developed had an R^2 of 93% implying that 93% of the variations in the *o*-diphenols of the cotyledons could be explained by the model. This also suggests that 7% of the variations were due to other factors that were not investigated in this work (Table 16.5).

16.4.3.3 Changes in Anthocyanins

Anthocyanins which are responsible for the characteristic purple colour of unfermented cocoa beans (Ziegleder 2009) are located in the specialised vacuoles within the cotyledon and are hydrolysed by glycosidase to anthocyanidins during cocoa fermentation (Afoakwa 2010). Thompson et al. (2001) observed that the enzyme cleaves the sugar moieties galactose and arabinose attached to the anthocyanins. These result in the bleaching of the purple colour of the beans as well as the release of reducing sugars that can participate in flavour precursor reactions during roasting. Anthocyanins usually disappear rapidly during the fermentation process.

The response surface plot (Figure 16.13) showed that the anthocyanin content of the beans decreased significantly ($p < 0.05$) with increasing fermentation time for all pod storage periods. It decreased from 9.53 mg/kg at the start of fermentation to 5.35 mg/kg at the end of fermentation for the unstored pods. It also decreased from 8.95–4.07 mg/kg, 6.57–3.36 mg/kg and 5.90–3.03 mg/kg at the end of fermentation for pods stored for 3, 7 and 10 days, respectively. The observed reduction might be due to the activities of glycosidase, which hydrolysed the anthocyanins in the cotyledons to anthocyanidins (Afoakwa 2010).

Increasing pod storage consistently reduced the anthocyanin levels at all fermentation times (Figure 16.13). The anthocyanin levels of the unfermented cocoa beans decreased significantly ($p < 0.05$) from 9.53 mg/kg for the unstored pods to 5.90 mg/kg after 10 days of pod storage. The observed reduction in the anthocyanin levels during pod storage suggests possible

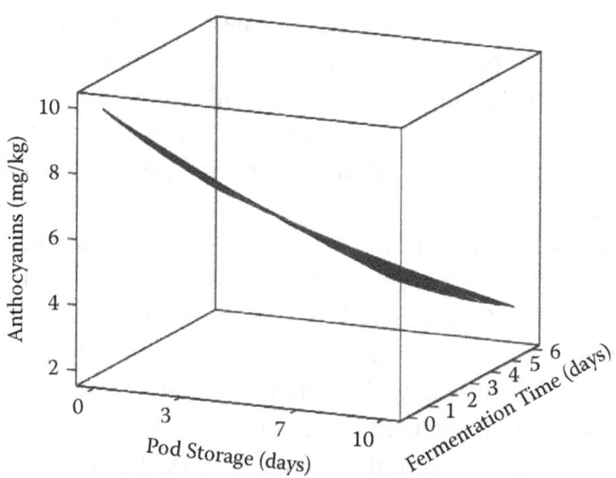

FIGURE 16.13 Response surface plot displaying anthocyanin content of cocoa beans as affected by pod storage and fermentation.

breakdown of anthocyanins in the cotyledons to anthocyanidins during the storage period. At the end of fermentation (6 days), anthocyanins decreased by 44% for the unstored pods, 55% for pods stored for 3 days and 49% for pods stored for 7 and 10 days. The breakdown of anthocyanins during pod storage and fermentation is important as this would lead to the formation of more condensation products of anthocyanin, such as cyanidin-3-β-D-galactosid and cyanidin-3-α-L arabinosid (Kim and Keeney 1984) and thus change the colour of the beans from slaty over purple to brown (Kealey et al. 1998).

Regression analysis of the data also showed significant ($p < 0.05$) influence of the linear factor of fermentation time and pod storage on the anthocyanin content of the cotyledons. There was also significant ($p < 0.05$) influence of the interaction between PS and FT on the anthocyanin content of the cotyledons. The quadratic factor of PS and FT was, however, not significant ($p > 0.05$). The model developed had an R^2 of 93%. This implies the model could explain about 93% of the variations in the anthocyanin content of the cotyledons, and that 7% of the variations were due to other factors not investigated in this work (Table 16.5).

16.4.3.4 Changes in Free Fatty Acids (FFAs)

Free fatty acids are carboxylic acids released from triglycerides (Selamat et al. 1996) through the effect of a lipase (E.C. 3.1.1.3) or an oxidation (Guehi et al. 2008). The quality of raw cocoa beans depends widely on their FFA content as it gives the measure of rancidity of cocoa beans (Afoakwa et al. 2011b) and high FFA content is reported to be a serious quality defect which reduces the technical and economic value of the cocoa beans (Guehi et al. 2008). Again, the cocoa fat (butter) hardness is reported to depend on the saturated and unsaturated fatty acid contents bound in triglycerides and on free fatty acid content (Guehi et al. 2008), and a high FFA content leads to a decrease in hardness of cocoa butter which reduces the commercial value for both processors and chocolate manufacturers.

There were general increases in FFA levels with increasing fermentation (Figure 16.14). The FFAs increased from 0.26% at the start of fermentation to 0.42% at the end of fermentation for the unstored pods. A similar trend of increase was observed for all pod storage treatments. The FFAs also increased with increasing pod storage (Figure 16.14). The FFAs for the unfermented beans increased from 0.26% for both the unstored pods and pods stored for 3 days, to 0.31% and 0.52% for pods stored for 7 and 10 days, respectively. At the end of fermentation, the FFA levels were 0.51%, 0.52% and 0.52% for pods stored for 3, 7 and 10 days, respectively. The model developed to predict the effect of pod

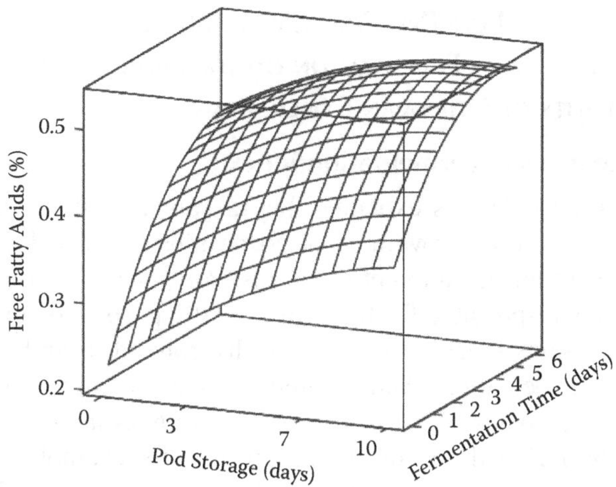

FIGURE 16.14 Response surface plot showing effect of pod storage on the free fatty acids of cocoa beans during fermentation.

storage and fermentation of the free fatty acids of cocoa beans had an R^2 of 88% (Table 16.5). This implies that the model developed could explain about 88% of the variations in the FFAs of the cotyledons, whereas 12% of the variations were due to other factors not investigated in this work. Regression coefficients revealed that both the linear and quadratic factors of fermentation time and pod storage as well as the interaction between PS and FT had significant ($p < 0.05$) influence on the FFAs of the cotyledons.

The gradual increase in FFAs in the cocoa beans during both pod storage and fermentation could be attributed to the activity of the lipase enzyme present in the natural cocoa beans which acts to break down the triglycerides into separate groups of the fatty acids and glycerol thereby freeing the fatty acids (Dand 1997). The European parliament and European council directive 73/241/EEC (EEC 1973) limits the maximum FFA content to 1.75% oleic acid equivalent in cocoa butter. To be able to meet the acceptable level, Dand (1997) reported that the FFA levels should be less than 1% in fresh cocoa beans and less than 1.75% in dried cocoa beans. Even though the FFA levels in the cocoa beans increased with both fermentation time and pod storage, the levels were all, however, below the acceptable limits of 1.75% oleic acid equivalent in cocoa butter. Results from this study suggest that cocoa pods can be stored up to 10 days and beans fermented for 6 days without adversely affecting the FFA levels in the fermented beans.

16.4.4 Effects of Pulp Pre-Conditioning (Pod Storage) and Fermentation on the Fermentative Quality of Ghanaian Cocoa Beans

16.4.4.1 Changes in Fermentation Index

The fermentation index, as explained by Takrama, Aculey, and Aneani (2006), is a measure of brownness of cocoa nibs and it helps to ascertain the degree of fermentation of the beans. Polyphenol compounds such as anthocyanins responsible for the characteristic purple colour of unfermented cocoa beans (Ziegleder 2009) are hydrolysed to anthocyanidins during cocoa fermentation. Anthocyanidins then polymerise along with simple catechins to form complex tannins. Anthocyanins usually disappear rapidly during the fermentation process; for example, 93% were reportedly lost after 4 days fermentation (Wollgast and Anklam 2000) and the colour of the beans changes from slaty over purple to brown (Kealey et al. 1998). Thus, anthocyanin content has been considered as a good index for determination of the degree of cocoa bean fermentation (Pettipher 1986; Shahidi and Naczk 1995). Fermented cocoa beans with FI values below one indicate under-fermentation whereas fermented beans with FI values of one and above are considered to be well-fermented (Gourieva and Tserrevitinov 1979).

The response surface plot (Figure 16.15) showed the fermentation index of the beans during fermentation for all pod storage treatments. FI increased from 0.674 at the start of fermentation to 1.390 for the unstored pods. It

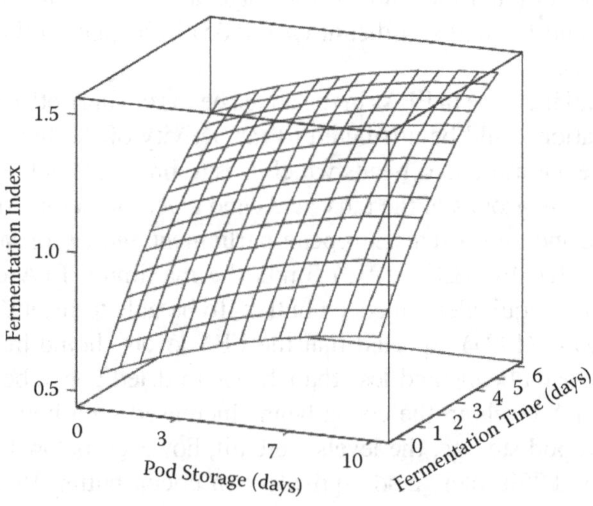

FIGURE 16.15 Response surface plot showing effect of pod storage and fermentation time on the fermentation index of cocoa beans.

also increased from 0.675–1.389 for pods stored for 3 days, 0.675–1.423 for pods stored for 7 days and from 0.763–1.424 for pods stored for 10 days.

The increase in FI as fermentation progressed could be due to the formation of more and more condensation products of anthocyanin, such as cyanidin-3-β-D-galactosid and cyanidin-3-α-L arabinosid (Kim and Keeney 1984) as a result of the breakdown of anthocyanin pigments. Afoakwa et al. (2011a) also observed a similar trend of increasing fermentation index with fermentation time for all pod storage treatments except for pods stored for 21 days which recorded a decrease in FI by the sixth day of fermentation.

Pod storage, however, caused only marginal increases in FI at all fermentation times. FI of the unfermented beans increased slightly from 0.674 for the unstored pods to 0.763 after 10 days of pod storage. The FI of the fermented beans (6 days fermentation) also increased from 1.390 for the unstored pods to 1.424 for pods stored for 10 days. Pod storage reduces the pulp volume per seed due to water evaporation and inversion of sucrose, and increases micro-aeration within the pulp as well as within the fermenting mass. This might serve to enhance the activity of polyphenol oxidase leading to oxidation of polyphenolic compounds resulting in the formation of brown pigment, thus increasing the FI of the cocoa nibs.

Regression analysis of the data showed a significant ($p < 0.05$) influence of the linear of fermentation time and pod storage and quadratic factor of FT on the FI of the nibs. There was no significant ($p > 0.05$) interaction between PS and FT on the FI of the nibs. The model developed could explain about 87% of the variations in the FI of the nibs, suggesting that 13% of the variations were due to other factors not investigated in this work (Table 16.5).

16.4.4.2 Cut Test of Unfermented and Fermented Cocoa Beans

The cut test is the simplest, and still the most widely used method to assess the quality of a random sample of beans from a batch by visual evaluation of the cut beans (Lopez and Dimick 1995). It is used for the evaluation of the sanitary quality of the beans (Guehi et al. 2007) and also assesses the degree of fermentation by counting the fully brown, brown/purple and purple coloured beans (Fowler 2009). This test identifies beans that are visibly mouldy, slaty (i.e., unfermented), infested, germinated, flat (i.e., containing no nib or cotyledon), purple or brown (Fowler 2009).

Results of the cut test of unfermented beans as well as beans fermented for 3 and 6 days for all the pod storage treatments are shown in Table 16.6. Generally, there were increases in brown beans with increasing fermentation time for all the pod storage treatments. The brown beans increased from 0.33 to 57% and from 5 to 66% for the unstored pods and pods stored for 3 days, respectively, at the end of fermentation. It also increased from 23 to

TABLE 16.6
Cut Test Result for Unfermented and Fermented Cocoa Beans

Pod Storage (Days)	Fermentation Time (Days)	Deep Purple (%)	Pale Purple (%)	Brown (%)	Slaty (%)	Mouldy (%)	Germinated (%)	Other Defects (%)
0	0	6.67	1.67	0.33	91.33	0	0	0
	3	7.67	59.0	22.33	11.0	0	0	0
	6	3.33	40.0	56.67	0	0	0	0
3	0	10.67	3.33	5.33	89.67	0	0	0
	3	6.34	74.33	14.33	5.0	0	0	0
	6	2.37	31.67	65.96	0	0	0	0
7	0	25.0	19.0	22.67	33.33	0	0	0
	3	0.67	20.0	78.33	1.0	0	0	0
	6	0	6.33	93.67	0	0	0	0
10	0	10.68	23.0	11.33	48.33	3.33	3.33	0
	3	0	44.67	52.0	0	0	3.33	0
	6	0	27.34	72.33	0	0	0.33	0

94% and 11 to 72% for pods stored for 7 and 10 days, respectively, at the end of fermentation. Changes in anthocyanins and oxidation products of the polyphenol oxidase might have contributed to the brown pigment formation in the cocoa beans during the fermentation period (Afoakwa et al. 2012). Anthocyanins are rapidly hydrolysed to anthocyanidins and sugars (galactose and arabinose) by glycosidase during cocoa fermentation (Camu et al. 2008). The decreasing of polyphenols and anthocyanins content normally leads to the changes of cocoa bean colour from purple to brown (Guehi et al. 2007). Again, Shamsuddin and Dimmick (1986) suggested that the brown pigments might also be produced from complexation of condensed tannin, a high molecular weight product of flavonoid polymerisation, with protein, via hydrogen bonding.

Results also showed that increasing pod storage increased the percentage of brown beans. Afoakwa et al. (2012) reported that cocoa flavours develop best when the degree of fermentation (percentage of fully brown beans) is above 60%. Beans stored for 3, 7 and 10 days produced brown beans above 60%. This is because pod storage reduced the pulp volume per seed and increased micro-aeration within the pulp and the fermenting mass. This is probably due to the activity of polyphenol oxidase resulting in the oxidation of polyphenols. There were also reductions in the deep purple beans with increasing fermentation for all pod storage times. Purple beans are reported to contain high polyphenol and anthocyanin content (Guehi et al. 2007) and reduction in purple beans with fermentation suggested that polyphenols and anthocyanins were being degraded. The percentage of pale purple beans at the end of fermentation for all pod storage treatments ranged from 6% to 40%.

Pale purple beans are not defective beans as they change to brown upon storage (Takrama et al. 2006) and the trade accepts up to 30–40%, but samples containing over 50% are unacceptable (Wood and Lass 1985). Results showed that the proportion of pale purple beans did not exceed 50% for all the pod storage days and this gives an indication that the beans were adequately fermented.

The percentage of slaty beans was higher in the unfermented beans for all pod storage treatments but reduced drastically at the end of the fermentation. Germination occurred in beans stored for 10 days prior to fermentation (Table 16.6) and this was as a result of the prolonged storage of the pods which resulted in the rotting of pods and consequently penetration of oxygen into the pods creating optimum conditions for growth of the beans. However, during fermentation heat was produced and, coupled with the diffusion of some metabolites (ethanol and acetic acid) into the beans, resulted in the death of the beans, hence arresting germination. This was evident as the percentage of germinated beans reduced from 3.33–0.33% by the end of the fermentation process. Earlier work by Afoakwa et al.

(2012) also reported the occurrence of germinated beans of 2% and 11% in the pods stored for 14 and 21 days, respectively, prior to fermentation due to prolonged storage. Germinated beans are considered a defect because the hole left by the emerging radical provides an easy entrance for insects and moulds (Afoakwa et al. 2012). They are also considered to lack good chocolate flavour (Wood and Lass, 1985).

About 3% mouldy beans were detected in the beans whose pods were stored for 10 days before fermentation and this could be ascribed to the invasion of mould species *Phytophthora palminovora* and *Botryodiplodia theobrommae* during the prolonged pod storage. Wood and Lass (1985) reported that internal moulds are the major causes of off-flavours during cocoa processing, and samples of beans with as little as 4% of internal moulds can produce off-flavours in their finished products. The beans of the pods stored for 10 days, however, did not exceed this limit. Moulds inside the beans can also increase the free fatty acid content of the cocoa butter (Wood and Lass 1985) and this might have accounted for the high FFA value of 0.52% for pods stored for 10 days at the end of fermentation.

16.5 CONCLUSION

The pH of unfermented cocoa pulp was acidic ranging from 3.88–4.02. Pod storage and fermentation increased the pH of the pulp with a consequent decrease in titratable acidity. At the end of fermentation, the pH of the pulp increased from 3.88–3.96 for the unstored pods, 3.98–5.04 for pods stored for 3 days, 4.01–5.23 for 7 days pod storage and 4.02–5.24 for pods stored for 10 days. The gradual increase in pH during fermentation was probably due to the breakdown of citric acid in the pulp by yeasts and lactic acid bacteria. Increasing pod storage and fermentation time also decreased the reducing sugars and total solids in the pulp. Changes in mineral composition of cocoa pulp during fermentation of pulp pre-conditioned cocoa beans were noted. The most abundant minerals in the freshly harvested and unfermented cocoa pulp are calcium, followed by potassium and sodium with values of 316.92 mg/100 g, 255.12 mg/100 g and 103.26 mg/100 g, respectively. Zinc is the mineral with the least concentration of 1.04 mg/100 g, and iron and magnesium had appreciable values of 4.26 mg/100 g and 32.52 mg/100 g, respectively. Pod storage and fermentation, however, showed variable trends in the studied mineral content. Pod storage led to decreases in sodium and iron concentrations at all fermentation times. Increasing fermentation time increased the composition of calcium, magnesium, zinc and potassium at all pod storage periods

The pH of unfermented cocoa beans was slightly acidic, ranging from 6.09–6.27. Increasing pod storage caused a consistent increase in pH of the nibs at the end of fermentation with a consequent decrease in titratable acidity. Pod storage between 3–7 days produced fermented cocoa nibs (6 days fermentation) with pH between 5.10–5.36 which falls within reported pH (5.0–5.5) needed to produce cocoa beans of higher flavour potentials. Pod storage for 10 days, however, produced cocoa nibs with very high pH (7.01) at the end of fermentation. Fermenting freshly harvested cocoa beans (unstored pods) produced cocoa nibs with low pH (4.80) and high titratable acidity (0.268 meq NaOH/100 g) at the end of fermentation. Pod storage between 3–7 days and 6 days of fermentation could be used to produce cocoa beans with higher concentrations of flavour precursors and also reduced nib acidification. Increasing fermentation consistently decreased the non-reducing sugars, total sugars and protein content of the beans, whereas reducing sugars increased. The reductions in total and non-reducing sugars with a consequential increase in reducing sugars might be due to the activities of seed invertase, which hydrolysed sucrose into glucose and fructose during the fermentation period, whereas the reductions in protein might be due to endogenous seed proteases which could lead to the formation of flavour precursors such as hydrophilic oligopeptides and hydrophobic amino acids. Pod storage caused marginal reductions in total and non-reducing sugars with consequential increase in reducing sugars, whereas protein content was reduced significantly.

Total polyphenols, *o*-diphenols and anthocyanins content of freshly harvested unfermented cocoa beans were 171.54 mg/g, 21.47 mg/g and 9.53 mg/kg, respectively. Pod storage and fermentation caused significant reductions in total polyphenol, *o*-diphenol and anthocyanin content of the cocoa beans and thus will reduce the astringency and bitterness in cocoa and cocoa products as well as develop the brown colour of the beans. The reductions in total polyphenols and *o*-diphenols during fermentation was due to the action of polyphenol oxidase whose activities were further enhanced by the pod storage due to the reductions in seed pulp volume and increased aeration within the fermenting mass. Anthocyanin breakdown might be due to the activities of glycosidase, which hydrolysed the anthocyanins in the cotyledons to anthocyanidins. Pod storage up to 10 days with 6 days fermentation produced beans with 0.52% FFAs which was below the acceptable limits of 1.75% oleic acid equivalent. Cocoa pods could be stored up to 10 days and beans fermented for 6 days without adversely affecting the FFA levels in the fermented beans.

The fermentation index of the beans increased with increasing pod storage and fermentation time due to the breakdown of anthocyanins. The fermentation index of all the beans at the end of fermentation were above 1.0 indicating that all the beans were well fermented. Increasing pod storage and fermentation caused reductions in the percentage of slaty beans and deep purple beans but increased the percentage of brown beans. Beans from pods stored for 3, 7 and 10 days produced brown beans above 60%. Prolonged pod storage (10 days) caused 3% mouldy beans and 3% germinated beans.

Appendix A: Abbreviations and Acronyms

ABBREVIATIONS, ACRONYMS AND THEIR MEANINGS

ADI: Acceptable Daily Intake
AI, ai: Active Ingredient
ALARA: As Low As Reasonably Achievable
AOEL: Acceptable Operator Exposure Level
ARfD: Acute Reference Dose as Active Substance
CBE: Cocoa Butter Equivalent
CBR: Cocoa Butter Replacer
CCRD: Central Composite Rotatable Design
cDNA: Complementary Deoxyribonucleic Acid
CMR: Substances that are carcinogenic, mutagenic or toxic to reproduction
CNS: Central Nervous System
CSD: Crystal Size Distribution
CTU: Chocolate Temper Units
CXL: Codex Maximum Residue Limit (Codex MRL)
DSC: Differential Scanning Calorimetry
DT50: Period required for 50% dissipation (define method of estimation)
EC: European Commission
EU: European Union
FID: Flame Ionisation Detection
g: Gram
GAP: Good Agricultural Practice(s)
GC: Gas Chromatography
GC-MS: Gas Chromatography-Mass Spectrometry
GC-O: Gas Chromatography-Olfactometry
GHP: Good Hygienic Practice(s)
GLC: Gas Liquid Chromatography
GLP: Good Laboratory Practice
GMO: Genetically Modified Organism
GMP: Good Manufacturing Practice(s)
GMS: Glycerol MonoStearates
GSP: Good Storage Practice
GU: Gloss Units
GWP: Good Warehouse Practice(s)
ha: Hectare

HACCP: Hazard Analysis Critical Control Point (usually food processing)
HPLC: High Performance Liquid Chromatography (sometimes high pressure ~)
HV: High Volume
ICA: International Confectionery Association
ICCO: International Cocoa Organization
IOCCC: International Office of Cocoa, Chocolate and Confectionery
IPM: Integrated Pest Management
IRM: Insecticide Resistance Management
JECFA: Joint FAO/WHO Meeting on Contaminants and Food Additives
JMPR: Joint FAO/WHO Meeting on Pesticide Residues (*Codex Alimentarius*)
k: Kilo (10^3)
kg: Kilogram (10^3 g)
K: Organic carbon adsorption coefficient
Koc: Hydroxyl radical rate constant
KOH: Organic matter adsorption coefficient
Kom: Octanol water partition coefficient
L: Litre
LCOW: Lethal concentration, median
LD_{50}: Median lethal dose; *dosis letalis media*
LOAEL: Lowest Observable Adverse Effect Level
LOD_{50}: Limit of Determination, has also been used for "limit of detection"
LOEC: Lowest Observable Effect Concentration
LOEL: Lowest Observable Effect Level
LOQ: Limit of Quantification
LV: Low Volume
µg: Microgram
µm: Micrometre (micron)
m: Metre
M: Molar (g molecular weight),
mega~: (10^6)
MC: Moisture Content
mg: milligram
milli~: (10^{-3})
mL: Millilitre
MLD: Minimum Lethal Dose
MLT: Median Lethal Time
mm: Millimetre
mM: Millimolar
MoA: Mode of Action
mol: Mole (usually g molecular weight)
MRL: Maximum Residue Level

MSDS: Material Safety Data Sheet
NCA/CMA: National Confectioners Association/Chocolate Manufacturers Association
nd: Not detected
NEDI: National Estimated Daily Intake
NEL: No Effect Level
ng: Nanogram
NOAEC: No Observed Adverse Effect Concentration
NOAEL: No Observed Adverse Effect Level
NOED: No Observed Effect Dose
NOEL: No Observed Effect Level
OP: Organophosphorous Pesticide
p: Pico~ (10^{-12})
P: Partition coefficient between n-octanol and water
Pa: Pascal (1 bar = 100 kPa)
PBT: Persistent Bioaccumulative Toxic chemicals
pH: pH-value
PHI: Pre-Harvest Interval
PIC: Prior Informed Consent po by mouth (*per os*)
POP: Persistent Organic Pollutants
ppb: Parts per billion (10^{-9})
PPE: Personal Protective Equipment
ppm: Parts per million (10^{-6})
QPS: Quarantine Pre-Shipment (Fumigation)
QSAR: Quantitative Structure-Activity Relationship
RfD: Reference Dose
RH: Relative Humidity
SI: Système International, International standard units for measurement
SOP: Standard Operating Procedures sp species (only after a generic name)
TAGs: Triacylglycerols
TLC: Thin Layer Chromatography
TMDI: Theoretical Maximum Daily Intake
tMRL: Temporary Maximum Residue Limit
ULV: Ultra Low Volume
UV: Ultraviolet
VAR: Volume Application Rate
vPvB: Very Persistent, very Bioaccumulative

Appendix B: Websites

WEBSITES OF ORGANIZATIONS RELATED TO COCOA AND CHOCOLATE INDUSTRY

Association of the Chocolate, Biscuit & Confectionery Industries of the EU (CAOBISCO) http://www.caobisco.com/english/main.asp

CAB International: http://www.cabi.org/index.asp

Certification bodies involved with cocoa traceability and GAP:

1. The Fairtrade Foundation: http://www.fairtrade.net
2. The Rainforest Alliance: http://www.rainforest-alliance.org
3. UTZ CERTIFIED: http://www.utzcertified.org

Cocoa Merchants Association of America (CMAA): http://www.cocoamerchants.com/

Codex Alimentarius Commission: http://www.codexalimentarius.net/pesticide MRLs: http://www.codexalimentarius.net/mrls/pestdes/jsp/pest_q-e.jsp

COLEACP (horticultural GAP project): http://www.coleacp.org/

CropLife International: http://www.croplife.org/

Distribution and use of pesticides: http://www.fao.org/ag/AGP/AGPP/Pesticid/p.htm

European Cocoa Association (ECA): www.eurococoa.com

European Commission (Directorate General for Development and Directorate General for Health and Consumer Affairs [DG SANCO]) EU Food safety: http://ec.europa.eu/food/index_en.htm

EU Agriculture: http://ec.europa.eu/dgs/agriculture/index_en.htm

EU legislation on MRLs: http://ec.europa.eu/food/plant/protection/pesticides/index_en.htm

European Food Safety Agency: http://www.efsa.eu.int/

European Initiative for the Sustainable Development in Agriculture (EISA): http://www.sustainable-agriculture.org/start.html

European and Mediterranean Plant Protection Organization (EPPO): http://www.eppo.org/

Federation of Cocoa Commerce (FCC): http://www.cocoafederation.com/

Food and Agriculture Organisation (FAO): http://www.fao.org/

Global Forum on Agricultural Research (GFAR): (enhancing national capacities to adapt and transfer knowledge: hosted by FAO): http://www.egfar.org/

Health & Safety Executive (UK - formerly PSD): http://www.pesticides.gov.uk/food_safety.asp?id = 726

International Cocoa Organisation (ICCO): http://www.icco.org/

JMPR: technical monographs http://www.inchem.org/pages/jmpr.html

JMPR Reports: http://www.fao.org/agriculture/crops/core-themes/theme/pests/jmpr/jmpr-rep/en/

QC procedure: http://ec.europa.eu/food/plant/protection/resources/qualcontrol_en.pdf

Understanding the Codex: http://www.fao.org/docrep/w9114e/W9114e04.htm

Appendix C: Glossary

GLOSSARY OF COCOA AND CHOCOLATE TERMINOLOGIES

Adulteration: Alteration of the composition of graded cocoa by any means whatsoever so that the resulting mixture or combination is either not of the grade prescribed, or its quality or flavour is injuriously affected, or its bulk or mass is altered.

Bean cluster: Two or more beans joined together which cannot be separated by finger and thumb.

Bean count: The total number of whole beans per 100 g as graded in accordance with Table 12.1.

Bittersweet chocolate: Also referred to as 'dark chocolate', this is chocolate manufactured by blending a minimum amount of 35% cocoa liquor with variations of sugar, cocoa butter, emulsifiers and flavourings.

Bloom: The appearance of fat or sugar on the surface of chocolate giving it a white sheen or sometimes individual white blobs.

Broken bean: A cocoa bean of which a fragment is missing, the remaining part more than half of the whole bean.

Casson equation: $\sqrt{\tau} = \sqrt{\tau CA} + \sqrt{\mu CA}.\sqrt{\dot{\gamma}}$. (Variable definitions; τ: yield stress; τCA: Casson yield stress; μCA: Casson viscosity; and $\dot{\gamma}$; shear rate.

Cocoa bean: The seed of the cocoa tree (*Theobroma cacao* Linnaeus); commercially, and for the purpose of this international standard (ISO), the term refers to the whole seed, which has been fermented and dried.

Cocoa butter: A natural fat that is present in cocoa beans and obtained by pressing cocoa liquor.

Cocoa Butter Equivalent (CBE): These are vegetable fats that are totally compatible with cocoa butter and can be mixed with it in proportions stipulated by regulation.

Cocoa Butter Replacer (CBR): These are vegetable fats that may be mixed with cocoa butter but only in a limited proportion by regulation.

Cocoa liquor, cocoa mass: Also known as chocolate liquor, this is composed of roasted and ground cocoa nibs.

Cocoa nibs: Similar to cocoa cotyledons, these are cocoa beans with shells removed.

Cocoa powder: A product obtained by grinding or pulverizing pressed cocoa cake and available in different fat levels. It can be natural or manufactured by the Dutch process.

Contamination: The presence of a smoky, hammy or other smell not typical of cocoa, or substance not natural to cocoa which is revealed during the cut test or physical inspection of a sample.

Compound: This is a confectionery product in which vegetable oil has been substituted for cocoa butter.

Cut test: The procedure by which the cotyledons of cocoa beans are exposed for the purpose of determining the incidence of defective or slaty cocoa beans or violet or purple beans or the presence of contamination within a sample.

Defective beans: An internally mouldy or insect-damaged bean.

Dextrose: Also known as glucose or corn starch, it is a sweetener which is commercially made from starch by the action of heat and acids or enzymes, resulting in the complete hydrolysis of corn starch. It is a reducing sugar that produces high-temperature browning effects in baked foods. Industrially, it is used in ice cream, bakery products, confections and chocolate cookie drops. The sugar helps maintain the shape of the cookie drop during baking and reduces smearing of the chocolate after baking.

Dry cocoa: A commercial term designating cocoa beans which have been evenly dried throughout and of which the moisture content corresponds to the requirements of this international standard (ISO).

Dutching process: This is an alkaline treatment of cocoa nibs prior to grinding, or the liquor prior to pressing. It facilitates darkening of the resultant cocoa liquor, modifies the chocolate flavour and also helps keep the cocoa solids in uniform suspension in chocolate beverages.

Emulsifier: A surface-active agent that promotes the formation and stabilization of an emulsion. Examples are lecithin and polyglycerol polyricinolete (PGPR) that are used in chocolate manufacturing to help control flow properties.

Enrobing: The act of coating a candy center by covering it with chocolate. This could be either done by hand or mechanical means.

Fair fermented: Cocoa beans that are not more than 10% slaty and 10% defective by count.

Fat bloom: This is the visually undesirable white cast that appears on chocolate products as a result of poor or insufficient tempering or exposure of the chocolate to high temperatures without re-tempering.

Fermentation: A process by which complex microbial interaction naturally modifies the composition of cocoa beans so that upon roasting, they yield characteristic chocolate flavour.

Flat beans: A cocoa bean which is too thin to be cut to give a complete surface of the cotyledons.

Foreign matter: Any substance other than cocoa beans and residue.

Fragment: A piece of cocoa bean equal to or less than half the original bean.

Germinated bean: A cocoa bean the shell of which has been pierced, split or broken by the growth of the seed-germ.

Good fermented: Cocoa beans that are not more than 5% slaty and 5% of all other defectives by count.

Grinding: A mechanical process by which roasted cocoa bean nib is reduced to a smooth liquid known as cocoa liquor.

Hard butter: This is a class of specialty fats with physical properties similar to cocoa butter. They are typically solid to semi-solid at ambient temperatures and melt relatively rapidly at higher temperatures depending upon application.

Insect-damaged or infested bean: A cocoa bean the internal parts of which are found to contain insects or mites at any stage of development or to show signs of damage caused thereby, which are visible to the naked eye.

Lauric fat: A vegetable fat typically containing 40–50% lauric fat acid and mainly obtained from coconut and palm-kernel origin. Compound coatings containing lauric fats usually require appropriate tempering.

Lecithin: A natural food additive which acts as an emulsifier and surface-active agent. Most commercial lecithin products are derived from soybean. In chocolate manufacture, it controls flow properties by reducing viscosity and is typically used in ranges between 0.1% and 0.5%.

Milk chocolate: A chocolate product made by the combination of about 10% cocoa liquor, 12% milk with cocoa butter, sugar or sweeteners, emulsifiers and some flavourings.

Mouldy bean: A cocoa bean on the internal parts of which mould is visible to the naked eye.

Natural process: Non-alkalized cocoa liquor processed into cocoa powder without alkalizing treatment.

Non-lauric fat: This is an edible fat which does not contain lauric fatty acids. Examples are cottonseed oil, soybean oil and palm oil. Manufacture of confectionery products containing non-lauric fat typically requires no tempering and will possess a higher melting point.

Non-Newtonian liquid: A liquid such as molten chocolate whose viscosity varies according to rate of stirring (shear).

Origin liquor: Cocoa mass manufactured in country of bean origin.

Particle fineness: This is the measurement of average particle size of component solids in a chocolate mix and are expressed in ten-thousandths of an inch or in microns.

Piece of shell: Part of the shell without the kernel.

Plastic viscosity: Amount of energy required to keep a non-Newtonian liquid moving once motion has been initiated.

Press cake: The product that remains after most of the cocoa butter has been expressed from the cocoa liquor. Press cake is pulverized for making cocoa powder.

Pressing: The process of partially removing cocoa butter from cocoa liquor by means of hydraulic presses. The two products obtained after pressing are cocoa butter and pressed cake.

Residue: Any cocoa element other than whole cocoa beans and flat beans (broken beans, fragments and pieces of shell).

Roasting: A cooking or heating process applied to cocoa beans using dry heat at high temperatures to facilitate winnowing of the beans into nibs and also help develop the chocolate flavour.

Semi-sweet: See 'Bittersweet', another name for semi-sweet.

Sieve: Means a screen with round holes, the diameter of which shall be 5.0 mm minimum or maximum.

Sievings: The material which will pass through the sieve.

Slaty bean: A cocoa bean which shows a slaty appearance on at least half of the surface of the cotyledons exposed by the cut test.

Sweet chocolate: A chocolate product prepared by blending a minimum of 15% cocoa liquor with varying amounts of sweeteners and cocoa butter. Flavourings may sometimes be added.

Tempering: The process of fat crystallization during chocolate manufacture, so that the finished product solidifies in a stable crystal form. Proper tempering, when followed, provides good contraction from moulds, good setting properties, good surface gloss and shelf-life characteristics. Tempering is a critical step in chocolate manufacture and certain confectionary products.

Unsweetened baking chocolate: This is a consumer term for cocoa or chocolate liquor.

Vanillin: An artificial substitute for vanilla.

Violet or purple bean: A cocoa bean which shows a violet or purple colour on at least half of the surface of the cotyledons exposed by the cut test.

Viscosity: A measure of the resistance to flow of molten chocolate and determines its ability to be pumped through pipes during industrial manufacture, and the extent to which the chocolate could be used to cover the center of confectionery, cake, cookie or ice cream.

Chocolate viscosity is influenced by process, solids, particle size, distribution and formulation variations.

White chocolate: A chocolate product composed of sugar, cocoa butter, whole milk and flavourings. In the United States, this product cannot be called chocolate because it does not contain cocoa solids. It is sometimes referred to as 'white cocoa butter based confectionery coating'.

Winnowing: The process of cracking and removing the cocoa bean shell to reveal the inner part of the bean 'the nibs'.

Yield value: Amount of energy required to initiate motion in a non-Newtonian liquid, for example, molten chocolate.

Chocolate viscosity is influenced by process conditions, particle size distribution and formulation variations.

White chocolate. A chocolate product composed of cocoa butter, sugar, whole milk and flavourings. In the United States, this product used to be called confectionery coating because it does not contain cocoa solids. It is sometimes referred to as [illegible] confectionery coating.

Winnowing. The process of cracking and removing the cocoa bean shells to leave the inner part of the bean, the nibs.

W[illegible] [illegible]

References

Abarca, M.L., M.R. Bragulat, G. Castella, and F. J. Cabanes. 1994. Ochratoxin A production by strains of *Aspergillus niger* var. niger. *Applied and Environmental Microbiology* 60: 2650–2652.

Abrahams, P.W. 2002. Soils: Their implications to human health. *Science of the Total Environment* 291: 1–32.

Abrunhosa, L., R.R.M. Paterson, Z. Kozakiewicz, N. Lima, and A. Venancio. 2001. Mycotoxin: Production from fungi isolated from grapes. *Letters in Applied Microbiology* 32: 240–242.

Adjinah, K.O. and I.Y. Opoku. 2010. The national cocoa diseases and pest control (CODAPEC): Achievements and challenges. http://www.myjoyonline.com (accessed April 28, 2010).

Afoakwa, E.O. 2010. *Chocolate Science and Technology*. Oxford, UK: Wiley-Blackwell. pp 3–82.

Afoakwa, E.O. and A. Paterson. 2010. Cocoa fermentation: Chocolate flavour quality. In: *Encyclopedia of Biotechnology in Agriculture and Food*. Oxford: Taylor & Francis, pp. 457–468.

Afoakwa, E.O., A. Paterson, and M. Fowler (2007). Factors influencing rheological and textural qualities in chocolate - A review. *Trends in Food Science & Technology*, 18 (2007) 290–298.

Afoakwa, E.O., A. Paterson, M. Fowler, and A. Ryan. 2008. Flavour formation and character in cocoa and chocolate: A critical review. *Critical Reviews in Food Science and Nutrition* 48:840–857.

Afoakwa, E.O., A. Paterson, M. Fowler, and A. Ryan. 2009. Matrix effects of flavour volatiles release in dark chocolates varying in particle size distribution and fat content. *Food Chemistry* 113(1):208–215.

Afoakwa, E.O., J. Quao, J. Takrama, A.S. Budu, and F.K. Saalia. 2011a. Effect of pulp preconditioning on acidification, proteolysis, sugars and free fatty acids concentration during fermentation of cocoa (*Theobroma cacao*) beans. *International Journal of Food Sciences and Nutrition* 62(7):755–764.

Afoakwa, E.O., J. Quao, J. Takrama, A.S. Budu, and F.K. Saalia. 2011b. Chemical composition and physical quality characteristics of Ghanaian cocoa beans as affected by pulp pre-conditioning and fermentation. *Journal of Food Science and Technology* 47(1):3–11.

Afoakwa, E.O., J. Quao, J. Takrama, A.S. Budu, and F.K. Saalia. 2012. Influence of pulp preconditioning and fermentation on fermentative quality and appearance of Ghanaian cocoa (*Theobroma cacao*) beans. *International Food Research Journal* 19(1): 127–133.

Afoakwa, E.O., J.E. Kongor, J.F. Takrama, A.S. Budu, and H. Mensah-Brown. 2013a. Effects of pulp preconditioning on total polyphenols, *o*-diphenols and anthocyanin concentrations during fermentation and drying of cocoa (*Theobroma cacao*) beans. *Journal of Food Science and Engineering* 3 (2013) 235–245.

Afoakwa, E.O., J.E. Kongor, J.F. Takrama, and A.S. Budu, 2013b. Changes in acidification, sugars and mineral composition of cocoa pulp during fermentation of pulp pre-conditioned cocoa (*Theobroma cacao*) beans. *International Food Research Journal* 20 (3): 1215–1222.

Aikpokpodion, P.E., and L.N. Dongo. 2010. Effects of fermentation intensity on polyphenols and antioxidant capacity of cocoa beans. *International Journal of Sustainable Crop Production* 5(4): 66–70.

Aikpokpodion, P.E., L. Lajide, and A. Aiyesanmi. 2010. Heavy metals contamination in fungicide treated cocoa plantations in cross river state, Nigeria. *American-Eurasian Journal of Agriculture and Environmental Science* 8(3): 268–274.

Akrofi, A.Y., and F. Baah. (Eds.). 2007. *Proceedings of the 5th INCOPED International Seminar on Cocoa Pests and Diseases, International Permanent Working Group for Cocoa Pests and Diseases (INCOPED) Secretariat,* Cocoa Research Institute, Tafo, Ghana.

Allen, J.B. 1987. London Cocoa Trade Amazon Project. Final Rep. Phase two. *Cocoa Growers' Bulletin* 30:1–94.

Alloway, B.J. 1995. Soil processes and the behavior of heavy metals. In *Heavy Metals in Soils*, 2nd edition, B.J. Alloway (Ed.). New York: Blackie, pp. 11–37.

Almela, L., V. Rabe, B. Sánchez, F. Torrella, J. López-Pérez, J.A. Gabaldón, and L. Guardiola. 2007. Ochratoxin A in red paprika: Relationship with the origin of the raw material. *Food Microbiology* 24:319–327.

Alva, A.K., B. Huang, and S. Paramasivam. 2000. Soil pH affects copper fractionation and phytotoxicity. *Soil Science Society of America Journal* 64: 955–962.

Alvim, P.D.T. 1984. Flowering of cacao. *Cocoa Growers' Bulletin*, 35:23–31.

Amézqueta, S., E. Gonzalez-Penas, M. Murillo, and A. Lopez de Cerain. 2004. Validation of a high-performance liquid chromatography analytical method for ochratoxin A quantification in cocoa beans. *Food Additives and Contaminants* 22:590–596.

Amézqueta, S., E. Gonzalez-Penas, M. Murillo, and A. Lopez de Cerfin. 2005. Occurrence of ochratoxin A in cocoa beans: Effect of shelling. *Food Additives and Contaminants* 22: 590 –595.

Amoa-Awua, W., M. Madsen, J. Takrama, A.O. Olaila, L. Ban-Koffi, and M. Jakobsen. 2006. *Quality Manual for Production & Primary Processing of Cocoa*. Ghana: Food Research Institute, Council for Scientific and Industrial Research. pp 2–18.

Amoye, S. 2006. Cocoa sourcing, world economics and supply. *The Manufacturing Confectioner* 86(1):81–85.

Ampuero, S. and J.O. Bosset. 2003. The electronic nose applied to dairy products: A review. *Sensors and Actuators* B 94:1–12.

Anon. 2002. Why buy ethically? An introduction to the philosophy behind ethical purchasing. http://www.ethicalconsumer.org (accessed June 2011).

Anon. 2008. History of chocolates. http://www.barry-callebaut.com/1589 (accessed October 20, 2011).

Anon. 2011. Cocoa. http://www.theice.com/publicdocs/ICE_Cocoa_Brochure.pdf (accessed August 7, 2011).

AOAC. 1990. *Official Methods of Analysis*. 15th edition. Washington DC: Association of Official Analytical Chemists.

AOAC. 2005. *Official Methods of Analysis*. 18th edition. Washington DC: Association of Official Analytical Chemists.

Ardhana, M.M., and G.H. Fleet. 2003. The microbial ecology of cocoa bean fermentations in Indonesia. *International Journal of Food Microbiology*, 86:87–99.

Aremu, C.Y., M.A. Agiang, and J.O.I. Ayatse. 1995. Nutrient and anti-nutrient profiles of raw and fermented cocoa beans. *Plant Foods for Human Nutrition* 48:221–223.

Arinze E.A., J.S. Sokhansanj, G.J. Schoenu and F.G. Trauttmansdorff. 1996. Experimental evaluation, simulation and optimization of a commercial heated-air batch hay dryer: Part 1, Dryer functional performance, product quality, and economic analysis of drying. *Journal of Agricultural Engineering Research*, 63:301–314.

Aroyeun, S.O., J.O. Ogunbayo, and A.O. Olaiya. 2006. Effect of modified packaging and storage time of cocoa pods on the commercial quality of cocoa beans. *British Food Journal* 108:141–151.

Asante, E.G. 1995. The economic relevance of plant diseases and pest management in the Ghana cocoa industry. A paper presented at the *International Cocoa Pests and Diseases Seminar*. Accra, November 6th–10th. Tafo: CRIG

Augier, F., J. Nganhou, M. Barel, J.C. Benet, and G. Berthomieu. 1998. Reducing cocoa acidity during drying. *Plantations, Recherche, Developpement* 5:127–133.

Awua, P.K. 2002. *Cocoa Processing and Chocolate Manufacture in Ghana*. Essex, UK: David Jamieson and Associates Press.

Baah, F., and V. Anchirinah. 2011. A review of Cocoa Research Institute of Ghana extension activities and the management of cocoa pests and diseases in Ghana. *American Journal of Social and Management Sciences* 2(1):196–201.

Baker, D.M., K.I. Tomlins, and C. Gray, 1994. Survey of Ghanaian cocoa farmer fermentation practices and their influence on cocoa flavour. *Food Chemistry* 51:425–431.

Bakker, J. H., and F.H.J. Bunte. 2009. Biologische internationale handel. LEI, Wageningen UR. http://www.lei.dlo.nl/publicaties/PDF/2009/2009-003.pdf (accessed May, 15, 2009).

Barel, M.A. 1986. Delai d'kcabossage. Influence sur les rendements et la qualite du cacao marchand et du cacao torrefie. *Café, Cacao, Thé* 31:141–149.

Barel, M. 1987. Pod breaking delay. Influence on the yields and the quality of raw and roasted cocoa. *Café, Cacao, Thé* 31:141–150.

Barel, M.A. 1997. Première transformation du cacao. Formation de l'arôme cacao. In *Cacao, chocolat, production, utilisation, caractéristiques*, J. Pontillon (Ed.). Paris: Techniques et documentation, Lavoisier, pp. 96–115.

Barry Callebaut. 2013. Organic cocoa farming: Expanding opportunities for cocoa farmers. http://www.barry-callebaut.com (accessed July 2, 2013).

BCCCA, Biscuit, Cake, Chocolate and Confectionery Alliance. 1996. *Cocoa Beans: Chocolate Manufacturers' Quality Requirements*, 4th edition. London: BCCCA.

Beckett, S.T. 2000. *The Science of Chocolate*. London: Royal Society of Chemistry.

Beckett, S.T. 2008. *The Science of Chocolate*, 2nd edition. London: Royal Society of Chemistry.

Beckett, S.T. 2009. *Industrial Chocolate Manufacture and Use*, 4th edition. Oxford, Blackwell, pp. 238–240, 277.

Belitz, H.D. and W. Grosch. 2004. Cocoa and chocolate. In *Food Chemistry*, 2nd edition. Berlin: Springer-Verlag, pp. 894–903.

Belitz, H.D., W. Grosch, and P. Schieberle. 2009. *Food Chemistry*, 4th edition. Springer-Verlag, Berlin, Heidelberg, pp. 959–967.

Bempah, C.K. et al. 2011. A preliminary assessment of consumer's exposure to organochlorine pesticides in fruits and vegetables and the potential health risk in Accra Metropolis, Ghana. *Food Chemistry* 128:1058–1065.

Benkouider, C. 2005. Going organic: Cadbury acquires Green and Black's. Euromonitor. http://www.marketresearchworld.net/index.php?option=com_content&task=view&id=182&Itemid=77 (accessed February 19, 2005).

Berbert, P.R.F. 1976. Revista. *Theobroma* 6:67.

Berbert, P.R.F. 1979. Contribuicao para o conhecimento dos acucares componentes da amendoa e do mel de cacao. *Revista Theobroma (Brasil)*, 9:55–61.

Betina, V. 1989. *Mycotoxins, Chemical, Biological and Environmental Aspects*. Amsterdam: Elsevier.

Biehl, B. 1984. Cocoa fermentation and the problem of acidity, over fermentation and low cocoa flavour. In *International Conference on Cocoa and Coconut*, 15–17 October, Kuala Lumpur, Malaysia, pp. 1–8.

Biehl, B., and D. Passern. 1982. Proteolysis during fermentation-like incubation of cocoa seeds. *Journal of Science of Food & Agriculture*, 33:1280–1290.

Biehl, B., and J. Voigt, 1994. Biochemical approach to raw cocoa quality improvement: Comparison of seed proteins and proteases in their ability to produce cocoa aroma precursors. Paper presented at the Malaysian International Cocoa Conference, 20–21 October 1994, Kuala Lumpur, Malaysia.

Biehl, B. and J. Voigt. 1996. Biochemistry of chocolate flavour precursors. In *Proceedings of International Cocoa Conference*. Salvador, Bahia.

Biehl, B., and J. Voigt. 1999. Biochemistry of cocoa flavour precursors. In *Proceedings of the 12th International Cocoa Research Conference, Salvador, Brazil, 1996*. Lagos, Nigeria: Cocoa Producers Alliance, pp. 929–938.

Biehl, B., and G. Ziegleder. 2003. Cocoa: Chemistry of processing. In *Encyclopaedia of Food Sciences and Nutrition*, 2nd edition, Vol. 3. New York: Academic Press, pp. 1436–1448.

Biehl, B., E. Brunner, D. Passern, V.C. Quesnel, and D. Adomako. 1985. Acidification, proteolysis and flavour potential in fermenting cocoa beans. *Journal of Agriculture and Food Chemistry* 36:583–598.

Biehl, B., B. Meyer, G. Crone, L. Pollmann, and M.B. Said. 1989. Chemical and physical changes in the pulp during ripening and post-harvest storage of cocoa pods. *Journal of the Science Food and Agriculture* 48:189–208.

Biehl, B., B. Meyer, M.B. Said, and R.J. Samarakhody. 1990. Bean spreading: A method for pulp preconditioning to impair strong nib acidification during cocoa fermentation in Malaysia. *Journal of Science of Food & Agriculture* 51:35–45.

Biehl, B., D. Passern, and W. Sagemann. 1982b. Effect of acetic acid on subcellular structures of cocoa bean cotyledons. *Journal of Agriculture and Food Chemistry* 33:1101–1109.

Biehl, B., C. Wewetzer, and D. Passern. 1982a. Vacuolar (storage) proteins of cocoa seeds and their degradation during germination and fermentation. *Journal of Agriculture and Food Chemistry* 33:1291–1304.

Birch, A.J., and F.W. Donovan. 1953. Studies in relation to biosynthesis I. Some possible routes to derivatives of Orcinol and Phloroglucinol. *Australian Journal of Chemistry* 6:360–368.

Boal, C. 2006. 'Green' consumers push for organic cocoa. Foodnavigator. http://www.foodnavigator.com/Product-Categories/Chocolate-and-confectionery-ingredients/Greenconsumers-push-for-organic-cocoa.

Bonvehi, J.S., and F.V. Coll. 1997. Evaluation of bitterness and astringency of polyphenolic compounds in cocoa powder. *Food Chemistry* 60:365–370.

Bonvehi, S.J. 2004. Occurrence of ochratoxin A in cocoa products and chocolate. *Journal of Agricultural and Food Chemistry* 52:6347 e6352.

Borges, G., F. Tomaś-Barbeŕan, and A. Crozier. 2006. Phytochemicals in cocoa and flavan-3-ol bioavailability. In *Teas, Cocoa and Coffee, Plant Secondary Metabolites and Health*. A. Crozier, H. Ashihara, and F. Tomás-Barbéran (Eds.), Blackwell Publishing Ltd., pp: 193–194.

Bourn, D. and J. Prescott. 2002. A comparison of the nutritional value, sensory qualities, and food safety of organically and conventionally produced foods. *Critical Reviews in Food Science and Nutrition* 1:1–34.

Brera, C., R. Caputi, M. Miraglia, I. Iavicoli, A. Salerno, and G. Carelli. 2002. Exposure assessment to mycotoxins in workplaces: Aflatoxins and ochratoxin A occurrence in airborne dusts and human sera. *Microchemical Journal* 73(1–2):167–173.

Brera, C., F. Debegnach, B. De Santis, et al. 2011. Ochratoxin A in cocoa and chocolate products from the Italian market: Occurrence and exposure assessment. *Food Control* 22:1663–1667.

Brown, M.B. 1993. *Fairtrade; Reform and Realities in the International Trading System*. London: Zed.

Brummer, Y., and S.W. Cui. 2005. Understanding carbohydrate analysis. In *Food Carbohydrates: Chemistry, Physical Properties and Applications*, S.W. Cui (Ed.). Boca Raton, FL: Taylor & Francis Group, Chapter 2, pp. 1–38.

Buamah, R., V.P. Dzogbefia, and J.H. Oldham. 1997. Pure yeast culture fermentation of cocoa (*Theobroma cacao* L.): Effect on yield of sweating and cocoa beans quality. *World Journal of Microbiology & Biotechnology* 13:457–462.

Bucheli, P., G. Rousseau, M. Avarez, M. Laloi, and J. McCarthy. 2001. Developmental variations of sugars, carboxylic acids, purine alkaloids, fatty acids and endoprotease activity during maturation of *Theobroma cacao* L. seeds. *Journal of Agriculture and Food Chemistry* 49:5046–5051.

Burdaspal, P. A., and T.M. Legarda. 1998. Ochratoxin A in beer produced in Spain and other European countries. *Alimentaria* 291:115–122.

Burdaspal, P.A., and T.M. Legarda. 2003. Occurrence of ochratoxin A in samples of different types of chocolate and cocoa powder, marketed in Spain and fifteen foreign countries. *Alimentaria* 10:143–153.

Camu, N.T., K.S. De Winter, J.S. Addo, J.F. Takrama, H. Bernaert, and L. De Vuyst. 2008. Fermentation of cocoa beans: Influence of microbial activities and polyphenol concentrations on the flavour of chocolate. *Journal of the Science of Food and Agriculture* 88:2288–2297.

Carpenter, D.R., J.F. Hammerstone, L.J. Romanezyk, and W.M. Aitken. 1994. Lipid composition of Herrania and Theobroma seeds. *Journal of American Oil Chemistry Society* 71:845–851.

Carr, J.G., P.A. Davies, and J. Dougan. 1979. Cocoa fermentation in Ghana and Malaysia. In *Proceedings of 7th International Cocoa Research Conference*, Douala, Cameroon, Vol. 4, pp. 573–576.

Carr, J.G., and P.A. Davies. 1980. *Cocoa Fermentation in Ghana and Malaysia (Part 2): Further Microbiological Methods and Results*. Bristol: Long Ashton Research Station.

Carrigan, M., and A. Attalla. 2001. The myth of the ethical consumer—Do ethics matter in purchase behaviour? *Journal of Consumer Marketing* 18:560–578.

CBI 2007. The (organic) coffee, tea and cocoa market in the EU. http://www.cbi.eu

Chaiseri, S., and P.S. Dimick. 1989. Lipid and hardness characterististics of cocoa butter from different geographic regions. *Journal of the American Oil Chemists Society* 66:1771–1780.

Chapin, F.S. 1983. Patterns of nutrient absorption and use by plants from natural and man-modified environments. In *Disturbance and Ecosystems*. H.A. Mooney and M. Gordron (Eds.). Ecological Studies 44, New York: Springer Verlag, pp. 175–187.

Chavan, J.K., and S.S. Kadam. 1989. Critical reviews in food science and nutrition. *Food Science* 28:348–400.

Chiodini, A.M., P. Scherpenisse, and A.A. Bergwerff. 2006. Ochratoxin A contents in wine: Comparison of organically and conventionally produced products. *Journal of Agricultural and Food Chemistry* 54:7399–7404.

Clapperton, J., J.F. Hammerstone, L.J. Romanczyk, S. Yow, J. Chan, and D. Lim. 1992. Polyphenols and cocoa flavour. In: *Proceedings of the Meeting of GroupPolyphenol, JIEP 92*, Lisbon.

Clapperton, J., R. Lockwood, L. Romanczyk, and J.F. Hammerstone. 1994. Contribution of genotype to cocoa (*Theobroma cacao* L.) flavour. *Tropical Agriculture (Trinidad)* 71:303–308.

Clapperton, J.F. 1994. A review of research to identify the origins of cocoa flavour characteristics. *Cocoa Growers' Bulletin* 48:7–16.

Codex Alimentarius Commission. 2013. Proposed draft code of practice for the prevention and reduction of ochratoxin A contamination in cocoa. In *Joint FAO/WHO Food Standards Programme Codex Committee on Contaminants in Foods*. Seventh Session Moscow, Russian Federation, 8–12 April 2013. http://www.codexalimentarius.org (accessed July 16, 2013).

Coe, S.D., and M.D. Coe. 1996. *The True History of Chocolate*. London: Thames and Hudson.

Cooper, B. 2008. Premium chocolate set to grow despite tough times. http://www.just-food.com/article.aspx?id = 104298 (accessed June 20, 2011).

COPAL 2008. Cocoa Producers' Alliance. COPAL homepage [online]. http://www.copal-cpa.org (accessed December 5, 2009).

Copetti, M.V., J.L. Pereira, B.T. Iamanaka, J.I. Pitt, and M.H. Taniwaki. 2010. Ochratoxigenic fungi and ochratoxin A in cocoa during farm processing. *International Journal of Food Microbiology* 143:67–70.

Counet, C., and S. Collin. 2003. Effect of the number of flavanol units on the antioxidant activity of procyanidin fractions isolated from chocolate. *Journal of Agriculture and Food Chemistry* 51:6816–6822.

Counet, C., C. Ouwerx, D. Rosoux, and S. Collin. 2004. Relationship between procyanidin and flavour contents of cocoa liquors from different origins. *Journal of Agriculture and Food Chemistry* 52:6243–6249.

Craig, W.J. and T.T. Nguyen. 1984. Caffeine and theobromine levels in cocoa and carob products. *Journal of Food Science* 49:302–303.

CRIG. 2008. *Guide to the Control of Black Pod Disease in Ghana.* Tafo: Cocoa Research Institute of Ghana (CRIG).

Crouzillat, D., E. Lerceteau, J. Rogers, and V. Petiard, 1999. Evolution of cacao bean proteins during fermentation: A study by two-dimensional electrophoresis. *Journal of the Science of Food and Agriculture* 79:619–625.

Crozier, J.S., G.A. Preston, W.J. Hurst, J.M. Payne, J. Mann, L. Hainly, and L.D. Miller. 2011. Cacao seeds are a "Super Fruit": A comparative analysis of various fruit powders and products. *Chemistry Central Journal* 5:2.

Cunha, J.D. 1990. Performance of Burairo 3×3m dryer for cocoa. *Agrotropica*, 2(3):157–160.

Cunningham, R.K., and P.W. Arnold. 1962. The shade and fertilizer requirements of cacao (Theobroma cacao L) in Ghana. *Journal of the Science of Food and Agriculture*, 13:213–221.

Dahiya, S., R. Karpe, A.G. Hegde, and R.M. Sharma. 2005. Lead, cadmium and nickel in chocolate and candies from suburban areas of Mumbai, India. *Journal of Food Composition and Analysis* 18(6): 517–522.

Dalcero, A., C. Magnoli, C. Hallak, S.M. Chiacchiera, G. Palacio, and C.A. Rosa. 2002. Detection of ochratoxin A in animal feeds and capacity to produce this mycotoxin by *Aspergillus* section Nigri in Argentina. *Food Additives and Contamination* 19:1065–1072.

Dalman, O., A. Demirak, and A. Balci. 2006. Determination of heavy metals (Cd, Pb) and trace elements (Cu, Zn) in sediments and fish of the Southeastern Aegean Sea (Turkey) by atomic absorption. *Food Chemistry* 95: 157–162.

Damiani, O. 2002. *Small Farmers and Organic Agriculture: Lessons Learned from Latin America and the Caribbean.* Rome: International Fund for Agricultural Development.

Dand, R. 1996. *The International Cocoa Trade*, 2nd edition. New York: Woodhead.

Dand, R. 1997. *The International Cocoa Trade.* New York: John Wiley & Sons.

David, S. 2005. *Learning About Sustainable Cocoa Production: A Guide for Participatory Farmer Training.* Yaoundé, Cameroon: International Institute of Tropical Agriculture.

de Brito, E.S., N.H.P. Garcia, M.I. Gallão, A.L. Cortelazzo, P.S. Fevereiro, and M.R Braga. 2000. Structural and chemical changes in cocoa (*Theobroma cacao L.*) during fermentation, drying and roasting. *Journal of the Science of Food and Agriculture* 81:281–288.

De La Cruz, M., R. Whitkus, A. Gomez-Pompa, and L. Mota-Bravo. 1995. Origins of cacao cultivation. *Science* 375:542–543.

de Magalhães, J.T., G.A. Sodré, H. Viscogliosi, and M. Grenier-Loustalot. 2011. Occurrence of ochratoxin A in Brazilian cocoa beans. *Food Control* 22:744–748.

De Zaan Cocoa Manual. 2009. ADM Cocoa International, Switzerland, pp 9–83.

Delonga, K., R.I. Redovniković, S. Mazor, V. Dragović-Uzelac, M. Carić, and J. Vorkapić-Furač. 2009. Polyphenolic content and composition and antioxidative activity of different cocoa liquors. *Czech Journal of Food Science* 27(5):330–335.

Denny, C. and L. Elliott. 2003. Agriculture: Farming's double standards, Trade Supplement. *The Guardian*. September 2003.

Despreaux, D. 1998. Le cacaoyer et la cacaoculture. In *Cacao et chocolates production, utilisation, caracteristiques*, J. Pontillon (Ed.). Paris: Technique & Documentation Lavoisier, pp. 44–93.

Dhoedt, A. 2008. Food of the gods – The rich history of chocolates. *Agro-Foods Industry Hi-Tech Journal* 19(3):4–6.

Dias, D.R., R.F. Schwan, E.S. Freire, and R. dos Santos Serodio. 2007. Elaboration of a fruit wine from cocoa (*Theobroma cacao* L.) pulp. *International Journal of Food Science and Technology* 42:319–329.

Dillinger, T.L., P. Barriga, S. Escarcega, M. Jimenez, D.S. Lowe, and L.E. Grivetti. 2000. Food of the gods: Cure for humanity? A cultural history of the medicinal and ritual use of chocolate. *Journal of Nutrition* 130:2057S–2072S.

Dimick, P.S., and J.C. Hoskin. 1999. The chemistry of flavour development in chocolate. In *Industrial Chocolate Manufacture and Use*, 3rd edition. S.T. Beckett (Ed.). Oxford: Blackwell Science, pp. 137–152.

Dimick, P.S., and J.M. Hoskin. 1981. Chemico-physical aspects of chocolate processing – A review. *Canadian Institute of Food Science and Technology Journal* 4:269–282.

Dos Santos, W.N., E.G. da Silva, M.S. Fernandes et al. 2005. Determination of copper in powdered chocolate samples by slurry sampling flame atomic-absorption spectrometry. *Analytical and Bioanalytical Chemistry* 382: 1099–1102.

Dubois, M., K.A. Gilles, J.K. Hamilton, P.A. Rebers, and F. Smith. 1979. Colorimetric method for the determination of sugars and related substances. *Analytical Chemistry* 28:350–356.

Duncan, R.J.E. 1984. A survey of Ghanaian cocoa farmers' fermentation and drying practices and their implications for Malaysian practices. In *Proceedings of International Conference on Cocoa and Coconut,* 15–17 October, Malaysia.

Duncan, R.J.E., G. Godfrey, T.N. Yap, G.L. Pettiphar, and T. Tharumarajah. 1989. Improvement of Malaysian cocoa bean flavour by modification of harvesting, fermentation and drying methods – The Sime Cadbury Process. *The Planter, Kuala Lumpur* 65:157–173.

Dzahini-Obaitey, H., O. Domfah, and M.F. Amoah. 2010. Over seventy years of viral disease of cocoa in Ghana: From researchers' perspective. *African Journal of Agricultural Research* 5(7):476–479.

EEC (1973). Directive 73/241/EEC by European Parliament and the European Council relating to cocoa and chocolate products intended for human consumption. *Official Journal of the European Communities* L 228 of 16/08/1973, pp. 0023–0035.

EEC. 1990. Folin-Ciocalteu index. *Official Journal of European Communities* 41:178–179.

EFTA. 2005. *Fairtrade in Europe 2005 – Facts and Figures on the Fairtrade Sector in 25 European Countries*. Brussels: European Fair Trade Association.

EFTA. 2006. *EFTA Surveillance Authority Annual Report 2006*. Retrieved from http://www.eftasurv.int/media/esa-docs/physical/11156/data.pdf on 20/05/12.

Egan, H., R.S. Kirk, and R. Sawyer. 1990. *Pearson's Chemical Analysis of Foods*, 8th edition. London: Churchill Livingstone.

End, M.J., R.M. Wadsworth, and P. Hadley. 1990. The primitive cocoa germplasm database. *Cocoa Grower's Bulletin* 43:25–33.

Faborode, M.O., J.F. Favier, and O.A. Ajayi. 1995. On the effects of forced air drying on cocoa quality. *Journal of Food Engineering* 25:455-472.

Fairtrade Federation. 1999. What does Fairtrade really mean? http://www.fairtrade-federation.com (accessed March 15, 2013).

Fairtrade Foundation. 2004. Fairtrade Facts and Figures. *Fairtrade Resources, SP6*, April 2004.

Fairtrade Foundation. 2006. *The Fairtrade Mark*. Fairtrade Foundation Report. http://www.fairtrade.org.uk (accessed June 12, 2008)

FAO/WHO Codex Alimentarius Commission. 1999. Organic agriculture. http://www.fao.org/organicag/oa-faq/oa-faq1/en/(accessed on July 7, 2013).

FAO/WHO/UNEP. 1999. Minimising risks posed by mycotoxins utilising the HACCP Concept. In *Third Joint FAO/WHO/UNEP International Conference on Mycotoxins*, 8b, 1–13. ftp://ftp.fao.org/es/esn/food/myco8b.pdf (accessed March 10, 2008).

Figueira, A., S.R.C. Silva, G.A.L. Goncalves, and V.M. Gilabert-Escriva. 2002. Fatty acid and triglycerol composition and thermal behaviour of fats from seeds of Brazilian Amazonian *Theobroma* species. *Journal of the Science of Food and Agriculture* 82: 1425–1431.

FLO. 2005. *Fairtrade Cocoa Standards for Small Farmer Organisations*. Fairtrade Labelling Organisation International Report. http://www.fairtrade.org.uk (accessed December 9, 2005).

FLO. 2006. *Fairtrade 2006 Annual Report*. Fairtrade Labelling Organisation International. http://www.fairtrade.org.uk (accessed December 17, 2006).

Forsyth, W.G.C. 1952. Cocoa polyphenolic substances. 2. Changes during fermentation. *Biochemical Journal* 51:516–520.

Forsyth, W.G.C. and V.C. Quesnel. 1957. Cacao glycosidase and colour changes during fermentation. *Journal of Science of Food and Agriculture* 8:505–509.

Fowler, M.S. 1995. Quality of cocoa beans for chocolate manufacturers. In *Proceedings of International Research Conference*, Montpellier. pp 143–153.

Fowler, M.S. 1999. Cocoa beans: From tree to factory. In *Industrial Chocolate Manufacture and Use*, 3rd edition. S.T. Beckett (Ed.). Oxford: Blackwell Science. pp. 8–35.

Fowler, M.S. 2009. Cocoa beans: From tree to factory. In *Industrial Chocolate Manufacture and Use*, 4th edition. S.T. Beckett (Ed.). Oxford: Blackwell. pp. 10–33, 137–152.

Fuleki, T., and F.J. Francis. 1968. Quantitative methods for anthocyanin. 1. Extraction and determination of total anthocyanin in cranberries. *Journal of Food Science* 33:72–77.

Galtier, P. 1991. Pharmacokinetics of ochratoxin A in animals. *IARC Sci. Publ.* 115:187–200.

Galvez, S.L., G. Loiseau, J.L. Paredes, M. Barel, and J.P. Guiraud. 2007. Study on the microflora and biochemistry of cocoa fermentation in the Dominican Republic. *International Journal of Food Microbiology* 114:124–130.

Garcìa-Alamilla, P., M.A. Salgado-Cervantes, M. Barel, G. Berthomieu, G.C. Rodríguez-Jímenes, and M.A. García-Alvarado. 2007. Moisture, acidity and temperature evolution during cacao drying. *Journal of Food Engineering* 79:1159–1165.

Geisen, R., Z. Mayer, A. Karolewiez, and P. Färber. 2004. Development of a real time PCR system for detection of *Penicillium nordicum* and for monitoring ochratoxin A production in foods by targeting the ochratoxin polyketide synthase gene. *Systematic and Applied Microbiology* 27:501–507.

Geographical. 2004. *The Trouble with Global Trade*. London: Royal Geographical Society.

Gilmour, M., and M. Lindblom. 2008. Management of ochratoxin A in the cocoa supply chain: A summary of work by the CAOBISCO/ECA/FCC working group on ochratoxin A. In *Mycotoxins: Detection Methods, Management, Public Health and Agricultural Trade*. J. F. Leslie, R. Bandyopadhyay, and A. Visconti (Eds.). Wallingford, UK: CABI, pp. 231–243.

Ghana Cocoa Board (COCOBOD). 2010. *The History of Cocoa and its Production in Ghana*. Retrieved from http://www.cocobod.gh on June 2011.

Global Research on Cocoa. 2008. Working with and for farmers. *GRO-Cocoa*, Issue 13. June 2008. Retrieved from http://www.cabi.org/Uploads/CABI/projects/gro-cocoa-issue-13-jun-2008.pdf (Accessed on 22/08/12).

Gotsch, N. 1997. Cocoa biotechnology: Status, constraints and future prospects. *Biotechnology Advances* 15:333–352.

Gourieva, K.B., and O.B. Tserrevitinov. 1979. Method of evaluating the degree of fermentation of cocoa beans. USSR Patent no. 646254.

Greenberg, R. 1998. Biodiversity in the cacao agro-ecosystems: Shade management and landscape considerations. Proceedings of the Smithsonian Migratory Bird Center Cacao Conf. Retrieved from http://nationalzoo.si.edu/conservationandscience/migratorybirds/research/cacao/papers.cfm (Accessed Nov. 2003).

The Green Guide. 2004. Flavour Perception. Taylor, A.J., and D.D. Roberts. Oxford, U.K: Blackwell. s.d. Organic and Fair Trade Chocolate.

Guehi, S.T., M. Dingkuhn, E. Cros, G. Fourny, R. Ratomahenina, G. Moulin, and A.C Vidal. 2008. Impact of cocoa processing technologies in free fatty acids formation in stored raw cocoa beans. *African Journal of Agricultural Research* 3(3):174–179.

Guehi, T.S., Y.M. Konan, R. Koffi-Nevry, N.D. Yao, and N.P. Manizan. 2007. Enumeration and identification of main fungal isolates and evaluation of fermentation's degree of Ivorian raw cocoa beans. *Australian Journal of Basic and Applied Sciences*, 1(4):479–486.

Guehi, T.S., I.B. Zahouli, L. Ban-Koffi, M.A. Fae, and J.G. Nemlin. 2010. Performance of different drying methods and their effects on the chemical quality attributes of raw cocoa material. *International Journal of Food Science and Technology* 45:1564–1571.

Halliday, J. 2008. ADM Cocoa starts supply of organic chocolate ingredients in UK. Foodnavigator. http://www.foodnavigator.com/Financial-Industry/ADM-Cocoa-starts-supply-of-organicchocolate-ingredients-in-UK (accessed August 5, 2008).

Hansen, C.E., M. del Olmo, and C. Burri. 1998. Enzyme activities in cocoa beans during fermentation. *Journal of the Science of Food and Agriculture* 77:273–281.

Hansen, C.E., A. Manez, C. Burri, and A. Bousbaine. 2000. Comparison of enzyme activities involved in flavour precursor formation in unfermented beans of different cocoa genotypes. *Journal of the Science of Food and Agriculture* 80:1193–1198.

Hartel, R.W. 2001. *Crystallization in Food.* Gaithersburg, MD: Aspen.

Hashim, P., S. Jinap, S.K.S. Mohammad, and A. Ali. 1998. Changes in free amino acid, peptide-N, sugar and pyrazine concentration during cocoa fermentation. *Journal of Science of Food and Agriculture* 78:535–550.

Hebbar, P., H.C. Bittenbender, and D. O'Doherty. 2011. Farm and forestry production and marketing profile for cacao (Theobroma cacao). Specialty crops for Pacific Island agroforestry. http://agroforestry.net/scps (accessed August10,2011)

Holm, C.S., J.W. Aston, and K. Douglas. 1993. The effects of the organic acids in cocoa on the flavour of chocolate. *Journal of Science of Food and Agriculture* 61:65–71.

Hoskin, J.C., and P.S. Dimick. 1994. Chemistry of flavour development in chocolate. In *Industrial Chocolate Manufacture and Use*, 2nd edition. S.T. Beckett (Ed.). New York: Van Nostrand Reinhold, pp. 102–115.

Howell, G.M., S.E. Jorge Villar, L.F.C. De Oliveira, and H. Mireille. 2005. Analytical Raman spectroscopic study of cacao seeds and their chemical extracts. *Analytica Chimica Acta* 538:175.

Hurst, W.J., and R.A. Martin. 1998. High-performance liquid chromatography determination of Ochratoxin A in artificially contaminated cocoa beans using automated clean-up. *Journal of Chromatography A* 810:89–94.

ICCO. 2005. Facts and figures on Fairtrade cocoa. *Consultative Board Report on the World Cocoa Economy.* Fifth Meeting, London.

ICCO. 2006. A study on the market for organic cocoa. http://www.icco.org/economics/market.aspx (accessed September 22, 2006).

ICCO. 2007a. *ICCO Document: Assessment of the Movements of Global Supply and Demand.* London: International Cocoa Organization.

ICCO. 2007b. Sustainable cocoa economy: A comprehensive and participatory approach. *Consultative Board Report on World Cocoa Economy.* Twelfth Meeting, Kuala Lumpur.

ICCO. 2007c. *Supply chain management for total quality cocoa in Africa.* Executive Committee Report. London.

ICCO. 2007d. *Progress Report Action Programme on Pesticides*, ICCO Executive Committee 133rd Meeting, EBRD Offices. 5–7 June. London.

ICCO. 2008. *Annual report 2006/2007*. International Cocoa Organisation, London, UK. Retrieved from www.icco.org on 05-06-2010.

ICCO. 2010a. *ICCO Quarterly Bulletin of Cocoa Statistics*, XXXVI: 4, Cocoa year 2009/2010. http://www.icco.org (accessed March, 2012)

ICCO. 2010b. *Functioning and Transparency of the Terminal Markets for Cocoa - An Overview and Analysis of Recent Events on the London Terminal Market.* Report presented to the Executive Committee at the 142nd Meeting, London.

ICCO. 2010c. *Pesticide Use in Cocoa - A Guide for Training Administrative and Research Staff.* London, UK. 3rd edition. Roy Bateman (Ed.), London.

ICCO. 2011. ICCO *Quarterly Bulletin of Cocoa Statistics*, Vol. XXXVII, No. 4, Cocoa year 2010/11. Published: 30-11-2011. Retrieved from www.icco.org on 05-04-2012.

ICCO. 2012a. *Annual Report of the International Cocoa Organization for 2010/2011.*

ICCO. 2012b. *The World Cocoa Economy: Past and Present.* Report presented to the Executive Committee at the 146th Meeting, London.

ICCO. 2013a. Production of cocoa beans. *ICCO Quarterly Bulletin of Cocoa Statistics*, XXXIX: 1, Cocoa year 2012/13.

ICCO (2013b). Grinding of cocoa beans. *ICCO Quarterly Bulletin of Cocoa Statistics*, XXXIX: 1, Cocoa year 2012/13.

ICCO. 2008. International Cocoa Organization Report of Cocoa Statistics. *The Manufacturing Confectioner* 88(3):39–40.

ICCO. 2011. ICCO *Quarterly Bulletin of Cocoa Statistics*, XXXVII: 4, Cocoa year 2010/11. http://www.icco.org (accessed April 5, 2012).

ICA (International Confectionery Association). 2000. *Viscosity of Cocoa and Chocolate Products. Analytical Method 46.* Available from CAOBISCO, rue Defacqz 1, B-1000 Bruxelles, Belgium.

IFOAM. 2006. *The World of Organic Agriculture: Statistics & Emerging Trends* 2006. Bonn Germany & Research Institute of Organic Agriculture FiBL, Frick, Switzerland. Retrieved from http://orgprints.org/5161/1/yussefi-2006-overview.pdf (Accessed on 14/10/11).

IFOAM. 2008. http://www.ifoam.org (accessed April, 2010).

Institute of Statistical Social and Economic Research (ISSER) (2008). *The State of the Ghanaian Economy in 2007.* Legon: University of Ghana.

IOCCC (International Office of Cocoa, Chocolate and Sugar Confectionery). 1996. Determination of free fatty acids (FFA) content of cocoa fat as a measure of cocoa nib acidity. *Analytical Methods* 42:130–136.

ISO 1998. Quality Management Systems—Concepts and vocabulary ISO 9000:2000. ISO/TC 176/SC1/N 185.

Javaid, S.B., J.A. Gadahi, M. Khaskeli, M.B. Bhutto, S. Kumbher, and A.H. Panhwar. 2009. Physical and chemical quality of market milk sold at Tandojam, Pakistan. *Pakistan Veterinary Journal*, 29(1):27–31.

Jayas, D.S. and J.S. Sokhansanj. 1989. Thin layer drying of barley at low temperatures. *Canadian Agricultural Engineering*, 31:21–23.

Jespersen, L., D.S. Nielsen, S. Hønholt, and M. Jakobsen. 2005. Occurrence and diversity of yeasts involved in fermentation of West African cocoa beans. *FEMS Yeast Research* 5:441–453.

Jinap, S. 1994. Organic acids in cocoa beans, a review. *ASEAN Food Journal* 9:3–12.

Jinap, S., and P.S. Dimick. 1990. Acidic characteristics of fermented and dried cocoa beans from different countries of origin. *Journal of Food Science* 55(2):547–550.

Jinap, S., P.S. Dimick, and R. Hollender. 1995. Flavour evaluation of chocolate formulated from cocoa beans from different countries. *Food Control* 6(2):105–110.

Jinap, S., M.H. Siti, and M.G. Norsiati. 1994. Formation of methyl pyrazine during cocoa bean fermentation. *Pertanika Journal of Tropical Agricultural Science* 17(1):27–32.

Jinap, S., Y. Ikrawan, J. Bakar, N. Saari, and H.N. Lioe. 2008. Aroma precursors and methylpyrazines in under fermented cocoa beans induced by endogenous carboxypeptidase. *Journal of Food Science* 73(7):141–147.

Johns, N.D. 1999. Conservation in Brazil's chocolate forest: The unlikely persistence of the traditional cocoa agro-ecosysytem. *Environmental Management* 23:31–47.

Kabata-Pendias, A. and H. Pendias. 1997. *Trace Elements in Soils and Plants*, 2nd edition. New York: CRC Press.

Kao, C.M., L. Katz, and C. Khosla. 1994. Engineered biosynthesis of a complete macrolactone in a heterologous host. *Science* 265:509–512.

Kaphueakngam, P., A. Flood, and S. Sonwai. 2009. Production of cocoa butter equivalent from mango seed, almond fat and palm oil mid-fraction. *Asian Journal of Food Agro-Industry* 2(4):442.

Kattenberg, H., and A. Kemming. 1993. The flavour of cocoa in relation to the origin and processing of the cocoa beans. In *Food Flavours, Ingredients and Composition*. G. Charalambous (Ed.). Amsterdam: Elsevier Science, pp 1–22.

Kealey, K.S., R.M. Snyder, L.J. Romanczyk, H.M. Geyer, M.E. Myers, E.J. Withcare, J.F. Hammerstone, and H.H. Schmitz. 1998. *Cocoa Components, Edible Products Having Enhanced Polyphenol Content, Methods of Making Same and Medical Uses*. Patent Cooperation Treaty (PCT) WO 98/09533. USA: Mars Inc.

Kilian, B., L. Pratt, C. Jones, and A. Villalobos. 2006. Is sustainable agriculture a viable strategy to improve farm income in Central America? A case study on coffee. *Journal of Business Research* 59:322–330.

Kim, H., and P.G. Keeney. 1983. Method of analysis for (–) epicatechin content in cocoa beans by high performance liquid chromatography. *Journal of Food Science* 48:548–551.

Kim, H., and P.G. Keeney. 1984. Epicatechin content in fermented and unfermented cocoa beans. *Journal of Food Science* 49:1090–1092.

Kirchhoff, P.M., B. Biehl, and G. Crone. 1989. Peculiarity of the accumulation of free amino acids during cocoa fermentation. *Food Chemistry* 31:295–311.

Knezevic, G. 1979. Heavy metals in food. Part 1. Content of cadmium in raw cocoa beans and in semi finished and finished chocolate products. *Dtsch. Lebensm - Rundsch* 75(10): 305–309.

Knezevic, G. 1980. Heavy metals in food stuff. The copper content of raw cocoa, intermediate and finished cocoa products, *GGB* 5(2): 24–26.

Knezevic, G. 1982. Heavy metals in food. Part 2. Lead content in unrefined cocoa and in semi finished and finished cocoa products. *Dtsch, Lebensm-Rundsch* 78(5): 178–180.

Knight, I. (Ed.). 2000. *Chocolate and Cocoa: Health and Nutrition*. Oxford, UK: Blackwell Science.

Koekoek, F.J. 2003. *The Organic Cocoa Market in Europe. Summary of a Market Study*. EPOPA.

Konczak, I., and W. Zhang, 2004. Anthocyanins—More than nature's colours. *Journal of Biomedicine and Biotechnology* (5):239–240.

Kris-Etherton, P.M., J.A. Derr, and D.C. Mitchell. 1993. The role of fatty acid saturation on plasma lipids, lipoproteins and apolipoproteins: I. Effects of whole food diets high in cocoa butter, olive oil, soybean oil, dairy butter and milk chocolate on the plasma lipids of young men. *Metabolism* 42:121–129.

Kuiper-Goodman, T., and P.M. Scott. 1989. Review: Risk assessment of the mycotoxin ochratoxin A. *Biomedical and Environmental Science* 2:179–248.

Kyi, T.M., W.R.W. Daud, A.B. Mohammad, M.W. Samsudin, A.A.H. Kadhum, and M.Z.M. Talib, 2005. The kinetics of polyphenol degradation during the drying of Malaysian cocoa beans. *International Journal of Food Science and Technology* 40:323–331.

Lange, H., and A. Fincke. 1970. Kakao und Schokolade. In *Handbuch der Lebensmittel Band VI: Alkaloidhaltige Genussmittel, Gewu` Erze, Kochsalz*. L. Acker, K.G. Bergner, W. Diemair (Eds.). Berlin, Heidelberg: Springer-Verlag, pp. 210–309.

Lannes, S.C.S., M.L. Medeiros, and L.A. Gioielli. 2003. Physical interactions between cupuassu and cocoa fats. *Grasasy Aceites* 54:253–258.

Lawless, H.T., and H. Heymann. 1998. *Sensory Evaluation of Food*; *Principles and Practices*. New York: Chapman & Hall.

Laws, E.A. 2000. *Aquatic Pollution*: *An Introductory Text,* 3rd edition. New York: John Wiley and Sons, p. 672.

Leake, L.L. 2009. Electronic noses and tongues. *Food Technology* 6:96–102.

Lehrian, D.W., and P.G. Keeney. 1980. Changes in lipid component of seed during growth and ripening of cocoa fruit. *Journal of the American Oil Chemists' Society* 57:61–65.

Lehrian, D.W., and G.R. Patterson. 1983. Cocoa fermentation. In *Biotechnology, a Comprehensive Treatise, Vol. 5*. G. Reed and H.J. Rehm (Eds.). Basel: Verlag Chemie, pp. 529–575.

Liu, P. (Ed). 2008. Value-adding standards in the North-American food market. Trade opportunities in certified products for developing countries. FAO. Rome. http://www.fao.org/docrep/010/a1585e/a1585e00.htm (accessed August, 2011).

Lopez, A., and V.C. Quesnel. 1973. Volatile fatty acid production in cacao fermentation and the effect on chocolate flavour. *Journal of the Science of Food & Agriculture* 24:319–326.

Lopez, A.S. 1986. Chemical changes occurring during the processing of cocoa. In *Proceedings of the cocoa biotechnology symposium.* P.S. Dimick (Ed.). Department of Food Science. Pennslyvania State University: University Park, pp. 19–53.

Lopez, A.S., and P.S. Dimick. 1995. Cocoa fermentation. In *Biotechnology: A Comprehensive Treatise, Vol. 9, Enzymes, Biomass, Food and Feed*, 2nd edition. G. Reed, and T.W. Nagodawithana (Eds.). Weinheim: VCH, pp. 563–577.

Lopez, A.S., D.W. Lehrian, and L.V. Lehrian. 1978. Optimum temperature and pH of invertase of the seeds of *Theobroma cacao L. Revista Theobroma* 8:105–112.

Loska, K., and D. Wiechuła. 2003. Application of principal component analysis for the estimation of source of heavy metal contamination in surface sediments from the Rybnik Reservoir. *Chemosphere* 51: 723–733.

Lucia, R.C., M.B.C. Neuza, and P.R. Fernando. 2005. Copper-iron metabolism interaction in rats. *Nutrition Research* 25(1): 79–92.

Luna, F., D. Crouzillat, L. Cirou, and P. Bucheli. 2002. Chemical composition and flavour of Ecuadorian cocoa liquor. *Journal of Agricultural and Food Chemistry* 50:3527–3532.

MacFarlane, G.R., and M.D. Burchett. 2001. Photosynthetic pigments and peroxidase activity as indicators of heavy metal stress in the grey mangrove, *Avicennia marina* (Forsk.) Vierh. *Marine Pollution Bulletin* 42(3): 233–240.

Medeiros, M.L., A. Ayrosa, R.N.M. Pitombo, and S.C.S. Lannes. 2006. Sorption isotherms of cocoa and cupuassu products. *Journal of Food Engineering* 73:402–406.

Menter, L.L. 2005. *The Sustainable Cocoa Trade. An Analysis of US Market and Latin American Trade Prospects.* EcoMercados Project. http://econegociosagricolas.com/files/The_Sustainable_Cocoa_Trade_An_Analysis_of_US_Market_1.pdf (Accessed on 20/07/10).

Merry, R.H., G.K. Tiller, and A.M. Alston. 1986. The effects of soil contamination with copper, lead and arsenic on the growth and composition of plants. *Plant and Soil* 95: 225–269.

Mesallam, A.S. 1987. Heavy metal content of canned orange juice as determined by direct current plasma atomic emission spectrophotometry (DCPAES). *Food Chemistry* 26(1): 47–58.

Meyer, B., B. Biehl, M.B. Said, and R.J. Samarakoddy. 1989. Post harvest pod storage: A method of pulp preconditioning to impair strong nib acidification during cocoa fermentation in Malaysia. *Journal of the Science of Food and Agriculture* 48:285–304.

Micco, C., A. Ambruzzi, M. Miraglia, C. Brera, R. Onori, and L. Benelli. 1991. Contamination of human milk by ochratoxin A. In *Mycotoxins Endemic Nephropathy and Urinary Tract Tumours, Vol. 115* Publications pp. 105–108. IARC Science.

Mikkelsen, L. 2010. *Quality Assurance Along the Primary Processing Chain of Cocoa Beans from Harvesting to Export in Ghana.* Student in Food Science at the University of Copenhagen, Faculty of Life Sciences, Frederiksberg, pp. 2–37.

Mingorance, M.D., B. Valdes, and S.R. Oliva. 2007. Strategies of heavy metal uptake by plants growing under industrial emissions. *Environmental International* 33: 514–20.

Minifie, B.W. 1989. *Chocolate, Cocoa and Confectionery – Science and Technology.* London: Chapman & Hall.

Miraglia, M., and C. Brera. 2002. Reports on tasks for scientific cooperation, task 3.2.7. http://europa.eu.int/comm/food/fs/scoop/3.2.7_en.pdf (accessed April 18, 2002)

Miraglia, M., A. de Dominicis, C. Brera, S. Corneli, E. Cava, E. Menghetti, and E. Miraglia. 1995. Ochratoxin A levels in human milk and related food samples: An exposure assessment. *Natural Toxins* 3:436–444.

Misnawi, S. 2008. Physico-chemical changes during cocoa fermentation and key enzymes involved. *Review Penelitian Kopi dan Kakao*, 24(1):54–71.

Misnawi, S., S. Jinap, B. Jamilah, and S. Nazamid. 2002. Oxidation of polyphenols in unfermented and partly fermented cocoa beans by cocoa polyphenol oxidase and tyrosinase. *Journal of the Science of Food Agriculture* 82:559–566.

Misnawi, S., S. Jinap, B. Jamilah, and S. Nazamid. 2003. Effects of incubation and polyphenol oxydase enrichment on colour, fermentation index, procyanidins and astringency of unfermented and partly fermented cocoa beans. *International Journal of Food Science & Technology* 38: 285–295.

Moran, M. 2008. 2008 Trend Report: Premium Chocolate. The Sweet Sales of Success. The Gourmet Retailer. Available at: <http://www.gourmetretailer.com/gourmetretailer/content_display/trends/e3ic1abd1883d2156371dc907ea114507eb> (accessed February, 2013).

Mossu, J. 1992. *Cocoa.* London: Macmillan Press.

Motamayor, C.J., P. Lachenaud, R. Loor, N.D. Kuhn, J.S. Brown, and R.J. Schnell. 2008. Geographic and genetic population differentiation of the Amazonian chocolate tree (Theobroma cacao L.). PLoS ONE 3(10): e3311. doi:10.1371/journal.pone.0003311.

Mounjouenpou, P., D. Gueule, A. Fontana-Tachon, B. Guyot, P.R. Tondje, and J.P. Guiraud, 2008. Filamentous fungi producing ochratoxin A during cocoa processing in Cameroon. *International Journal of Food Microbiology* 121:234–241.

Moy, G.G., and J.R. Wessel. 2000. Codex Standard for Pesticides Residues. In *International Standards for Food Safety*, N. Rees and D. Watson (Eds.). Gaithersburg, MD: Aspen.

Musche, R., and W. Lucas. 1973. Heavy metal contamination of foods I. Cocoa beans. *Dtsch. Lebensm-Rundsch* 69(8): 227–278.

Nair, K.P. 2010. Cocoa (*Theoboma cacao L.*). In *The Agronomy and Economy of Important Tree Crops of the Developing World.* The Hague, Netherlands: Elsevier, Chapter 5, pp. 131–180.

Nazaruddin, R., M.Y. Ayub, S. Mamot, and C.H. Heng. 2001. HPLC Determination of methylxanthines and polyphenols levels in cocoa and chocolate products, *Malaysian Journal Analytical Science* 7:377–386.

Nazaruddin, R., L. Seng, O. Hassan, and M. Said. 2006a. Effect of pulp preconditioning on the content of polyphenols in cocoa beans (*Theobroma cacao*) during fermentation. *Industrial Crops and Products* 24:87–94.

Nazaruddin, R., O. Hassan, M. Said, W. Samsudin, and I. Noraini. 2006b. Influence of roasting conditions on volatile flavour of roasted Malaysian cocoa beans. *Journal of Food Processing and Preservation* 30:283–284.

New Agriculturalist. 2007. Ghanian farmers' group deals with challenges of organic cocoa. http://africanagriculture.blogspot.com/2007/06/ghanian-farmers-group-deals-with.html (accessed June, 2013).

N'Goran, K. 1998. Reflections on a sustainable cacao production system: The situation in the Ivory Coast, Africa. Proceedings of the Smithsonian Migratory Bird Center cacao conf. Retrieved from http://nationalzoo.si.edu/conservationandscience/migratorybirds/research/cacaopapers.cfm (Accessed November, 2003).

Nielsen, D.S. 2006. *The microbiology of Ghanaian cocoa fermentations*. PhD diss., Department of Food Science, Food Microbiology. The Royal Veterinary and Agricultural University, Denmark, p. 7.

Nielsen, D.S., U. Schillinger, C.M.A. Franz, J. Bresciani, W. Amoa-Awua, W.H. Holzapfel, and M. Jakobsen. 2007a. *Lactobacillus ghanesis* sp. A motile lactic acid bacterium isolated from Ghanaian cocoa fermentations. *International Journal System Evolution Microbology* 57:1468–1472.

Nielsen, D.S., O.D. Teniola, L. Ban-Koffi, M. Owusu, T.S. Andersson, and W.H. Holzapfel. 2007b. The microbiology of Ghanaian cocoa fermentations analyzed using culture-dependent and culture-independent methods. *International Journal of Food Microbiology* 114:168–186.

Nout, M.J.R., and Y. Motarjemi. 1997. Assessment of fermentation as a household technology for improving food safety: A joint FAO/WHO workshop. *Food Control* 8:221–226.

O'Callaghan, J., M.X. Caddick, and A.D.W. Dobson. 2003. A polyketide synthase gene required for Ochratoxin A biosynthesis in *Aspergillus ochraceus*. *Microbiology* 149:3485–3491.

Oliveira, A.M., N.R. Pereira, A. Marsaioli Jr. and F. Augusto. 2004. Studies on the aroma of cupuassu liquor by headspace solid-phase micro extraction and gas chromatography. *Journal of Chromatography* A 1025:15–124.

Ollennu, L.A.A., G.K. Owusu, and J.M. Thresh. 1989. The control of cocoa swollen shoots disease in Ghana. C*rop Protection* 5(1):41–52.

Opoku, I.Y., and K. Owusu. 1995. A severe type of black pod disease of cocoa in Ghana. *Newsletter* 11:1–4. Tafo: Cocoa Research Institute of Ghana.

Opoku, I.Y., M.K. Assuah, and O. Domfeh. 2007a. Manual for the identification and control of diseases of cocoa. *Technical Bulletin* No. 16, Tafo: Cocoa Research Institute of Ghana.

Opoku, I.Y., A.Y. Akrofi, and A.A. Appiah. 2007b. Assessment of sanitation and fungicide application directed at cocoa tree trunks for the control of Phytophthora black pod infections in pods growing in the canopy. *European Journal of Plant Pathology* 117:167–175.

Osman, H., R. Nazaruddin, and S.L. Lee. 2004. Extracts of cocoa (Theobroma cacao L.) leaves and their antioxidation potential. *Food Chemistry* 86:41–45.

Ostovar, K., and P.G. Keeney. 1973. Isolation and characterization of microorganisms involved in the fermentation of Trinidad's cacao beans. *Journal of Food Science* 38:611–617.

Othman, A., A. Ismail, N.A. Ghani, and I. Adenan. 2007. Antioxidant capacity and phenolic content of cocoa beans. *Food Chemistry* 100(4):1523–1530.

Ouattara, H.G., B.L. Koffi, G.T. Karou, A. Sangare, S.L. Niamke, and J.K. Diopoh. 2008. Implication of *Bacillus spp.* in the production of pectinolytic enzymes during cocoa fermentation. *World Journal of Microbiology and Biotechnology* 24:1753–1760.

Palli, D., M. Miraglia, C. Saieva, G. Masala, E. Cava, M. Colatosti et al. 1999. Serum levels of ochratoxin A in healthy adults in Tuscany: Correlation with individual characteristics and between repeat measurements. *Cancer Epidemiology, Biomarkers and Prevention* 8(3):265–269.

Pettipher, G.L. 1986. Analysis of cocoa pulp and the formulation of a standardized artificial cocoa pulp medium. *Journal of the Science of Food and Agriculture* 37:297–309.

Pettipher, G.L. 1990. The extraction and partial purification of cocoa storage proteins. *Café, Cacao, Thé* XXXIV:23–26.

Petzinger, E., and K. Ziegler. 2000. Ochratoxin A from a toxicological perspective. *J. Vet. Pharmacology and Therapautics* 23:91–98.

Pitt, J.L., J.C. Basilico, M.L. Abarca, and C. Lopez. 2000. Mycotoxins and toxigenic fungi. *Medical Mycology* 38:41–46.

Pittet, A. 2001. Natural occurrence of mycotoxins in foods and feeds: A decade in review. In *Mycotoxins and Phycotoxins in Perspective at the Turn of the Millennium*, W.J. de Coe, R.A. Samson, and H.P. van Egmond (Eds.). Wageningen, The Netherlands, 153–172.

Poelman, A., J. Mojet, D. Lyon, and S. Sefa-Dedeh. 2007. The influence of information about organic production and fairtrade on preferences for and perception of pineapple, *Food Quality and Preference* 19(1):114–121. doi:10.1016/j.foodqual.2007.07.005.

Purdy, L.H., and R.A. Schmidt. 1996. Status of cacao witches' broom: Biology, epidemiology, and management. *Annual Review of Phytopathology* 34:573–594.

Puziah, H., S. Jinap, K.S.M. Sharifah, and A. Asbi. 1998a. Changes in free amino acid, peptide-N, sugar and pyrazine concentration during cocoa fermentation. *Journal of the Science of Food Agriculture* 78:535–542.

Puziah, H., S. Jinap, K.S.M. Sharifah, and A. Asbi. 1998b. Effect of mass and turning time on free amino acids, peptide-N, sugar and pyrazine concentration during cocoa fermentation. *Journal of the Science of Food Agriculture* 78:543–550.

Quast, B.L., V. Luccas, and G.T. Kieckbusch. 2011. Physical properties of precrystallized mixtures of cocoa butter and cupuassu fat. *Grasasy aceites* 62(1):62–67.

Rankin, C.W., J.O. Nriagu, J.K. Aggarwal et al. 2005. Lead contamination in cocoa and cocoa products: Isotopic evidence of global contamination. *Environmental Health Perspectives* 113(10) 1344–1348.

Rano, L. 2008. Organic chocolate fans unlikely to switch to cheaper options. http://www.foodanddrinkeurope.com/Product-Categories/Company-news/Organic-chocolate-fans-unlikelyto-switch-to-cheaper-options (accessed July 9, 2008).

Reineccius, G. 2006. *Flavour Chemistry and Technology*, 2nd edition. Boca Raton, FL: CRC Press.

Reineccius, G.A., D.A. Andersen, T.E. Kavanagh, and P.G. Keeney. 1972. Identification and quantification of the free sugars in cocoa beans. *Journal of Agriculture and Food Chemistry* 20:199–202.

Riiner, U. 1970. Investigation of the polymorphism of fats and oils by temperature programmed X-ray diffraction. Lebensmittel-Wissenschaft Und-Technologie (*Food Science and Technology*) 3(6):101–106.

Ringot, D., A. Chango, Y. Schneider, and Y. Larondelle. 2006. Toxicokinetics and toxicodynamics of ochratoxin A, an update. *Chemico-Biological Interactions* 159:18–46.

Rodrıguez, P., E. Péreza, and R. Guzmán. 2009. Effect of the types and concentrations of alkali on the color of cocoa liquor. *Journal of the Science of Food and Agriculture 89*:1186–1194.

Roelofsen, P.A. 1958. Fermentation, drying, and storage of cocoa beans. *Advances in Food Research* 8:225–296.

Rohan, T.A. 1963. Processing of raw cocoa for the market. *FAO Tech. Bull. No* 60.

Rohan, T.A., and T. Stewart. 1967. The precursors of chocolate aroma: Production of reducing sugars during fermentation of cocoa beans. *Journal of Food Science* 32:399–402.

Rössner, S. 1997. Chocolate – Divine food, fattening junk or nutritious supplementation? *European Journal of Clinical Nutrition* 51:341–345.

Rusconi, M., and A. Conti. 2010. Theobroma cacao L., the food of the gods: A scientific approach beyond myths and claims: Review. *Pharmacological Research* 61:5–13.

Said, M.B., B. Meyer, and B. Biehl. 1987. Pulp preconditioning–A new approach towards quality improvement of Malaysian cocoa beans. *Peruk Planters Association Journal* 63:35–47.

Said, M.B., M.J. Musa, and B. Biehl. 1988. An integrated approach towards quality improvement of Malaysian cocoa beans. In *Proceedings of the 10th International Conference on Cocoa, Santo Domingo*. pp. 767–773.

Said, M.B. 1989. Some methods to determine the degree of fermentation in cocoa beans. *Food Processing: Issues and Prospect*, Kuala Lumper 12–14 Sep., pp. 41–51.

Said, M.B., and R.J. Samarakhody. 1984. Cocoa fermentation: Effect of surface area, frequency of turning and depth of cocoa masses. In *International Conference on Cocoa and Coconut*, 15–17 October, 1984. Kuala Lumpur, Malaysia.

Sakharov, I.Y., and G.B. Ardila. 1999. Variations of peroxidase activity in cocoa (*Theobroma cacao* L.) beans during their ripening, fermentation and drying. *Food Chemistry* 65:51–54.

Saldaña, M.D.A., R.S. Mohamed, and P. Mazzafera. 2002. Extraction of cocoa butter from Brazilian cocoa beans using supercritical CO_2 and ethane. *Fluid Phase Equilibria* 197:886.

Sanagi, M.M., W.P. Hung, and S.M. Yasir. 1997. Supercritical fluid extraction of pyrazines in roasted cocoa beans: Effect of pod storage period. *Journal of Chromatography A* 785: 361–367.

Sanchez, J., G. Daquenet, J.P. Guiraud, J.C. Vincent, and P. Galzy. 1985. A study of the yeast flora and the effect of pure culture seedling during the fermentation process of cocoa beans. *Lebensm.- Wiss. Technol.* 18:69–75.

Sanchez-Hervas, M., J.V. Gil, F. Bisbal, D. Ramon, and P.V. Martínez-Culebras, 2008. Mycobiota and mycotoxin producing fungi from cocoa beans. *International Journal of Food Microbiology* 125:336–340.

Savithri, P., B.I.J.U. Joseph, and S. Poongothai. 2003. Effect of copper fungicide sprays on the status of micronutrient in soils of hot semi-arid region of India. Tamil Nadu Agricultural University, Coimbatore 641 003.

Schieberle, P. 2000. The chemistry and technology of cocoa. In *Caffeinated Beverages*. ACS Symposium Series *754:*262.

Schlichter, A., J.S. Sarig, and N. Garti. 1988. Reconsideration of polymorphic transformations in cocoa butter using the DSC. *Journal of the American Oil Chemists Society* 65(7):1140–1143.

Schwan, R.F. 1998. Cocoa fermentations conducted with a defined microbial cocktail inoculum. *Applied and Environmental Microbiology* 64:1477–1483.

Schwan, R.F., and A.E. Wheals. 2004. The microbiology of cocoa fermentation and its role in chocolate quality. *Critical Reviews in Food Science and Nutrition* 44:205–221.

Schwan, R.F., A.H. Rose, and R.G. Board. 1995. Microbial fermentation of cocoa beans, with emphasis on enzymatic degradation of the pulp. *Journal of Applied Bacteriology* 79:96S–107S (Suppl).

Selamat, J., M.A. Hamid, S. Mohamed, and C.Y. Man. 1996. Physical and chemical characteristics of Malaysian cocoa butter. In *Proceedings of the Malaysian International Cocoa Conference*. J. Selamat, B.C. Lian, T.K. Lai, W.R.W. Ishak, and M. Mansor (Eds.). Kuala Lumpur, pp. 351–357

Selinus, O., B. Alloway, J.A. Centeno et al. (Eds). 2005. Essentials of medical geology: Impact of natural environment on public health. *Geological Magazine* 144: 890-891.

Senesi, G.S., G. Baldassarre, N. Senesi, and B. Radina. 1999. Trace element inputs by anthropogenic activities and implications for human health. *Chemosphere* 39: 343–377.

Serna-Saldivar, S.O., M.H. Gomez, and L.W. Rooney. 1990. Technology, chemistry and nutritive value of alkaline-cooked corn products. In *Advances in Cereal Science and Technology, Vol. 10.* Y. Pomeranz (Ed.). St. Paul, MN: America Association of Cereal Chemists Inc., pp. 243–295.

Servais, C., H. Ranc, and I.D. Roberts. 2004. Determination of chocolate viscosity. *Journal of Texture Studies* 34:467–497.

Shahidi, F., and M. Naczk. 1995. *Food Phenolics: Sources, Chemistry, Effects, and Applications*. Lancaster, PA: Technomic.

Shamsuddin, S.B., and P.S. Dimmick. 1986. Qualitative and quantitative measurement of cacao bean fermentation. In *Proceedings of the Symposium Cacao Biotechnology*. P.S. Dimmick (Ed.). Pennsylvania State University, pp. 55–78.

Shaw, D.S. and I. Clarke. 1999. Belief formation in ethical consumer groups: An exploratory study. *Marketing Intelligence and Planning* 17(2):109–120.

SIPPO, 2002: *Organic Coffee, Cocoa and Tea*. Retrieved from http://www.sippo.ch/internet/osec/en/home/import/publications/food.-ContentSlot-49371-ItemList-85076-File.File.pdf/pub_food_org_coffee.pdf (Accessed on 11/12/2005)

Smart, G.A., and J.C. Sherlock. 1987. Nickel in foods and the diet. *Food Additives and Contaminants* 4(1): 61–71.

Spencer, M.A., and R. Hodge.1992. Cloning and sequencing of a cDNA encoding the major storage proteins of *Theobroma cacao*: Identification of the proteins as members of the vicilin class of storage proteins. *Planta* 186:567–576.

Steyn, P.S. 1971. Ochratoxin and other dihydroisocoumarins. In *Microbial Toxins, Vol. VI, Fungal Toxins*. A. Ciegler, S. Kadis, and S.J. Ajl (Eds.). New York: Academic Press, pp. 179–205.

Steyn, P.S. and C.W. Holzapfel. 1967. The synthesis of ochratoxin A and B metabolites of *Aspergillus ochraceus* Wilh, *Tetrahedron* 23:4449–4461.

Sukha, D.A. 2003. Primary processing of high quality Trinidad and Tobago cocoa beans: Targets, problems, and options. In *Proceedings on Revitalization of the Trinidad and Tobago Cocoa Industry (TTIC): Targets, Problems and Options (TPO)*, 20 Sep, 2003. Wilson, L.A. (Ed.). The Association of Professional Agricultural Scientists of Trinidad and Tobago (APASTT): St. Augustine, Trinidad, pp. 27–31.

Svanberg, U., and W. Lorri. 1997. Fermentation and nutrient availability. *Food Control* 8(5/6):319–327.

Tafuri, A., R. Ferracane, and A. Ritieni. 2004. Ochratoxin A in Italian marketed cocoa products. *Food Chemistry* 88:487–494.

Takrama, J.F., P.C. Aculey, and F. Aneani. 2006. Fermentation of cocoa with placenta: A scientific study. In *Proceedings of 15th International Cocoa Research Conference*; *Costa Rica. Volume II*, pp. 1373–1379.

Talbot, G. (1999). Chocolate temper. In *Industrial Chocolate Manufacture and Use*, 3rd edition. S.T. Beckett (Ed.), Oxford, UK: Blackwell Science, pp. 218–230.

Taylor, A.J. 2002. *Food Flavour Technology*. Sheffield, UK: Sheffield Academic Press.

Thamke, I., K. Dürrschmid, and H. Rohm. 2009. Sensory description of dark chocolates by consumers. *LWT-Food Science and Technology*, 42:534–539.

Thompson, S.S., K.B. Miller, and A.S. Lopez. 2001. Cocoa and coffee. In *Food Microbiology – Fundamentals and Frontiers*. M.J. Doyle, L.R. Beuchat, and T.J. Montville (Eds.). Washington DC: ASM Press, pp. 721–733.

Tomlins, K.I., D.M. Baker, P. Dapyln, and D. Adomako. 1993. Effect of fermentation and drying practices on the chemical and physical profiles of Ghana cocoa. *Food Chemistry* 46:257–263.

Toselli, M., E. Baldi, G. Marcolini et al. 2009. Response of potted grapevines to increasing soil copper concentration. *Australian Journal Grape Wine Research* 15: 85–92.

Valenta, H. 1998. Chromatographic methods for the determination of ochratoxin A in animal and human tissues and fluids. *Journal of Chromatography A* 815:75–92.

Valiente, C., E. Molla, M.M. Martin-Cabrejas et al. 1996. Cadmium binding capacity of cocoa and isolated total dietary fibre under physiological pH conditions. *Journal of the Science of Food and Agriculture* 72(4): 476–482.

Van der Merwe, K.J., P.S. Steyne, L.F. Fourie, D.B. Scott, and J.J. Theron. 1965. Ochratoxin A, a toxic metabolite produced by *Aspergillus ochraceus* Wilh. *Nature* 205:1112 –1113.

Varga, J., K. Rigo, S. Kocsube, B. Farkas, and K. Pal. 2003. Diversity of polyketide synthase gene sequences in *Aspergillus* species. *Res. Microbiol.* 154:593– 600.

Velazquez-Manoff, M. 2009. The challenge of fair-trade chocolate. *Christian Science Monitor.* http://features.csmonitor.com/economyrebuild/2009/06/09/the-challenge-of-fair-trade-chocolate (accessed June, 2013).

Venter, M.J., N. Schouten, R. Hink, N.J.M. Kuipers, and A.B. de Haan. 2007. Expression of cocoa butter from cocoa nibs. *Separation and Purification Technology* 55:256.

Voigt, J., B. Biehl, and S. Kamaruddin. 1993. The major seed proteins of *Theobroma cacao* L. *Food Chemistry* 47:145–171.

Voigt, J., B. Biehl, H. Heinrichs, S. Kamaruddin, G. Marsoner, and A. Hugi. 1994c. *In-vitro* formation of cocoa specific aroma precursors: Aroma-related peptides generated from cocoa seed protein by co-operation of an aspartic endoprotease and a carboxypeptidase. *Food Chemistry* 49:173–80.

Voigt, J., H. Heinrichs, G. Voigt, and B. Biehl. 1994a. Cocoa-specific aroma precursors are generated by proteolytic digestion of the vicilin like globulin of cocoa seeds. *Food Chemistry* 50:177–184.

Voigt, J., G. Voigt, H. Heinrichs, D. Wrann, and B. Biehl. 1994b. *In vitro* studies on the proteolytic formation of the characteristic aroma precursors of fermented cocoa seeds – the significance of endoprotease specificity. *Food Chemistry* 51:7–14.

White, A., P. Handler, E.L. Smith, R.L. Hill, and I.R. Lehman. 1978. Carbohydrate metabolism I. In *Principles of biochemistry*, 6th edition. A. White et al. (Eds.) New York: McGraw Hill, pp. 423–476.

Whitefield, R. 2005. *Making Chocolates in the Factory.* London, UK: Kennedy's.

WHO/FAO. 2009. Definition of Pesticide.

Wikipedia 2013. Theobroma. http://en.wikipedia.org/wiki/Theobroma (accessed July 29, 2013).

Willer, H., and M. Yussefi. 2006. *The World of Organic Agriculture: Statistics and Emerging Trends 2006.* Bonn: International Federation of Organic Agriculture Movements (IFOAM) and the Research Institute of Organic Agriculture (FiBL), Frick, Switzerland.

Willson, K.C. 1999. *Coffee, Cocoa and Tea.* Wallingford, UK: CABI.

Wollgast, J. and E. Anklam. 2000. Review on polyphenols in *Theobroma cacao*: Changes in composition during the manufacture of chocolate and methodology for identification and quantification. *Food Research International* 33:423–447.

Wood, G.A.R. 1983. From harvest to store. In *Cocoa*. G.A.R. Wood, and R.A. Lass (Eds.). New York: Longman, p. 444.

Wood, G.A.R., and R.A. Lass. 1985. *Cocoa*, 4th edition. London, UK: Longman Group.

World Cocoa Foundation. 2010. *Cocoa Market Update*, World Cocoa Foundation Published Reports and Resources. Retrieved from http://worldcocoafoundation.org on May 2012.

Wright, D.C., W.D. Park, N.R. Leopold, P.M. Hasegawa, and J. Janick. 1982. Accumulation of lipids, proteins, alkaloids and anthocyanins during embryo development *in vivo* of Theobroma cacao L. *Journal of American Oil Chemistry Society* 59:475–479.

Wright, L.T., and S. Heaton. 2006. Fair Trade marketing: An exploration through quantitative research. *Journal of Strategic Marketing* 14:411–426.

Yağar, H., and A. Sağiroğlu. 2002. Partial purification and characterization of polyphenol oxidase of quince. *Turkish Journal of Chemistry* 26:97–103.

Yusep, I., S. Jinap, B. Jamilah, and S. Nazamid. 2002. Influence of carboxypeptidases on free amino acid, peptide and methylpyrazine contents of underfermented cocoa beans. *Journal of the Science of Food and Agriculture* 82:1584–1592.

Zak, D.L., and P.G. Keeney. 1976. Changes in cocoa proteins during ripening of fruit, fermentation, and further processing of cocoa beans. *Journal of Agriculture and Food Chemistry* 24:483–486.

Ziegleder, G. 1991. Composition of flavour extracts of raw and roasted cocoas. *Z. Lebensm. Unters. Forsch* 193:32–35.

Ziegleder, G. 2009. Flavour development in cocoa and chocolate. In *Industrial Chocolate Manufacture and Use*, 4th edition. S.T. Beckett (Ed.), Oxford, UK: Blackwell, pp. 169–174.

ADDITIONAL READING

Asiedu, J.J. 1989. *Processing Tropical Crops: A Technological Approach.* London: Macmillan Press, pp. 24–41.

Bartley, T. 2003. Certifying forests and factories: States social movements, and the rise of private regulation in the apparel and forest products fields. *Politics and Society* 31(3):433–64.

Beckett, S.T. 1999. *Industrial Chocolate Manufacture and Use*, 3rd edition. Oxford: Blackwell, pp. 153–181, 201–230, 405–428, 460–465.

Beckett, S.T. 2003. Is the taste of British milk chocolate different? *International Journal of Dairy Technology* 56(3):139–142.

Boulstridge, E. and M. Carrigan. 2000. Do consumers really care about corporate responsibility? Highlighting the attitude-behaviour gap. *Journal of Communication Management* 4(4):355–68.

Creyer, E.H., and W.T. Ross. 1997. The influence of firm behaviour on purchase intention: Do consumers really care about business ethics. *Journal of Consumer Marketing* 14(6):421–433.

Dankers, C. 2003. Environmental and social standards, certification and labelling for cash crops. Technical Paper No. 2, Commodities and Trade Division. Food and Agriculture Organization of the United Nations, Rome.

De Vuyst, L., T. Lefeber, Z. Papalexandratou, and N. Camu. 2010. The functional role of lactic acid bacteria in cocoa bean fermentation. In *Biotechnology of Lactic Acid Bacteria: Novel Applications,* F. Mozzi, R.R. Raya, and G.M. Vignolo (Eds.), 301–326. Ames, IA: Wiley-Blackwell.

FLO. 2007. *Fairtrade 2007 Annual Report*. Fairtrade Labelling Organisation International Report. http://www.fairtrade.org.uk (accessed December 2, 2007).

Getz, C., and A. Shreck. 2006. What organic and Fair Trade labels do not tell us: Towards a place-based understanding of certification. *International Journal of Consumer Studies* 30(5):490–501.

Harrison, R., T. Newholm, and D. Shaw. 2005. *The Ethical Consumer*. London: Sage.

Hopkins, R. 2000. *Impact Assessment Study of Oxfam Fair Trade, Final Report*. Oxford: Oxfam.

Jaffee, D. 2007. *Brewing Justice–Fairtrade Coffee, Sustainability and Survival*. Berkeley: University of California Press.

Jaffee, D., J.R. Kloppenburg, and M.B. Monroy. 2004. Bringing the 'moral charge' home: Fairtrade within the North and within the South. *Rural Sociology* 69:169–196.

Kramer, A. and B.A. Twigg. 1970. *Quality Control for the Food Industry, Vol. 1–Fundamentals*. 3rd edition. Westport, CT: AVI, pp. 10–18.

LeClair, M.S. 2002. Fighting the tide: Alternative trade organizations in the era of global free trade. *World Development* 30(6):949–958.

Lindsey, B. 2003. *Grounds for complaint? Understanding the 'coffee crisis'*. Trade Briefing Paper No. 16. Cato Institute, Washington, DC.

Low, W., and E. Davenport. 2005. Has the medium (roast) become the message? The ethics of marketing Fairtrade in the mainstream. *International Marketing Review,* 22(5):494–511.

Low, W., and E. Davenport. 2007. To boldly go...exploring ethical spaces to re-politicise ethical consumption and fair trade. *Journal of Consumer Behaviour* 6:336–348.

Maseland R., and A. de Vaal. 2002. How fair is fairtrade? *De Economist,* 150(3):251–272.

Nicholls, A.J. 2004. Fairtrade product development. *The Services Industries Journal* 24(2):102–107.

Parrish, B.D., V.A. Luzadis, and W.R. Bentley. 2005. What Tanzania's coffee farmers can teach the world: A performance-based look at the fair trade–free trade debate. *Sustainable Development* 13:177–189.

Pohland, A.E., P.L. Schuller, P.S. Steyn, and H.P. Van Egmond. 1982. Physicochemical data for some selected mycotoxins. *Pure and Applied Chemistry* 54:2219–2284.

Ronchi L. 2002a. The impact of fair trade on producers and their organizations: A case study with Coocafé in Costa Rica. *PRUS Working Paper No. 11*. Poverty Research Unit at Sussex.

Ronchi, L. 2002b. *Monitoring Impact of Fairtrade Initiatives: A Case Study of Kuapa Kokoo and the Day Chocolate Company*. London, UK: Twin.

Shreck, A. 2005. Resistance, redistribution, and power in the Fairtrade banana initiative. *Agriculture and Human Values* 22:17–29.

Strong, C. 1997. The problems of translating fair trade principles into consumer purchase behaviour. *Marketing Intelligence and Planning* 15(1):32–37.

Vreeland, C. 2000. Organic chocolate market skyrockets. Candy Industry. http://www.allbusiness.com/wholesale-trade/merchant-wholesalers-nondurable/676051-1.html (accessed June, 2013).

Watkins, K. 1998. 'Green dream turns turtle.' *The Guardian*, September 9, 1998: 4.

Zehner, D. 2002. An economic assessment of 'fairtrade' in coffee. *Chazen Web Journal of International Business*, Fall 2002. New York: Columbia University.

Zehner, D. 2003. Response to Paul Rice's rebuttal of 'An economic assessment of "fair trade" in coffee'. *Chazen Web Journal of International Business.* New York: Columbia University.

Index

B

C

D

E

F

G

H

M

N

P

Q

R

S

T

Y